D1091473

*Structural Analysis Of
Laminated Anisotropic Plates*

Structural Analysis
Of Laminated
Anisotropic
Plates

WITHDRAWN

JAMES M. WHITNEY
Materials Laboratory
Air Force Wright Aeronautical Laboratories
Wright-Patterson Air Force Base, Ohio
and
University of Dayton
Dayton, Ohio

WITHDRAWN

TECHNOMIC
PUBLISHING CO., INC.
LANCASTER · BASEL

Tennessee Tech. Library
Cookeville, Tenn.

Published in the Western Hemisphere by
Technomic Publishing Company, Inc.
851 New Holland Avenue
Box 3535
Lancaster, Pennsylvania 17604 U.S.A.

Distributed in the Rest of the World by
Technomic Publishing AG

© 1987 by Technomic Publishing Company, Inc.
All rights reserved

No part of this publication may be reproduced, stored in a
retrieval system, or transmitted, in any form or by any means,
electronic, mechanical, photocopying, recording, or otherwise,
without the prior written permission of the publisher.

Printed in the United States of America
10 9 8 7 6 5 4 3 2 1

Main entry under title:
  Structural Analysis of Laminated Anisotropic Plates

A Technomic Publishing Company book
Bibliography: p.
Includes index p. 339

Library of Congress Card No. 87-50430
ISBN No. 87762-518-2

*To my wife Phyllis and to my children for their
patience, understanding, and encouragement*

Table of Contents

Preface

This book is a revision and major expansion of *Theory of Laminated Plates* by J. E. Ashton and J. M. Whitney published in 1970. In the original book both the theoretical development and pertinent solutions for plates fabricated of thin layers of anisotropic material were presented. With the expanded structural use of advanced composite materials comes a continued need for a textbook which addresses the structural behavior of laminated plates. Five of the original seven chapters are contained in the present book with minor revision. The subject matter of the remaining two chapters is contained in the new book with major revision. In addition, the new book contains four additional chapters which include material on laminated beams, expansional strain effects, curved plates, and free-edge effects.

The objective of this book is to provide a clear foundation in the theory of laminated anisotropic plates, including the problems of bending under transverse load, stability, and free-vibration. Although the theoretical development is complete, the principal demonstration of the behavior of laminated plates is made through the presentation of a large number of actual solutions. In particular, the effects of bending anisotropy, stacking sequence, and bending-stretching coupling are illustrated through numerous solutions with comparison to the simpler cases of orthotropic plates. The solutions presented by J. E. Ashton in Chapters 4 and 5 of the original book are contained in Chapters 5 and 6 of the new book with some revision, including new material. These solutions have become a classic in laminated plate analysis and form an important part of the new book.

The book contains eleven chapters. Chapter 1 presents fundamental information from anisotropic elasticity; Chapters 2 and 3 provide a development of the governing partial differential equations and boundary conditions, including variational forms, for thin laminated anisotropic plates subject to the assumption of non-deformable normals. Chapter 4 treats one-dimensional theories associated with cylindrical bending and laminated beams. Chapter 5 treats the simplified form of the laminated plate equations equivalent to homogeneous orthotropic plates. This form of the equations is rarely applicable to real laminated plates except as an approximation, and Chapters 6 and 7 indicate the effect of an assumption of orthotropic behavior by comparing solutions, including bending anisotropy (Chapter 6) and bending-stretching coupling (Chapter 7) to these orthotropic solutions. In Chapter 8 the effect of expansional strains on the behavior of laminated plates is presented. Example problems include the effects of thermal expansion and dimensional changes induced by matrix swelling associated with moisture absorption. The basic theory is extended to cylindrical plates in Chapter 9. In Chapter 10 a higher order theory applicable to laminated anisotropic plates which includes the effects of transverse shear deformation is developed. Solutions involving the higher order theory are compared to results obtained from classical laminated plate theory in which transverse shear deformation is neglected. A discussion of sandwich plates is also included in Chapter 10. Free-edge effects are discussed in Chapter 11 along with the development of a higher order laminated plate theory which includes a thickness-stretch mode in addition to transverse shear deformation. The new theory is then applied to an approximate free-edge analysis of cross-ply laminates.

This book is intended to combine theoretical development with solutions to the governing equations in order to indicate the importance of stacking sequence, degree of bending anisotropy, bending-extensional coupling, expansional strains, transverse shear deformation, and free-edge effects. A software program called LAMPCAL is available with the book to perform many of these calculations. The appendix of this book provides a full description of LAMPCAL and can serve as the users' guide. It is hoped that engineers and materials scientists will find both the book and software useful in developing an understanding of laminated structural elements.

JAMES M. WHITNEY

Dayton, Ohio
March 1987

C HAPTER 1

Theory of an Anisotropic
Elastic Continuum

1.1 INTRODUCTION

W ITH THE INCREASED use of composite materials in structural applications
has come a need for the analysis of laminated anisotropic plates. This
chapter provides the fundamental principles of anisotropic elasticity from which
laminated plate theory is developed in the following two chapters. Much of the
presentation on anisotropic elasticity is based on the works of Lekhnitskii [1] and
Hearmon [2]. A detailed derivation of the theory of finite deformations can be
found in Fung [3].

1.2 STRESS AND STRAIN IN AN ANISOTROPIC CONTINUUM

Figure 1.1 shows the stress nomenclature in cartesian coordinates. In linear
mechanics little or no distinction is made between the stresses with respect to the
deformed and undeformed coordinates since the difference is a second order effect. However in the development of a plate theory which includes inplane force
effects it is useful to relate stresses on the deformed body to the initial configuration.

Consider a force vector $d\bar{F}$ acting on a deformed surface dS and a corresponding force vector $d\bar{F}_o$ acting on the same surface in the undeformed state $d\bar{S}_o$. The
stress components in the deformed state are given by the Cauchy relationship:

$$dF_i = \sum_{j=1}^{3} \tau_{ij} n_j dS \tag{1.1}$$

where τ_{ij} are components of the Eulerian stress tensor and n_j are direction
cosines of the outward normal to the deformed surface. The Kirchhoff stress tensor refers to the original configuration and its components are defined as follows:

$$dF_{oi} = \sum_{j=1}^{3} \sigma_{ij} n_{oj} dS_o = \sum_{j=1}^{3} \frac{\partial x_i}{\partial \bar{x}_j} dF_j \tag{1.2}$$

1

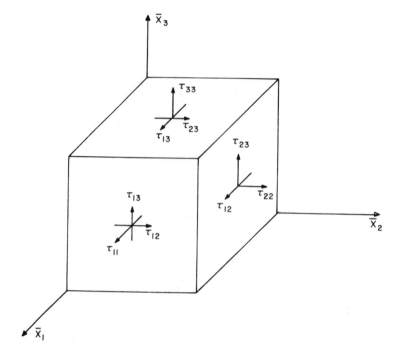

Figure 1.1. *Stress nomenclature on body in deformed state.*

where σ_{ij} are the components of the Kirchhoff stress tensor, n_{oj} are direction cosines of the outward normal to the undeformed surface, x_i are coordinates of the undeformed surface, and $\bar{x}_j$ are coordinates of the deformed surface.

In order to determine the principal stresses in a plate it is often necessary to consider the stresses with respect to an axis system rotated in the plane of the plate. Consider a rotation through an angle θ from x_1 and x_2 about the x_3 axis (see Figure 1.2). The rotated axes are denoted by x_1' and x_2'. The transformed stresses σ_{ij}' are given by

$$
\begin{bmatrix} \sigma_{11}' \\ \sigma_{22}' \\ \sigma_{33}' \\ \sigma_{23}' \\ \sigma_{13}' \\ \sigma_{12}' \end{bmatrix} = \begin{bmatrix} m^2 & n^2 & 0 & 0 & 0 & 2mn \\ n^2 & m^2 & 0 & 0 & 0 & -2mn \\ 0 & 0 & 1 & 0 & 0 & 0 \\ 0 & 0 & 0 & m & -n & 0 \\ 0 & 0 & 0 & +n & m & 0 \\ -mn & mn & 0 & 0 & 0 & (m^2 - n^2) \end{bmatrix} \begin{bmatrix} \sigma_{11} \\ \sigma_{22} \\ \sigma_{33} \\ \sigma_{23} \\ \sigma_{13} \\ \sigma_{12} \end{bmatrix} \quad (1.3)
$$

where $m = \cos\theta,\ n = \sin\theta$.

For finite deformation the Green strain tensor is used. The strain displacement relations are given by

$$\epsilon_{11} = \frac{\partial u_1}{\partial x_1} + \frac{1}{2}\left[\left(\frac{\partial u_1}{\partial x_1}\right)^2 + \left(\frac{\partial u_2}{\partial x_1}\right)^2 + \left(\frac{\partial u_3}{\partial x_1}\right)^2\right] \tag{1.4}$$

$$\epsilon_{22} = \frac{\partial u_2}{\partial x_2} + \frac{1}{2}\left[\left(\frac{\partial u_1}{\partial x_2}\right)^2 + \left(\frac{\partial u_2}{\partial x_2}\right)^2 + \left(\frac{\partial u_3}{\partial x_2}\right)^2\right] \tag{1.5}$$

$$\epsilon_{33} = \frac{\partial u_3}{\partial x_3} + \frac{1}{2}\left[\left(\frac{\partial u_1}{\partial x_3}\right)^2 + \left(\frac{\partial u_2}{\partial x_3}\right)^2 + \left(\frac{\partial u_3}{\partial x_3}\right)^2\right] \tag{1.6}$$

$$\epsilon_{23} = \frac{\partial u_3}{\partial x_2} + \frac{\partial u_2}{\partial x_3} + \frac{\partial u_1}{\partial x_2}\frac{\partial u_1}{\partial x_3} + \frac{\partial u_2}{\partial x_2}\frac{\partial u_2}{\partial x_3} + \frac{\partial u_3}{\partial x_2}\frac{\partial u_3}{\partial x_3} \tag{1.7}$$

$$\epsilon_{13} = \frac{\partial u_3}{\partial x_1} + \frac{\partial u_1}{\partial x_3} + \frac{\partial u_1}{\partial x_1}\frac{\partial u_1}{\partial x_3} + \frac{\partial u_2}{\partial x_1}\frac{\partial u_2}{\partial x_3} + \frac{\partial u_3}{\partial x_1}\frac{\partial u_3}{\partial x_3} \tag{1.8}$$

$$\epsilon_{12} = \frac{\partial u_2}{\partial x_1} + \frac{\partial u_1}{\partial x_2} + \frac{\partial u_1}{\partial x_1}\frac{\partial u_1}{\partial x_2} + \frac{\partial u_2}{\partial x_1}\frac{\partial u_2}{\partial x_2} + \frac{\partial u_3}{\partial x_1}\frac{\partial u_3}{\partial x_2} \tag{1.9}$$

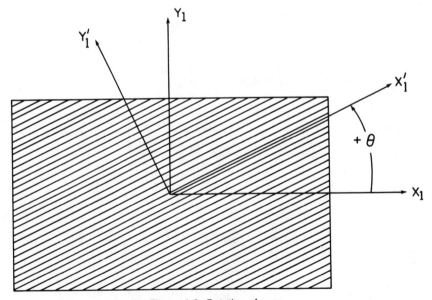

Figure 1.2. Rotation of axes.

where ϵ_{ij} are engineering strains and u_i are displacements along the x_i coordinate. For small displacement theory Equations (1.4–1.9) become

$$\left.\begin{array}{ccc} \epsilon_{11} = \dfrac{\partial u_1}{\partial x_1} & \epsilon_{22} = \dfrac{\partial u_2}{\partial x_2} & \epsilon_{33} = \dfrac{\partial u_3}{\partial x_3} \\[2ex] \epsilon_{23} = \dfrac{\partial u_3}{\partial x_2} + \dfrac{\partial u_2}{\partial x_3} \quad \epsilon_{13} = \dfrac{\partial u_3}{\partial x_1} + \dfrac{\partial u_1}{\partial x_3} \quad \epsilon_{12} = \dfrac{\partial u_1}{\partial x_2} + \dfrac{\partial u_2}{\partial x_1} \end{array}\right\} (1.10)$$

For a rotation about the x_3 axis we have the following transformation:

$$\begin{bmatrix} \epsilon'_{11} \\ \epsilon'_{22} \\ \epsilon'_{33} \\ \epsilon'_{23} \\ \epsilon'_{13} \\ \epsilon'_{12} \end{bmatrix} = \begin{bmatrix} m^2 & n^2 & 0 & 0 & 0 & mn \\ n^2 & m^2 & 0 & 0 & 0 & -mn \\ 0 & 0 & 1 & 0 & 0 & 0 \\ 0 & 0 & 0 & m & -n & 0 \\ 0 & 0 & 0 & n & m & 0 \\ -2mn & 2mn & 0 & 0 & 0 & (m^2 - n^2) \end{bmatrix} \begin{bmatrix} \epsilon_{11} \\ \epsilon_{22} \\ \epsilon_{33} \\ \epsilon_{23} \\ \epsilon_{13} \\ \epsilon_{12} \end{bmatrix} \quad (1.11)$$

1.3 EQUATIONS OF MOTION AND COMPATIBILITY

The Kirchhoff stress tensor must satisfy the following nonlinear equations of motion:

$$\frac{\partial}{\partial x_1}\left[\sigma_{11}\left(1 + \frac{\partial u_1}{\partial x_1}\right) + \sigma_{12}\frac{\partial u_1}{\partial x_2} + \sigma_{13}\frac{\partial u_1}{\partial x_3}\right]$$

$$+ \frac{\partial}{\partial x_2}\left[\sigma_{12}\left(1 + \frac{\partial u_1}{\partial x_1}\right) + \sigma_{22}\frac{\partial u_1}{\partial x_2} + \sigma_{23}\frac{\partial u_1}{\partial x_3}\right] \quad (1.12)$$

$$+ \frac{\partial}{\partial x_3}\left[\sigma_{13}\left(1 + \frac{\partial u_1}{\partial x_1}\right) + \sigma_{23}\frac{\partial u_1}{\partial x_2} + \sigma_{33}\frac{\partial u_1}{\partial x_3}\right] + X_1 = \varrho_\circ \frac{\partial^2 u_1}{\partial t^2}$$

$$\frac{\partial}{\partial x_1}\left[\sigma_{11}\frac{\partial u_2}{\partial x_1} + \sigma_{12}\left(1 + \frac{\partial u_2}{\partial x_2}\right) + \sigma_{13}\frac{\partial u_2}{\partial x_3}\right]$$

$$+ \frac{\partial}{\partial x_2}\left[\sigma_{12}\frac{\partial u_2}{\partial x_1} + \sigma_{22}\left(1 + \frac{\partial u_2}{\partial x_2}\right) + \sigma_{23}\frac{\partial u_2}{\partial x_3}\right] \quad (1.13)$$

$$+ \frac{\partial}{\partial x_3}\left[\sigma_{13}\frac{\partial u_2}{\partial x_1} + \sigma_{23}\left(1 + \frac{\partial u_2}{\partial x_2}\right) + \sigma_{33}\frac{\partial u_2}{\partial x_3}\right] + X_2 = \varrho_\circ \frac{\partial^2 u_2}{\partial t^2}$$

$$\frac{\partial}{\partial x_1}\left[\sigma_{11}\frac{\partial u_3}{\partial x_1} + \sigma_{12}\frac{\partial u_3}{\partial x_2} + \sigma_{13}\left(1 + \frac{\partial u_3}{\partial x_3}\right)\right]$$

$$+ \frac{\partial}{\partial x_2}\left[\sigma_{12}\frac{\partial u_3}{\partial x_1} + \sigma_{22}\frac{\partial u_3}{\partial x_2} + \sigma_{23}\left(1 + \frac{\partial u_3}{\partial x_3}\right)\right] \qquad (1.14)$$

$$+ \frac{\partial}{\partial x_3}\left[\sigma_{13}\frac{\partial u_3}{\partial x_1} + \sigma_{23}\frac{\partial u_3}{\partial x_2} + \sigma_{33}\left(1 + \frac{\partial u_3}{\partial x_3}\right)\right] + X_3 = \varrho_o\frac{\partial^2 u_3}{\partial t^2}$$

where t denotes time, ϱ_o is the density, and X_i are body forces. For linear small deformation theory, Equations (1.12–1.14) become

$$\frac{\partial \sigma_{11}}{\partial x_1} + \frac{\partial \sigma_{12}}{\partial x_2} + \frac{\partial \sigma_{13}}{\partial x_3} + X_1 = \varrho_o\frac{\partial^2 u_1}{\partial t^2}$$

$$\frac{\partial \sigma_{12}}{\partial x_1} + \frac{\partial \sigma_{22}}{\partial x_2} + \frac{\partial \sigma_{23}}{\partial x_3} + X_2 = \varrho_o\frac{\partial^2 u_2}{\partial t^2} \qquad (1.15)$$

$$\frac{\partial \sigma_{13}}{\partial x_1} + \frac{\partial \sigma_{23}}{\partial x_2} + \frac{\partial \sigma_{33}}{\partial x_3} + X_3 = \varrho_o\frac{\partial^2 u_3}{\partial t^2}$$

Given a strain field the question arises as to how Equations (1.10) can be integrated to determine the displacements. Since there are six strain equations in three unknown displacements, solutions will not be single-valued or continuous unless certain relations are satisfied. The following compatibility equations from linear theory of elasticity are well known.

$$\frac{\partial^2 \epsilon_{12}}{\partial x_1 \partial x_2} = \frac{\partial^2 \epsilon_{11}}{\partial x_2^2} + \frac{\partial^2 \epsilon_{22}}{\partial x_1^2} \qquad (1.16)$$

$$\frac{\partial^2 \epsilon_{23}}{\partial x_2 \partial x_3} = \frac{\partial^2 \epsilon_{22}}{\partial x_3^2} + \frac{\partial^2 \epsilon_{33}}{\partial x_2^2} \qquad (1.17)$$

$$\frac{\partial^2 \epsilon_{13}}{\partial x_1 \partial x_3} = \frac{\partial^2 \epsilon_{11}}{\partial x_3^2} + \frac{\partial^2 \epsilon_{33}}{\partial x_1^2} \qquad (1.18)$$

$$2\frac{\partial^2 \epsilon_{11}}{\partial x_2 \partial x_3} = \frac{\partial}{\partial x_1}\left(-\frac{\partial \epsilon_{23}}{\partial x_1} + \frac{\partial \epsilon_{13}}{\partial x_2} + \frac{\partial \epsilon_{12}}{\partial x_3}\right) \qquad (1.19)$$

$$2 \frac{\partial^2 \epsilon_{22}}{\partial x_1 \partial x_3} = \frac{\partial}{\partial x_2} \left(\frac{\partial \epsilon_{23}}{\partial x_1} - \frac{\partial \epsilon_{13}}{\partial x_2} + \frac{\partial \epsilon_{12}}{\partial x_3} \right) \tag{1.20}$$

$$2 \frac{\partial^2 \epsilon_{33}}{\partial x_1 \partial x_2} = \frac{\partial}{\partial x_3} \left(\frac{\partial \epsilon_{23}}{\partial x_1} + \frac{\partial \epsilon_{13}}{\partial x_2} - \frac{\partial \epsilon_{12}}{\partial x_3} \right) \tag{1.21}$$

1.4 GENERALIZED HOOKE'S LAW

Consider the following contracted stresses and strains:

$$\sigma_1 = \sigma_{11} \quad \sigma_2 = \sigma_{22} \quad \sigma_3 = \sigma_{33} \quad \sigma_4 = \sigma_{23} \quad \sigma_5 = \sigma_{13} \quad \sigma_6 = \sigma_{12} \tag{1.22}$$

$$\epsilon_1 = \epsilon_{11} \quad \epsilon_2 = \epsilon_{22} \quad \epsilon_3 = \epsilon_{33} \quad \epsilon_4 = \epsilon_{23} \quad \epsilon_5 = \epsilon_{13} \quad \epsilon_6 = \epsilon_{12} \tag{1.23}$$

Using Equations (1.22) and (1.23), the generalized Hooke's law can be written in the following matrix form:

$$\begin{bmatrix} \sigma_1 \\ \sigma_2 \\ \sigma_3 \\ \sigma_4 \\ \sigma_5 \\ \sigma_6 \end{bmatrix} = \begin{bmatrix} c_{11} & c_{12} & c_{13} & c_{14} & c_{15} & c_{16} \\ c_{12} & c_{22} & c_{23} & c_{24} & c_{25} & c_{26} \\ c_{13} & c_{23} & c_{33} & c_{34} & c_{35} & c_{36} \\ c_{14} & c_{24} & c_{34} & c_{44} & c_{45} & c_{46} \\ c_{15} & c_{25} & c_{35} & c_{45} & c_{55} & c_{56} \\ c_{16} & c_{26} & c_{36} & c_{46} & c_{56} & c_{66} \end{bmatrix} \begin{bmatrix} \epsilon_1 \\ \epsilon_2 \\ \epsilon_3 \\ \epsilon_4 \\ \epsilon_5 \\ \epsilon_6 \end{bmatrix} \tag{1.24}$$

where c_{ij} is the stiffness matrix. Equation (1.24) can be written in the inverted form

$$\epsilon_i = \sum_{j=1}^{6} S_{ij} \sigma_j \tag{1.25}$$

where s_{ij} is the compliance matrix. Obviously the compliance matrix is the inverse of the stiffness matrix.

For the general case there are 21 independent elastic constants. If, however, there are any planes of elastic symmetry this number is reduced. Assume that the x_3-axis is perpendicular to a plane of elastic symmetry. Then

$$c_{14} = c_{15} = c_{24} = c_{25} = c_{34} = c_{35} = c_{46} = c_{56} = 0$$

and the stiffness matrix in Equation (1.24) becomes

$$
\begin{bmatrix}
c_{11} & c_{12} & c_{13} & 0 & 0 & c_{16} \\
c_{12} & c_{22} & c_{23} & 0 & 0 & c_{26} \\
c_{13} & c_{23} & c_{33} & 0 & 0 & c_{36} \\
0 & 0 & 0 & c_{44} & c_{45} & 0 \\
0 & 0 & 0 & c_{45} & c_{55} & 0 \\
c_{16} & c_{26} & c_{36} & 0 & 0 & c_{66}
\end{bmatrix}
\tag{1.26}
$$

The number of independent elastic stiffness coefficients is reduced to 13.
If there are two mutually perpendicular planes of elastic symmetry then

$$
c_{16} = c_{26} = c_{36} = c_{45} = 0
$$

and (1.26) becomes

$$
\begin{bmatrix}
c_{11} & c_{12} & c_{13} & 0 & 0 & 0 \\
c_{12} & c_{22} & c_{23} & 0 & 0 & 0 \\
c_{13} & c_{23} & c_{33} & 0 & 0 & 0 \\
0 & 0 & 0 & c_{44} & 0 & 0 \\
0 & 0 & 0 & 0 & c_{55} & 0 \\
0 & 0 & 0 & 0 & 0 & c_{66}
\end{bmatrix}
\tag{1.27}
$$

Thus the number of stiffness constants is reduced to 9. Two mutually perpendicular planes of elastic symmetry imply the existence of a third mutually perpendicular plane of symmetry and the material is referred to as orthotropic.

Further reduction can be made if one of the planes is isotropic, i.e. the properties in the plane are independent of direction. Consider the case where the x_1 axis is perpendicular to a plane of isotropy. Then

$$
c_{33} = c_{22}, \; c_{13} = c_{12}, \; c_{55} = c_{66}
$$

$$
c_{44} = \tfrac{1}{2}(c_{22} - c_{23})
$$

and (1.27) becomes

$$
\begin{bmatrix}
c_{11} & c_{12} & c_{12} & 0 & 0 & 0 \\
c_{12} & c_{22} & c_{23} & 0 & 0 & 0 \\
c_{12} & c_{23} & c_{22} & 0 & 0 & 0 \\
0 & 0 & 0 & \tfrac{1}{2}(c_{22}-c_{23}) & 0 & 0 \\
0 & 0 & 0 & 0 & c_{66} & 0 \\
0 & 0 & 0 & 0 & 0 & c_{66}
\end{bmatrix}
\tag{1.28}
$$

Such a material is often referred to as transversely isotropic, and its stiffness properties are determined by 5 independent constants. For complete isotropy

$$c_{22} = c_{11}, \; c_{23} = c_{12}, \; c_{66} = \tfrac{1}{2}(c_{11} - c_{12})$$

and the number of independent constants is reduced to 2 which leads to the isotropic stiffness matrix

$$\begin{bmatrix} c_{11} & c_{12} & c_{12} & 0 & 0 & 0 \\ c_{12} & c_{11} & c_{12} & 0 & 0 & 0 \\ c_{12} & c_{12} & c_{11} & 0 & 0 & 0 \\ 0 & 0 & 0 & \tfrac{1}{2}(c_{11} - c_{12}) & 0 & 0 \\ 0 & 0 & 0 & 0 & \tfrac{1}{2}(c_{11} - c_{12}) & 0 \\ 0 & 0 & 0 & 0 & 0 & \tfrac{1}{2}(c_{11} - c_{12}) \end{bmatrix} \tag{1.29}$$

1.5 TRANSFORMATION OF ELASTIC STIFFNESSES UNDER ROTATION OF COORDINATE AXES

In engineering applications it is often necessary to determine the elastic properties with respect to x_1, x_2, x_3 coordinate system when the elastic stiffnesses are known relative to a rotated x_1', x_2', x_3' coordinate system, i.e., in matrix notation

$$\sigma' = C' \epsilon' \tag{1.30}$$

where in most cases C' is a known orthotropic stiffness matrix. We seek a relationship of the form

$$\sigma = C \epsilon \tag{1.31}$$

where C is to be determined from C', m, and n. Substituting Equations (1.3) and (1.11) into Equation (1.30), we obtain the result

$$T_\sigma \, \sigma = C' \, T_\epsilon \, \epsilon \tag{1.32}$$

where T_σ and T_ϵ are the stress and strain transformations as defined by Equations (1.3) and (1.11), respectively. Pre-multiplying Equation (1.32) by the inverse of the stress transformation matrix, denoted by T_σ^{-1}, we obtain the relationship

$$\sigma = T_\sigma^{-1} \, C' \, T_\epsilon \, \epsilon \tag{1.33}$$

Thus,

$$C = T_\sigma^{-1} \, C' \, T_\epsilon \tag{1.34}$$

The inverse of the stress transformation matrix, T_σ^{-1}, can be obtained by substituting $-n$ for n in the stress transformation matrix, T_σ, as given by Equation (1.3). Performing the operations defined by Equation (1.34), we obtain the following stiffness transformation equations for an orthotropic C' matrix:

$$C_{11} = c'_{11}m^4 + 2m^2n^2(C'_{12} + 2C'_{66}) + C'_{22}n^4 \tag{1.35}$$

$$C_{12} = m^2n^2(C'_{11} + C'_{22} - 4C'_{66}) + C'_{12}(m^4 + n^4) \tag{1.36}$$

$$C_{13} = c'_{13}m^2 + C'_{23}n^2 \tag{1.37}$$

$$C_{16} = mn[C'_{11}m^2 - C'_{22}n^2 - (C'_{12} + 2C'_{66})(m^2 - n^2)] \tag{1.38}$$

$$C_{22} = C'_{11}n^4 + 2m^2n^2(C'_{12} + 2C'_{66}) + C'_{22}m^4 \tag{1.39}$$

$$C_{23} = C'_{13}n^2 + C'_{23}m^2 \tag{1.40}$$

$$C_{26} = mn[C'_{11}n^2 - C'_{22}m^2 + (C'_{12} + 2C'_{66})(m^2 - n^2)] \tag{1.41}$$

$$C_{33} = C'_{33} \tag{1.42}$$

$$C_{36} = (C'_{23} - C'_{13})mn \tag{1.43}$$

$$C_{44} = C'_{44}m^2 + C'_{55}n^2 \tag{1.44}$$

$$C_{45} = (C'_{44} - C'_{55})mn \tag{1.45}$$

$$C_{55} = C'_{44}n^2 + C'_{55}m^2 \tag{1.46}$$

$$C_{66} = (C'_{11} + C'_{22} - 2C'_{12})m^2n^2 + C'_{66}(m^2 - n^2)^2 \tag{1.47}$$

$$C_{14} = C_{15} = C_{24} = C_{25} = C_{34} = C_{35} = C_{46} = C_{56} = 0 \tag{1.48}$$

Note that the transformed matrix, C, is not orthotropic.

1.6 ENGINEERING CONSTANTS

Consider the case of a material with three mutually perpendicular planes of elastic symmetry. The constitutive relations in terms of the elastic compliances (1.25) are

$$
\begin{bmatrix} \epsilon_1 \\ \epsilon_2 \\ \epsilon_3 \\ \epsilon_4 \\ \epsilon_5 \\ \epsilon_6 \end{bmatrix} =
\begin{bmatrix}
S_{11} & S_{12} & S_{13} & 0 & 0 & 0 \\
S_{12} & S_{22} & S_{23} & 0 & 0 & 0 \\
S_{13} & S_{23} & S_{33} & 0 & 0 & 0 \\
0 & 0 & 0 & S_{44} & 0 & 0 \\
0 & 0 & 0 & 0 & S_{55} & 0 \\
0 & 0 & 0 & 0 & 0 & S_{66}
\end{bmatrix}
\begin{bmatrix} \sigma_1 \\ \sigma_2 \\ \sigma_3 \\ \sigma_4 \\ \sigma_5 \\ \sigma_6 \end{bmatrix}
\tag{1.49}
$$

These compliances can be expressed in terms of engineering constants as follows:

$$
S_{11} = \frac{1}{E_{11}}, \qquad S_{12} = \frac{-\nu_{12}}{E_{11}}, \qquad S_{13} = \frac{-\nu_{13}}{E_{11}}
$$

$$
S_{22} = \frac{1}{E_{22}}, \qquad S_{23} = \frac{-\nu_{23}}{E_{22}}, \qquad S_{33} = \frac{1}{E_{33}}
\tag{1.50}
$$

$$
S_{44} = \frac{1}{G_{23}}, \qquad S_{55} = \frac{1}{G_{13}}, \qquad S_{66} = \frac{1}{G_{12}}
$$

where E_{ii} are the Young's moduli in tension (compression) along the i directions; ν_{ij} are the Poisson's ratios as determined from the contraction in the x_j direction during a tensile test in the x_i direction; and G_{ij} are shear moduli in the $x_i - x_j$ planes. Due to symmetry of the compliances, the following relationships exist between the Young's moduli and the Poisson's ratios:

$$
E_{11}\nu_{21} = E_{22}\nu_{12}, \quad E_{22}\nu_{32} = E_{33}\nu_{23}, \quad E_{33}\nu_{13} = E_{11}\nu_{31}
\tag{1.51}
$$

Thus for an orthotropic material we have the 9 independent engineering constants: E_{11}, E_{22}, E_{33}, ν_{12}, ν_{13}, ν_{23}, G_{12}, G_{13}, and G_{23}. For this case Equation (1.49) takes the form

$$
\begin{bmatrix} \epsilon_1 \\ \epsilon_2 \\ \epsilon_3 \\ \epsilon_4 \\ \epsilon_5 \\ \epsilon_6 \end{bmatrix}
=
\begin{bmatrix}
\dfrac{1}{E_{11}} & \dfrac{-\nu_{12}}{E_{11}} & \dfrac{-\nu_{13}}{E_{11}} & 0 & 0 & 0 \\[2mm]
\dfrac{-\nu_{12}}{E_{11}} & \dfrac{1}{E_{22}} & \dfrac{-\nu_{23}}{E_{22}} & 0 & 0 & 0 \\[2mm]
\dfrac{-\nu_{13}}{E_{11}} & \dfrac{-\nu_{23}}{E_{22}} & \dfrac{1}{E_{33}} & 0 & 0 & 0 \\[2mm]
0 & 0 & 0 & \dfrac{1}{G_{23}} & 0 & 0 \\[2mm]
0 & 0 & 0 & 0 & \dfrac{1}{G_{13}} & 0 \\[2mm]
0 & 0 & 0 & 0 & 0 & \dfrac{1}{G_{12}}
\end{bmatrix}
\begin{bmatrix} \sigma_1 \\ \sigma_2 \\ \sigma_3 \\ \sigma_4 \\ \sigma_5 \\ \sigma_6 \end{bmatrix}
\tag{1.52}
$$

For a transversely isotropic material in which the $x_2 - x_3$ plane is the plane of isotropy, we find

$$
E_{33} = E_{22}, \ G_{13} = G_{12}, \ \nu_{13} = \nu_{12}, \tag{1.53}
$$

$$
G_{23} = \frac{E_{22}}{2(1 + \nu_{23})}
$$

and we have 5 independent engineering constants.

If we invert Equation (1.52), we obtain the elastic stiffness, C_{ij}, in terms of engineering constants with the following results:

$$
C_{11} = (1 - \nu_{23}^2 E_{33}/E_{22}) \frac{E_{11}}{V} \tag{1.54}
$$

$$
C_{12} = (\nu_{12} + \nu_{13}\nu_{23} E_{33}/E_{22}) \frac{E_{22}}{V} \tag{1.55}
$$

$$
C_{13} = (\nu_{13} + \nu_{12}\nu_{23}) \frac{E_{33}}{V} \tag{1.56}
$$

$$C_{23} = (\nu_{23} + \nu_{12}\nu_{13}E_{22}/E_{11}) \frac{E_{33}}{V} \tag{1.57}$$

$$C_{22} = (1 - \nu_{13}^2 E_{33}/E_{11}) \frac{E_{22}}{V} \tag{1.58}$$

$$C_{33} = (1 - \nu_{12}^2 E_{22}/E_{11}) \frac{E_{33}}{V} \tag{1.59}$$

$$C_{44} = G_{23}, \ C_{55} = G_{13}, \ C_{66} = G_{12} \tag{1.60}$$

where

$$V = [1 - \nu_{12}(\nu_{12}E_{22}/E_{11} + 2\nu_{23}\nu_{13}E_{33}/E_{11}) - \nu_{13}^2 E_{33}/E_{11} - \nu_{23}^2 E_{33}/E_{22}]$$

For a material having transverse isotropy relative to the $x_2 - x_3$ plane, we find

$$C_{11} = (1 - \nu_{23}^2) \frac{E_{11}}{V} \tag{1.61}$$

$$C_{13} = C_{12} = \nu_{12}(1 + \nu_{23}) \frac{E_{22}}{V} \tag{1.62}$$

$$C_{23} = (\nu_{23} + \nu_{12}^2 E_{22}/E_{11}) \frac{E_{22}}{V} \tag{1.63}$$

$$C_{33} = C_{22} = (1 - \nu^2{}_{12}E_{22}/E_{11}) \frac{E_{22}}{V} \tag{1.64}$$

$$C_{44} = G_{23} = \frac{E_{22}}{2(1 + \nu_{23})} \tag{1.65}$$

$$C_{55} = C_{66} = G_{12} \tag{1.66}$$

and

$$V = [(1 + \nu_{23})(1 - \nu_{23} - 2\nu_{12}^2 E_{22}/E_{11})]$$

For an isotropic material

$$C_{11} = C_{22} = C_{33} = \frac{(1 - \nu)E}{(1 + \nu)(1 - 2\nu)} \tag{1.67}$$

$$C_{12} = C_{13} = C_{23} = \frac{\nu E}{(1 + \nu)(1 - 2\nu)} \qquad (1.68)$$

$$C_{44} = C_{55} = C_{66} = \frac{E}{2(1 + \nu)} \qquad (1.69)$$

1.7 PLANE STRESS

In practical engineering problems one is often interested in a state of plane stress. For plane stress relative to the $x_1 - x_2$ axes we have

$$\sigma_3 = \sigma_4 = \sigma_5 = 0 \qquad (1.70)$$

and the constitutive relations for an orthotropic material, Equation (1.51), in terms of engineering constants become:

$$
\begin{bmatrix} \epsilon_1 \\ \epsilon_2 \\ \epsilon_3 \end{bmatrix} = \begin{bmatrix} \dfrac{1}{E_{11}} & \dfrac{-\nu_{12}}{E_{11}} & 0 \\ \dfrac{-\nu_{12}}{E_{11}} & \dfrac{1}{E_{22}} & 0 \\ 0 & 0 & \dfrac{1}{G_{12}} \end{bmatrix} \begin{bmatrix} \sigma_1 \\ \sigma_2 \\ \sigma_6 \end{bmatrix} \qquad (1.71)
$$

Inverting Equation (1.71) we arrive at the constitutive relations in terms of stiffness coefficients which are of the form:

$$
\begin{bmatrix} \sigma_1 \\ \sigma_2 \\ \sigma_6 \end{bmatrix} = \begin{bmatrix} Q_{11} & Q_{12} & 0 \\ Q_{12} & Q_{22} & 0 \\ 0 & 0 & Q_{66} \end{bmatrix} \begin{bmatrix} \epsilon_1 \\ \epsilon_2 \\ \epsilon_6 \end{bmatrix} \qquad (1.72)
$$

where

$$Q_{11} = \frac{E_{11}}{(1 - \nu_{12}E_{22}/E_{11})}, \quad Q_{12} = \frac{\nu_{12}E_{22}}{(1 - \nu_{12}E_{22}/E_{11})}$$

$$(1.73)$$

$$Q_{22} = \frac{E_{22}}{(1 - \nu_{12}E_{22}/E_{11})}, \quad Q_{66} = G_{12}$$

and Q_{ij} are referred to as the reduced stiffnesses for plane stress.

Equation (1.72) can also be derived directly from equation (1.27) by employing the definition of plane stress as given by Equation (1.70), which allows ϵ_3 to be eliminated from the inplane constitutive relations. This approach yields the Q_{ij}'s in terms of C_{ij}'s and is discussed in detail in Chapter 2.

A more detailed discussion of the engineering constants can be found in References [4,5]. In addition, both of these references show how one can predict the elastic constants of a unidirectional fiber reinforced composite material as a function of constituent material properties and volume content of reinforcement. Once these "micromechanical" properties are determined the unidirectional composite is treated as a homogeneous orthotropic sheet. If one considers a multidirectional composite as being composed of unidirectional layers with different fiber orientations, then the theory of laminated plates as developed in the following chapters is directly applicable.

1.8 PLY AND LAMINATE NOTATION

The principal material directions within each ply of a laminate will be denoted by an $x_L - x_T$ axes system. Thus, the engineering constants relative to this system are: E_L, E_T, ν_{LT}, and G_{LT}. For unidirectional layers this notation has a specific physical meaning. In particular, x_L denotes the axis parallel to the fibers, and x_T denotes the axis perpendicular to (transverse) to the fiber direction. Under the notation the elastic constants for a unidirectional material take on the following nomenclature:

E_L = Young's modulus parallel to the fibers.
E_T = Young's modulus transverse to the fibers.
ν_{LT} = Major Poisson's ratio as measured from the transverse contraction under uniaxial tension parallel to the fibers.
G_{LT} = Shear modulus relative to the $x_L - x_T$ plane.

Laminate stacking sequences can be easily described for composites composed of layers of the same material with equal ply thickness by simply listing the ply orientations from the top of the laminate to the bottom. Thus, the notation [0°/90°/0°] uniquely defines a three-layer laminate. The angle denotes the orientation of the principal material axis, x_L, within each ply. If a ply were repeated, a subscript would be utilized to denote the number of repeating plies. Thus, [0°/90$_3$/0°] indicates that the 90° ply is repeated three times.

Any laminate in which the ply stacking sequence below the midplane is a mirror image of the stacking sequence above the midplane is referred to as a symmetric laminate. For a symmetric laminate, such as a [0°/90$_2$/0°] plate, the notation can be abbreviated by using [0°/90°]$_s$, where the subscript s denotes that the stacking sequence is repeated symmetrically. Angle-ply laminates are denoted by $[0°/45°/-45°]_s$ which can be abbreviated to $[0°/\pm45°]_s$. For laminates with repeating sets of plies, e.g. $[0°/\pm45°/0°/\pm45°]_s$, an abbreviated notation is of the form $[0°/\pm45°]_{2s}$.

If a symmetric laminate contains a layer which is split at the centerline, then a bar is utilized to denote the split. Thus the laminate $[0°/90°/0°]$ can be abbreviated by $[0°/\overline{90}°]_s$.

REFERENCES

1. Lekhnitskii, S. G., *Theory of Elasticity of an Anisotropic Elastic Body*, Holden-Day (1963).
2. Hearmon, R. F. S., *An Introduction to Applied Anisotropic Elasticity*, Oxford University Press (1961).
3. Fung, Y. C., *Foundations of Solid Mechanics*, Prentice-Hall, Inc. (1965).
4. Jones, R. M., *Mechanics of Composite Materials*, McGraw-Hill Book Co. (1975).
5. Halpin, J. C., *Primer on Composite Materials: Analysis*, Second Edition, Technomic Publishing Co. (1984).

CHAPTER 2

Equations of a Laminated Anisotropic Plate

2.1 BASIC ASSUMPTIONS

IN MOST PRACTICAL applications of thin plates the magnitude of the stresses acting on the surface parallel to the middle plane are small compared to the bending and membrane stresses. Since the plate is thin, this implies that the tractions on any surface parallel to the midplane are relatively small. In particular, an approximate state of plane stress exists.

A standard, x, y, z coordinate system, as shown in Figure 2.1, is used in deriving the equations. The displacements in the x, y, z directions are denoted by u, v, w, respectively. The following basic assumptions are made:

1. The plate is constructed of an arbitrary number of layers of orthotropic sheets bonded together. However, the orthotropic axes of material symmetry of an individual layer need not coincide with the x-y axes of the plate.
2. The plate is thin, i.e., the thickness h is much smaller than the other physical dimensions.
3. The displacements u, v, and w are small compared to the plate thickness.
4. In-plane strains ϵ_x, ϵ_y, and ϵ_{xy} are small compared to unity.
5. In order to include inplane force effects, nonlinear terms in the equations of motion involving products of stresses and plate slopes are retained. All other nonlinear terms are neglected.
6. Transverse shear strains ϵ_{xz} and ϵ_{yz} are negligible.
7. Tangential displacements u and v are linear functions of the z coordinate.
8. The transverse normal strain ϵ_z is negligible.
9. Each ply obeys Hooke's law.
10. The plate has constant thickness.
11. Rotatory inertia terms are negligible.
12. There are no body forces.
13. Transverse shear stresses σ_{xz} and σ_{yz} vanish on the surfaces $z = \pm\, h/2$.

It should be noted that assumption 6 is a direct consequence of plane stress. Together 6 and 7 constitute the classical assumptions of Kirchhoff. Assumption

17

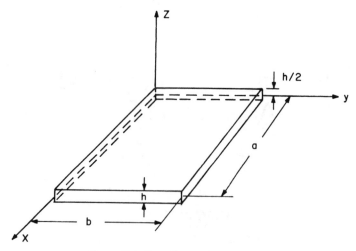

Figure 2.1. *Coordinate system of plate.*

8 allows the problem to be reduced to a two dimensional study of the middle plane.

2.2 STRAIN-DISPLACEMENT RELATIONS

Using assumption 7 the tangential displacements are of the form

$$u = u^o(x,y,t) + zF_1(x,y,t)$$

$$v = v^o(x,y,t) + zF_2(x,y,t)$$

(2.1)

where u^o and v^o are tangential displacements of the middle-plane. Substituting (2.1) into the strain-displacement relations, (1.10), and applying assumption 6 yields

$$\epsilon_{xz} = F_1(x,y,t) + \frac{\partial w}{\partial x} = 0$$

$$\epsilon_{yz} = F_2(x,y,t) + \frac{\partial w}{\partial y} = 0$$

(2.2)

Therefore,

$$F_1(x,y,t) = -\frac{\partial w}{\partial x} \quad F_2(x,y,t) = -\frac{\partial w}{\partial y}$$

(2.3)

Assumption 8 implies

$$w = w(x,y,t) \tag{2.4}$$

Thus the strain-displacement relations are of the form

$$\begin{aligned}
\epsilon_x &= \epsilon_x^o + z\varkappa_x \\
\epsilon_y &= \epsilon_y^o + z\varkappa_y \\
\epsilon_{xy} &= \epsilon_{xy}^o + z\varkappa_{xy}
\end{aligned} \tag{2.5}$$

where

$$\epsilon_x^o = \frac{\partial u^o}{\partial x} \qquad \epsilon_y^o = \frac{\partial u^o}{\partial y} \qquad \epsilon_{xy}^o = \frac{\partial u^o}{\partial y} + \frac{\partial v^o}{\partial x} \tag{2.6}$$

$$\varkappa_x = -\frac{\partial^2 w}{\partial x^2} \qquad \varkappa_y = -\frac{\partial^2 w}{\partial y^2} \qquad \varkappa_{xy} = -2\frac{\partial^2 w}{\partial x \partial y} \tag{2.7}$$

Equations (2.5–2.7) coincide with those of classical homogeneous plate theory [1].

2.3 EQUATIONS OF MOTION

Assumption 6 cannot be satisfied unless the resultant shear vanishes. Obviously, this is not physically correct. This apparent inconsistency in classical plate theory is recognized and accepted [1], and resultant shears are considered when developing the equations of motion.

Using the equations of motion (1.12, 1.13, 1.14) and retaining only those nonlinear terms which are consistent with assumption 5 yields the following relations for the kth layer of the laminate.

$$\frac{\partial \sigma_x^{(k)}}{\partial x} + \frac{\partial \sigma_{xy}^{(k)}}{\partial y} + \frac{\partial \sigma_{xz}^{(k)}}{\partial z} = \varrho_o^{(k)}\frac{\partial^2 u}{\partial t^2}$$

$$\frac{\partial \sigma_{xy}^{(k)}}{\partial x} + \frac{\partial \sigma_y^{(k)}}{\partial y} + \frac{\partial \sigma_{yz}^{(k)}}{\partial z} = \varrho_o^{(k)}\frac{\partial^2 v}{\partial t^2} \tag{2.8}$$

$$\frac{\partial}{\partial x}\left(\sigma_{xz}^{(k)} + \sigma_x^{(k)}\frac{\partial w}{\partial x} + \sigma_{xy}^{(k)}\frac{\partial w}{\partial y} \right) + \frac{\partial}{\partial y}\left(\sigma_{yz}^{(k)} + \sigma_{xy}^{(k)}\frac{\partial w}{\partial x} + \sigma_y^{(k)}\frac{\partial w}{\partial y} \right)$$

$$+ \frac{\partial}{\partial z}\left(\sigma_z^{(k)} + \sigma_{xz}^{(k)}\frac{\partial w}{\partial x} + \sigma_{yz}^{(k)}\frac{\partial w}{\partial y} \right) = \varrho_o^{(k)}\frac{\partial^2 w}{\partial t^2}$$

Stress and moment resultants are defined as follows:

$$(N_x, N_y, N_{xy}) = \int_{-h/2}^{h/2} (\sigma_x^{(k)}, \sigma_y^{(k)}, \sigma_{xy}^{(k)})\,dz$$

$$(Q_x, Q_y) = \int_{-h/2}^{h/2} (\sigma_{xz}^{(k)}, \sigma_{yz}^{(k)})\,dz \qquad (2.9)$$

$$(M_x, M_y, M_{xy}) = \int_{-h/2}^{h/2} (\sigma_x^{(k)}, \sigma_y^{(k)}, \sigma_{xy}^{(k)})z\,dz$$

These resultants are illustrated in Figure 2.2 and 2.3.

We now proceed by integrating the first of Equation (2.8) with respect to z:

$$\int_{-h/2}^{h/2} \frac{\partial \sigma_x^{k)}}{\partial x}\,dz + \int_{-h/2}^{h/2} \frac{\partial \sigma_{xy}^{(k)}}{\partial y}\,dz = \int_{-h/2}^{h/2} \varrho_o^{(k)} \frac{\partial^2 u}{\partial t^2}\,dz \qquad (2.10)$$

Interchanging the order of differentiation and integration, and using the definition of stress resultants along with Equations (2.1) and (2.3) yields

$$\frac{\partial N_x}{\partial x} + \frac{\partial N_{xy}}{\partial y} = \varrho\,\frac{\partial^2 u^o}{\partial t^2} \qquad (2.11)$$

where

$$\varrho = \int_{-h/2}^{h/2} \varrho_o^{(k)}\,dz \qquad (2.12)$$

It should be noted that the rotatory inertia term

$$-\frac{\partial^3 w}{\partial x \partial t^2} \int_{-h/2}^{h/2} \varrho_o^{(k)} z\,dz$$

is neglected in accordance with assumption 11.

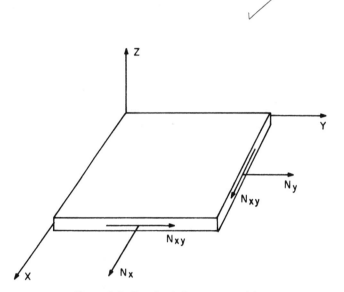

Figure 2.2. *Resultant stress nomenclature.*

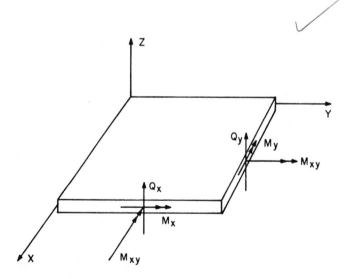

Figure 2.3. *Nomenclature for moment and transverse shear resultant.*

In a similar manner the second of Equations (2.8) integrated with respect to z becomes

$$\frac{\partial N_{xy}}{\partial x} + \frac{\partial N_y}{\partial y} = \varrho \frac{\partial^2 v^o}{\partial t^2} \tag{2.13}$$

Integrating the third of Equations (2.8) with respect to z and again interchanging the order of differentiation and integration yields

$$N_x \frac{\partial^2 w}{\partial x^2} + 2N_{xy} \frac{\partial^2 w}{\partial x \partial y} + N_y \frac{\partial^2 w}{\partial y^2} + \frac{\partial Q_x}{\partial x} + \frac{\partial Q_y}{\partial y}$$

$$+ \frac{\partial w}{\partial x}\left(\frac{\partial N_x}{\partial x} + \frac{\partial N_{xy}}{\partial y}\right) + \frac{\partial w}{\partial y}\left(\frac{\partial N_{xy}}{\partial x} + \frac{\partial N_y}{\partial y}\right) \tag{2.14}$$

$$+ q = \varrho \frac{\partial^2 w}{\partial t^2}$$

where

$$q = \sigma_z^{(k)}\left(\frac{h}{2}\right) - \sigma_z^{(k)}\left(-\frac{h}{2}\right) \tag{2.15}$$

A cursory examination of Equations (2.11) and (2.13) reveals that the last two expressions on the left side of (2.14) vanish for the static case and are second order in accordance with assumption 5 for the dynamic case. Thus (2.14) becomes

$$N_x \frac{\partial^2 w}{\partial x^2} + 2N_{xy} \frac{\partial^2 w}{\partial x \partial y} + N_y \frac{\partial^2 w}{\partial y^2}$$

$$+ \frac{\partial Q_x}{\partial x} + \frac{\partial Q_y}{\partial y} + q = \varrho \frac{\partial^2 w}{\partial t^2} \tag{2.16}$$

Now multiplying the first of Equations (2.8) by z and integrating with respect to z over the plate thickness yields

$$\frac{\partial M_x}{\partial x} + \frac{\partial M_{xy}}{\partial y} + \int_{-h/2}^{h/2} \frac{\partial \sigma_{xz}^{(k)}}{\partial z} z\,dz = \int_{-h/2}^{h/2} \varrho_o^{(k)} \frac{\partial^2 u}{\partial t^2} z\,dz \tag{2.17}$$

We note that

$$z \frac{\partial \sigma_{xz}^{(k)}}{\partial z} = \frac{\partial [z \sigma_{xz}^{(k)}]}{\partial z} - \sigma_{xz}^{(k)} \tag{2.18}$$

Taking (2.9) into account along with assumptions 11 and 13, Equation (2.17) becomes

$$\frac{\partial M_x}{\partial x} + \frac{\partial M_{xy}}{\partial y} - Q_x = 0 \tag{2.19}$$

Dynamic terms in (2.19) are neglected, as they lead to rotatory inertia terms in later operations. A similar operation with the second of Equations (2.8) yields

$$\frac{\partial M_{xy}}{\partial x} + \frac{\partial M_y}{\partial y} - Q_y = 0 \tag{2.20}$$

Differentiating Equations (2.19) and (2.20) with respect to x and y, respectively, yields

$$\frac{\partial Q_x}{\partial x} = \frac{\partial^2 M_x}{\partial x^2} + \frac{\partial^2 M_{xy}}{\partial x \partial y}$$

$$\frac{\partial Q_y}{\partial y} = \frac{\partial^2 M_{xy}}{\partial x \partial y} + \frac{\partial^2 M_y}{\partial y^2} \tag{2.21}$$

Putting (2.21) into (2.16) yields

$$\frac{\partial^2 M_x}{\partial x^2} + 2 \frac{\partial^2 M_{xy}}{\partial x \partial y} + \frac{\partial^2 M_y}{\partial y^2} + N_x \frac{\partial^2 w}{\partial x^2}$$

$$+ 2N_{xy} \frac{\partial^2 w}{\partial x \partial y} + N_y \frac{\partial^2 w}{\partial y^2} + q = \varrho \frac{\partial^2 w}{\partial t^2} \tag{2.22}$$

Equation (2.11), (2.13), and (2.22) constitute the equations of motion and are identical to those of classical homogeneous plate theory [1].

2.4 CONSTITUTIVE EQUATIONS

Assuming an approximate state of plane stress, the transverse normal strain ϵ_z

can be calculated in terms of plate stiffnesses through Hooke's law (1.24):

$$\epsilon_z = -\frac{c_{13}}{c_{33}}\epsilon_x - \frac{c_{23}}{c_{33}}\epsilon_y - \frac{c_{36}}{c_{33}}\epsilon_{xy} \qquad (2.23)$$

Using (2.23), the plane stress constitutive equation for the kth layer becomes

$$\begin{bmatrix} \sigma_x^{(k)} \\ \sigma_y^{(k)} \\ \sigma_{xy}^{(k)} \end{bmatrix} = \begin{bmatrix} Q_{11}^{(k)} & Q_{12}^{(k)} & Q_{16}^{(k)} \\ Q_{12}^{(k)} & Q_{22}^{(k)} & Q_{26}^{(k)} \\ Q_{16}^{(k)} & Q_{26}^{(k)} & Q_{66}^{(k)} \end{bmatrix} \begin{bmatrix} \epsilon_x \\ \epsilon_y \\ \epsilon_{xy} \end{bmatrix} \qquad (2.24)$$

where the reduced stiffness terms Q_{ij} are given by

$$Q_{ij} = c_{ij} - \frac{c_{i3}c_{j3}}{c_{33}} \qquad (2.25)$$

Using (2.25) in conjunction with (2.5) and the stress and moment resultant definitions yields the following constitutive relations for the plate:

$$\begin{bmatrix} N_x \\ N_y \\ N_{xy} \\ M_x \\ M_y \\ M_{xy} \end{bmatrix} = \begin{bmatrix} A_{11} & A_{12} & A_{16} & B_{11} & B_{12} & B_{16} \\ A_{12} & A_{22} & A_{26} & B_{12} & B_{22} & B_{26} \\ A_{16} & A_{26} & A_{66} & B_{16} & B_{26} & B_{66} \\ B_{11} & B_{12} & B_{16} & D_{11} & D_{12} & D_{16} \\ B_{12} & B_{22} & B_{26} & D_{12} & D_{22} & D_{26} \\ B_{16} & B_{26} & B_{66} & D_{16} & D_{26} & D_{66} \end{bmatrix} \begin{bmatrix} \epsilon_x^0 \\ \epsilon_y^0 \\ \epsilon_{xy}^0 \\ \varkappa_x \\ \varkappa_y \\ \varkappa_{xy} \end{bmatrix} \qquad (2.26)$$

where

$$(A_{ij}, B_{ij}, D_{ij}) = \int_{-h/2}^{h/2} Q_{ij}^{(k)}(1, z, z^2)dz \qquad (2.27)$$

The most important feature of (2.26) is the coupling phenomenon which exists between stretching and bending. If $Q_{ij}^{(k)}$ is an even function of z (symmetric layup of the laminate), $B_{ij} = 0$ and coupling is eliminated. One might conjecture that coupling could be eliminated by choosing the coordinate system to be other than the middle surface of the plate. However, it can be shown [2] that in the general case coupling cannot be completely eliminated.

It should be noted that variable thickness can be accounted for by considering variable limits of integration. Thus

$$A_{ij} = A_{ij}(x,y) \quad B_{ij} = B_{ij}(x,y) \quad D_{ij} = D_{ij}(x,y) \tag{2.28}$$

Using the transformation Equations (1.35 to 1.48), the transformation of the reduced stiffness matrix Q_{ij} can be established. For example

$$Q_{11} = c_{11} - \frac{c_{13}^2}{c_{33}} = m^4 c_{11}' + n^4 c_{22}' + 2m^2 n^2 c_{12}' + 4m^2 n^2 c_{66}'$$

$$- \frac{1}{c_{33}}(m^2 c_{13}' + n^2 c_{23}')^2 = m^4\left(c_{11}' - \frac{c_{13}'^2}{c_{33}}\right) + n^4\left(c_{22}'\right.$$

$$- \frac{c_{23}'^2}{c_{33}}\right) + 4m^2 n^2\left(c_{66}' - \frac{c_{36}'^2}{c_{33}}\right) = Q_{11}' m^4 + Q_{22}' n^4$$

$$+ 2(Q_{12}' + 2Q_{66}')m^2 n^2$$

(2.29)

Similar relationships can be obtained for other components of Q_{ij} leading to the conclusion that Q_{ij} transforms the same as c_{ij}. Multiplying Equation (2.29) by $(1,z,z^2)$ and integrating with respect to z yields

$$\int_{-h/2}^{h/2} Q_{11}^{(k)}\,(1,z,z^2)dz = (A_{11},B_{11},D_{11}) = m^4 \int_{-h/2}^{h/2} Q_{11}'^{(k)}(1,z,z^2)dz$$

$$+ n^4 \int_{-h/2}^{h/2} Q_{22}'^{(k)}\,(1,z,z^2)dz + 2m^2 n^2 \int_{-h/2}^{h/2} (Q_{12}'^{(k)} + 2Q_{66}'^{(k)})(1,z,z^2)dz$$

(2.30)

$$= (A_{11}',B_{11}',D_{11}')m^4 + (A_{22}',B_{22}',D_{22}')n^4$$

$$+ 2[(A_{12}',B_{12}',D_{12}') + 2(A_{66}',B_{66}',D_{66}')]m^2 n^2$$

Thus A_{ij}, B_{ij}, and D_{ij} transform as Q_{ij}.

It is sometimes useful to express stiffness properties in terms of invariants. The following invariants which are pertinent to laminated plate analysis have been previously established [3]:

$$L_1 = (Q_{11} + Q_{22} + 2Q_{12})$$

$$L_2 = (Q_{66} - Q_{12})$$

$$P_1 = A_{11} + A_{22} + 2A_{12} = hL_1$$

$$P_2 = A_{66} - A_{12} = hL_2$$

$$P_3 = B_{11} + B_{22} + 2B_{12} = 0 \qquad (2.31)$$

$$P_4 = B_{66} - B_{12} = 0$$

$$P_5 = D_{11} + D_{22} + 2D_{12} = L_1 \frac{h^3}{12}$$

$$P_6 = D_{66} - D_{12} = L_2 \frac{h^3}{12}$$

2.5 EQUATIONS OF MOTION IN TERMS OF DISPLACEMENTS

In the following derivation the effect of inplane forces on bending is neglected. In particular, it is assumed that large membrane forces are not applied externally to the plate. It will be shown in Chapter 7 that the inplane forces generated by coupling are not large enough to invalidate linear theory.

Substituting the constitutive relations (2.26) into the equations of motion (2.11), (2.13), (2.22), and using the strain-displacement relations (2.6) along with the curvature-displacement relations (2.7) yields the following equations of motion:

$$A_{11} \frac{\partial^2 u^o}{\partial x^2} + 2A_{16} \frac{\partial^2 u^o}{\partial x \partial y} + A_{66} \frac{\partial^2 u^o}{\partial y^2} + A_{16} \frac{\partial^2 v^o}{\partial x^2}$$

$$+ (A_{12} + A_{66}) \frac{\partial^2 v^o}{\partial x \partial y} + A_{26} \frac{\partial^2 v^o}{\partial y^2} - B_{11} \frac{\partial^3 w}{\partial x^3} \qquad (2.32)$$

$$- 3B_{16} \frac{\partial^3 w}{\partial x^2 \partial y} - (B_{12} + 2B_{66}) \frac{\partial^3 w}{\partial x \partial y^2} - B_{26} \frac{\partial^3 w}{\partial y^3} = \varrho \frac{\partial^2 u^o}{\partial t^2}$$

$$A_{16}\frac{\partial^2 u^o}{\partial x^2} + (A_{12} + A_{66})\,\frac{\partial^2 u^o}{\partial x \partial y} + A_{26}\,\frac{\partial^2 u^o}{\partial y^2} + A_{66}\,\frac{\partial^2 v^o}{\partial x^2}$$

$$+ 2A_{26}\,\frac{\partial^2 v^o}{\partial x \partial y} + A_{22}\,\frac{\partial^2 v^o}{\partial y^2} - B_{16}\,\frac{\partial^3 w}{\partial x^3} - (B_{12} + 2B_{66})\,\frac{\partial^3 w}{\partial x^2 \partial y}$$

$$(2.33)$$

$$- 3B_{26}\,\frac{\partial^3 w}{\partial x \partial y^2} - B_{22}\,\frac{\partial^3 w}{\partial y^3} = \varrho\,\frac{\partial^2 v^o}{\partial t^2}$$

$$D_{11}\frac{\partial^4 w}{\partial x^4} + 4D_{16}\,\frac{\partial^4 w}{\partial x^3 \partial y} + 2(D_{12} + 2D_{66})\,\frac{\partial^4 w}{\partial x^2 \partial y^2}$$

$$+ 4D_{26}\frac{\partial^4 w}{\partial x \partial y^3} + D_{22}\,\frac{\partial^4 w}{\partial y^4}$$

$$- B_{11}\,\frac{\partial^3 u^o}{\partial x^3} - 3B_{16}\,\frac{\partial^3 u^o}{\partial x^2 \partial y} - (B_{12} + 2B_{66})\,\frac{\partial^3 u^o}{\partial x \partial y^2}$$

$$(2.34)$$

$$- B_{26}\,\frac{\partial^3 u^o}{\partial y^3} - B_{16}\,\frac{\partial^3 v^o}{\partial x^3} - (B_{12} + 2B_{66})\,\frac{\partial^3 v^o}{\partial x^2 \partial y}$$

$$- 3B_{26}\,\frac{\partial^3 v^o}{\partial x \partial y^2} - B_{22}\,\frac{\partial^3 v^o}{\partial y^3} + \varrho\,\frac{\partial^2 w}{\partial t^2} = q$$

For symmetric laminates $B_{ij} = 0$ and Equations (2.32–2.34) reduce to the following:

$$A_{11}\frac{\partial^2 u^o}{\partial x^2} + 2A_{16}\,\frac{\partial^2 u^o}{\partial x \partial y} + A_{66}\,\frac{\partial^2 u^o}{\partial y^2} + A_{16}\,\frac{\partial^2 v^o}{\partial x^2}$$

$$(2.35)$$

$$+ (A_{12} + A_{66})\,\frac{\partial^2 v^o}{\partial x \partial y} + A_{26}\,\frac{\partial^2 v^o}{\partial y^2} = \varrho\,\frac{\partial^2 u^o}{\partial t^2}$$

$$A_{16}\frac{\partial^2 u^o}{\partial x^2} + (A_{12} + A_{66})\,\frac{\partial^2 u^o}{\partial x \partial y} + A_{26}\,\frac{\partial^2 u^o}{\partial y^2} + A_{66}\,\frac{\partial^2 v^o}{\partial x^2}$$

$$(2.36)$$

$$+ 2A_{26}\,\frac{\partial^2 v^o}{\partial x \partial y} + A_{22}\,\frac{\partial^2 v^o}{\partial y^2} = \varrho\,\frac{\partial^2 v^o}{\partial t^2}$$

$$D_{11} \frac{\partial^4 w}{\partial x^4} + 4D_{16} \frac{\partial^4 w}{\partial x^3 \partial y} + 2(D_{12} + 2D_{66}) \frac{\partial^4 w}{\partial x^2 \partial y^2}$$

$$+ 4D_{26} \frac{\partial^4 w}{\partial x \partial y^3} + D_{22} \frac{\partial^4 w}{\partial y^4} + \varrho \frac{\partial^2 w}{\partial t^2} = q \qquad (2.37)$$

Equations (2.35) and (2.36) are those of an anisotropic continuum in a state of plane stress [4], and Equation (2.37) coincides with the governing equations of a homogeneous anisotropic plate [5]. It should be noted, however, that this does not imply that the laminate can be considered homogeneous. The plate is homogeneous only if

$$D_{ij} = A_{ij} \frac{h^2}{12} \qquad (2.38)$$

If A_{ij} is orthotropic ($A_{16} = A_{26} = 0$) then Equations (2.35) and (2.36) become

$$A_{11} \frac{\partial^2 u^o}{\partial x^2} + A_{66} \frac{\partial^2 u^o}{\partial y^2} + (A_{12} + A_{66}) \frac{\partial^2 v^o}{\partial x \partial y} = \varrho \frac{\partial^2 u^o}{\partial t^2} \qquad (2.39)$$

$$(A_{12} + A_{66}) \frac{\partial^2 u^o}{\partial x \partial y} + A_{66} \frac{\partial^2 v^o}{\partial x^2} + A_{22} \frac{\partial^2 v^o}{\partial y^2} = \varrho \frac{\partial^2 v^o}{\partial t^2} \qquad (2.40)$$

For laminates composed of isotropic layers

$$A_{11} = A_{22} = A = \frac{Eh}{(1 - v^2)}$$

$$A_{12} = vA \qquad (2.41)$$

$$A_{66} = \frac{(1 - v)}{2} A$$

and the inplane equations become

$$\left. \begin{array}{c} \dfrac{\partial^2 u^o}{\partial x^2} + \dfrac{(1 - v)}{2} \dfrac{\partial^2 u^o}{\partial y^2} + (1 + v) \dfrac{\partial^2 v^o}{\partial x \partial y} = \dfrac{\varrho}{A} \dfrac{\partial^2 u^o}{\partial t^2} \\[4mm] (1 + v) \dfrac{\partial^2 u^o}{\partial x \partial y} + \dfrac{(1 - v)}{2} \dfrac{\partial^2 v^o}{\partial x^2} + \dfrac{\partial^2 v^o}{\partial y^2} = \dfrac{\varrho}{A} \dfrac{\partial^2 v^o}{\partial t^2} \end{array} \right\} \qquad (2.42)$$

For the case in which D_{ij} is orthotropic ($D_{16} = D_{26} = 0$) Equation (2.37) becomes

$$D_{11} \frac{\partial^2 w}{\partial x^4} + 2(D_{12} + 2D_{66}) \frac{\partial^4 w}{\partial x^2 \partial y^2} + D_{22} \frac{\partial^4 w}{\partial y^4} + \varrho \frac{\partial^2 w}{\partial t^2} = q \quad (2.43)$$

For isotropic layers

$$D_{11} = D_{22} = D = \frac{Eh^3}{12(1 - v^2)} \quad D_{12} = vD \quad D_{66} = \frac{(1 - v)}{2} D \quad (2.44)$$

and (2.43) becomes

$$D\nabla^4 w + \varrho \frac{\partial^2 w}{\partial t^2} = q \quad (2.45)$$

The force and moment resultants are obtained from the constitutive relations (2.26):

$$N_x = A_{11} \frac{\partial u^0}{\partial x} + A_{16} \left(\frac{\partial u^0}{\partial y} + \frac{\partial v^0}{\partial x} \right) + A_{12} \frac{\partial v^0}{\partial y} - B_{11} \frac{\partial^2 w}{\partial x^2}$$
$$- 2B_{16} \frac{\partial^2 w}{\partial x \partial y} - B_{12} \frac{\partial^2 w}{\partial y^2} \quad (2.46)$$

$$N_y = A_{12} \frac{\partial u^0}{\partial x} + A_{26} \left(\frac{\partial u^0}{\partial y} + \frac{\partial v^0}{\partial x} \right) + A_{22} \frac{\partial v^0}{\partial y} - B_{12} \frac{\partial^2 w}{\partial x^2}$$
$$- 2B_{26} \frac{\partial^2 w}{\partial x \partial y} - B_{22} \frac{\partial^2 w}{\partial y^2} \quad (2.47)$$

$$N_{xy} = A_{16} \frac{\partial u^0}{\partial x} + A_{66} \left(\frac{\partial u^0}{\partial y} + \frac{\partial v^0}{\partial x} \right) + A_{26} \frac{\partial v^0}{\partial y} - B_{16} \frac{\partial^2 w}{\partial x^2}$$
$$- 2B_{66} \frac{\partial^2 w}{\partial x \partial y} - B_{26} \frac{\partial^2 w}{\partial y^2} \quad (2.48)$$

$$M_x = B_{11}\frac{\partial u^0}{\partial x} + B_{16}\left(\frac{\partial u^0}{\partial y} + \frac{\partial v^0}{\partial x}\right) + B_{12}\frac{\partial v^0}{\partial y} - D_{11}\frac{\partial^2 w}{\partial x^2}$$

$$(2.49)$$

$$- 2D_{16}\frac{\partial^2 w}{\partial x \partial y} - D_{12}\frac{\partial^2 w}{\partial y^2}$$

$$M_y = B_{12}\frac{\partial u^0}{\partial x} + B_{26}\left(\frac{\partial u^0}{\partial y} + \frac{\partial v^0}{\partial x}\right) + B_{22}\frac{\partial v^0}{\partial y} - D_{12}\frac{\partial^2 w}{\partial x^2}$$

$$(2.50)$$

$$- 2D_{26}\frac{\partial^2 w}{\partial x \partial y} - D_{22}\frac{\partial^2 w}{\partial y^2}$$

$$M_{xy} = B_{16}\frac{\partial u^0}{\partial x} + B_{66}\left(\frac{\partial u^0}{\partial y} + \frac{\partial v^0}{\partial x}\right) + B_{26}\frac{\partial v^0}{\partial y} - D_{16}\frac{\partial^2 w}{\partial x^2}$$

$$(2.51)$$

$$- 2D_{66}\frac{\partial^2 w}{\partial x \partial y} - D_{26}\frac{\partial^2 w}{\partial y^2}$$

From Equations (2.19), (2.20), and (2.49–2.51) the following shear resultants are obtained

$$Q_x = B_{11}\frac{\partial^2 u^0}{\partial x^2} + 2B_{16}\frac{\partial^2 u^0}{\partial x \partial y} + B_{66}\frac{\partial^2 u^0}{\partial y^2} + B_{16}\frac{\partial^2 v^0}{\partial x^2}$$

$$+ (B_{12} + B_{66})\frac{\partial^2 v^0}{\partial x \partial y} + B_{26}\frac{\partial^2 v^0}{\partial y^2} - D_{11}\frac{\partial^3 w}{\partial x^3} \qquad (2.52)$$

$$- 3D_{16}\frac{\partial^3 w}{\partial x^2 \partial y} - (D_{12} + 2D_{66})\frac{\partial^3 w}{\partial x \partial y^2} - D_{26}\frac{\partial^3 w}{\partial y^3}$$

$$Q_y = B_{16}\frac{\partial^2 u^0}{\partial x^2} + (B_{12} + B_{66})\frac{\partial^2 u^0}{\partial x \partial y} + B_{26}\frac{\partial^2 u^0}{\partial y^2} + B_{66}\frac{\partial^2 v^0}{\partial x^x}$$

$$+ 2B_{26}\frac{\partial^2 v^0}{\partial x \partial y} + B_{22}\frac{\partial^2 v^0}{\partial y^2} - D_{16}\frac{\partial^3 w}{\partial x^3} \qquad (2.53)$$

$$- (D_{12} + 2D_{66})\frac{\partial^3 w}{\partial x^2 \partial y} - 3D_{26}\frac{\partial^3 w}{\partial x \partial y^2} - D_{22}\frac{\partial^3 w}{\partial y^3}$$

Using the constitutive Equation (2.24) the following stresses are obtained within each layer:

$$\sigma_x^{(k)} = Q_{11}^{(k)} \frac{\partial u^o}{\partial x} + Q_{16}^{(k)} \left(\frac{\partial u^o}{\partial y} + \frac{\partial v^o}{\partial x} \right) + Q_{12}^{(k)} \frac{\partial v^o}{\partial y}$$

$$- z \left(Q_{11}^{(k)} \frac{\partial^2 w}{\partial x^2} + 2Q_{16}^{(k)} \frac{\partial^2 w}{\partial x \partial y} + Q_{12}^{(k)} \frac{\partial^2 w}{\partial y^2} \right) \tag{2.54}$$

$$\sigma_y^{(k)} = Q_{12}^{(k)} \frac{\partial u^o}{\partial x} + Q_{26}^{(k)} \left(\frac{\partial u^o}{\partial y} + \frac{\partial v^o}{\partial x} \right) + Q_{22}^{(k)} \frac{\partial v^o}{\partial y}$$

$$- z \left(Q_{12}^{(k)} \frac{\partial^2 w}{\partial x^2} + 2Q_{26}^{(k)} \frac{\partial^2 w}{\partial x \partial y} + Q_{22}^{(k)} \frac{\partial^2 w}{\partial y^2} \right) \tag{2.55}$$

$$\sigma_{xy}^{(k)} = Q_{16}^{(k)} \frac{\partial u^o}{\partial x} + Q_{66}^{(k)} \left(\frac{\partial u^o}{\partial y} + \frac{\partial v^o}{\partial x} \right) + Q_{26}^{(k)} \frac{\partial v^o}{\partial y}$$

$$- z \left(Q_{16}^{(k)} \frac{\partial^2 w}{\partial x^2} + 2Q_{66}^{(k)} \frac{\partial^2 w}{\partial x \partial y} + Q_{26}^{(k)} \frac{\partial^2 w}{\partial y^2} \right) \tag{2.56}$$

The interlaminar shear stresses are determined from the first two equilibrium Equations in (2.8). Integration of these equations with respect to z, after taking into account (2.54–2.56), yields

$$\sigma_{xz}^{(k)} = \frac{z^2}{2} \left[Q_{11}^{(k)} \frac{\partial^3 w}{\partial x^3} + 3Q_{16}^{(k)} \frac{\partial^3 w}{\partial x^2 \partial y} + (Q_{12}^{(k)} + 2Q_{66}^{(k)}) \frac{\partial^3 w}{\partial x \partial y^2} \right.$$

$$+ Q_{26}^{(k)} \frac{\partial^3 w}{\partial y^3} \right] - z \left[Q_{11}^{(k)} \frac{\partial^2 u^o}{\partial x^2} \right.$$

$$\tag{2.57}$$

$$+ 2Q_{16}^{(k)} \frac{\partial^2 u^o}{\partial x \partial y} + Q_{66}^{(k)} \frac{\partial^2 u^o}{\partial y^2} + Q_{16}^{(k)} \frac{\partial^2 v^o}{\partial x^2}$$

$$\left. + (Q_{12}^{(k)} + Q_{66}^{(k)}) \frac{\partial^2 v^o}{\partial x \partial y} + Q_{26}^{(k)} \frac{\partial^2 v^o}{\partial y^2} \right] + f^{(k)}(x,y)$$

$$
\sigma_{yz}^{(k)} = \frac{z^2}{2} \left[Q_{16}^{(k)} \frac{\partial^3 w}{\partial x^3} + (Q_{12}^{(k)} + 2Q_{66}^{(k)}) \frac{\partial^3 w}{\partial x^2 \partial y} + 3Q_{26}^{(k)} \frac{\partial^3 w}{\partial x \partial y^2} \right.
$$

$$
\left. + Q_{22}^{(k)} \frac{\partial^3 w}{\partial y^3} \right] - z \left[Q_{16}^{(k)} \frac{\partial^2 u^o}{\partial x^2} + (Q_{12}^{(k)} + Q_{66}^{(k)}) \frac{\partial^2 u^o}{\partial x \partial y} \right.
$$

$$
+ Q_{26}^{(k)} \frac{\partial^2 u^o}{\partial y^2} + Q_{66}^{(k)} \frac{\partial^2 v^o}{\partial x^2} + 2Q_{26}^{(k)} \frac{\partial^2 v^o}{\partial x \partial y}
$$

$$
\left. + Q_{22}^{(k)} \frac{\partial^2 v^o}{\partial y^2} \right] + g^{(k)}(x,y)
$$

(2.58)

where $f^{k)}$ and $g^{k)}$ are functions of integration which are determined by continuity conditions between layers and by the vanishing of the shear tractions at the top and bottom surfaces of the plate.

2.6 GOVERNING EQUATIONS IN TERMS OF A STRESS FUNCTION AND TRANSVERSE DISPLACEMENT

For certain static problems, and dynamic problems in which the inplane inertia terms can be neglected, a stress function formulation of the inplane problem often proves useful [6,7]. We now define a stress function Φ such that

$$
N_x = \frac{\partial^2 \Phi}{\partial y^2} \quad N_y = \frac{\partial^2 \Phi}{\partial x^2} \quad N_{xy} = - \frac{\partial^2 \Phi}{\partial x \partial y}
$$

(2.59)

Equations (2.11) and (2.13) are exactly satisfied by (2.59).

Equation (2.26) can be written in the abbreviated form

$$
\left(\frac{N}{M} \right) = \left(\begin{array}{c|c} A & B \\ \hline B & D \end{array} \right) \left(\frac{\epsilon^o}{\varkappa} \right)
$$

(2.60)

or using matrix equations

$$
N = A\epsilon^o + B\varkappa
$$

$$
M = B\epsilon^o + D\varkappa
$$

(2.61)

Multiplying the first equation in (2.61) by A^{-1} (A inverse) yields

$$\epsilon^0 = A^{-1}N - A^{-1}B\varkappa \tag{2.62}$$

Putting (2.62) into the second of (2.61) yields

$$M = BA^{-1}N + (D - BA^{-1}B)\varkappa \tag{2.63}$$

Now (2.60) can be written in the semi-inverted form

$$\begin{pmatrix} \epsilon^0 \\ M \end{pmatrix} = \left(-\frac{A^*}{(-B^*)^T} \middle| \frac{B^*}{D^*} \right) \begin{pmatrix} N \\ \varkappa \end{pmatrix} \tag{2.64}$$

where the superscript T denotes a transpose matrix, and

$$A^* = A^{-1}, \quad B^* = -A^{-1}B, \quad D^* = D - BA^{-1}B$$

In the general case A^* and D^* are symmetric while B^* is not.
    Putting (2.64) into (2.22) and again neglecting inplane force effects yields

$$D_{11}^* \frac{\partial^4 w}{\partial x^4} + 4D_{16}^* \frac{\partial^4 w}{\partial x^3 \partial y} + 2(D_{12}^* + 2D_{66}^*) \frac{\partial^4 w}{\partial x^2 \partial y^2} + 4D_{26}^* \frac{\partial^4 w}{\partial x \partial y^3}$$

$$+ D_{22}^* \frac{\partial^4 w}{\partial y^4} + B_{21}^* \frac{\partial^4 \Phi}{\partial x^4} + (2B_{26}^* - B_{61}^*) \frac{\partial^4 \Phi}{\partial x^3 \partial y} \tag{2.65}$$

$$+ (B_{11}^* + B_{22}^* - 2B_{66}^*) \frac{\partial^4 \Phi}{\partial x^2 \partial y^2}$$

$$+ (2B_{16}^* - B_{62}^*) \frac{\partial^4 \Phi}{\partial x \partial y^3} + B_{12}^* \frac{\partial^4 \Phi}{\partial y^4} + \varrho \frac{\partial^2 w}{\partial t^2} = q$$

Equation (2.65) involves two unknowns; thus, a second relationship is necessary. A cursory examination of the compatibility Equations (1.16 to 1.21) reveals that they are exactly satisfied by the strain-displacement Equations (2.5) with the exception of the relationship

$$\frac{\partial^2 \epsilon_x^0}{\partial y^2} + \frac{\partial^2 \epsilon_y^0}{\partial x^2} - \frac{\partial^2 \epsilon_{xy}^0}{\partial x \partial y} = 0 \tag{2.66}$$

Substituting (2.64) into ,2.66) and taking into account (2.59) and (2.7) leads to the compatibility equation

$$
A_{22}^* \frac{\partial^4 \Phi}{\partial x^4} - 2A_{26}^* \frac{\partial^4 \Phi}{\partial x^3 \partial y} + (2A_{12}^* + A_{66}^*) \frac{\partial^4 \Phi}{\partial x^2 \partial y^2}
$$

$$
- 2A_{16}^* \frac{\partial^4 \Phi}{\partial x \partial y^3} + A_{11}^* \frac{\partial^4 \Phi}{\partial y^4} - B_{21}^* \frac{\partial^4 w}{x^4}
$$

$$
+ (B_{61}^* - 2B_{26}^*) \frac{\partial^4 w}{\partial x^3 \partial y} + (2B_{66}^* - B_{11}^* - B_{22}^*) \frac{\partial^4 w}{\partial x^2 \partial y^2}
$$

$$
+ (B_{62}^* - 2B_{16}^*) \frac{\partial^4 w}{\partial x \partial y^3} - B_{12}^* \frac{\partial^4 w}{\partial y^4} = 0
$$

(2.67)

For symmetric layups $B_{ij}^* = 0$, $D_{ij}^* = D_{ij}$, and Equation (2.65) reduces to (2.37), while (2.67) becomes

$$
A_{22}^* \frac{\partial^4 \Phi}{\partial x^4} - 2A_{26}^* \frac{\partial^4 \Phi}{\partial x^3 \partial y} + (2A_{12}^* + A_{66}^*) \frac{\partial^4 \Phi}{\partial x^2 \partial y^2}
$$

$$
- 2A_{16}^* \frac{\partial^4 \Phi}{\partial x \partial y^3} + A_{11}^* \frac{\partial^4 \Phi}{\partial y^4} = 0
$$

(2.68)

which is the stress function equation for an anisotropic plane stress elasticity problem. For specially orthotropic materials ($A_{16}^* = A_{26}^* = 0$), and (2.68) becomes

$$
A_{22}^* \frac{\partial^4 \Phi}{\partial x^4} + (2A_{12}^* + A_{66}^*) \frac{\partial^4 \Phi}{\partial x^2 \partial y^2} + A_{11}^* \frac{\partial^4 \Phi}{\partial y^4} = 0 \qquad (2.69)
$$

For isotropic layers

$$
A_{11}^* = A_{22}^* = \frac{1}{Eh} \quad A_{12}^* = - vA_{11}^* \quad A_{66}^* = 2(1 + v)A_{11}^* \qquad (2.70)
$$

and (2.69) reduces to the well known bi-harmonic equation

$$
\nabla^4 \Phi = 0 \qquad (2.71)
$$

The moment resultants are obtained from the constitutive relations (2.64):

$$M_x = -\left(B_{21}^* \frac{\partial^2 \Phi}{\partial x^2} - B_{61}^* \frac{\partial^2 \Phi}{\partial x \partial y} + B_{11}^* \frac{\partial^2 \Phi}{\partial y^2} + D_{11}^* \frac{\partial^2 w}{\partial x^2} \right.$$

$$\left. + 2D_{16}^* \frac{\partial^2 w}{\partial x \partial y} + D_{12}^* \frac{\partial^2 w}{\partial y^2} \right)$$

$$M_y = -\left(B_{22}^* \frac{\partial^2 \Phi}{\partial x^2} - B_{62}^* \frac{\partial^2 \Phi}{\partial x \partial y} + B_{12}^* \frac{\partial^2 \Phi}{\partial y^2} + D_{12}^* \frac{\partial^2 w}{\partial x^2} \right.$$

$$\left. + 2D_{26}^* \frac{\partial^2 w}{\partial x \partial y} + D_{22}^* \frac{\partial^2 w}{\partial y^2} \right)$$

(2.72)

$$M_{xy} = -\left(B_{26}^* \frac{\partial^2 \Phi}{\partial x^2} - B_{66}^* \frac{\partial^2 \Phi}{\partial x \partial y} + B_{16}^* \frac{\partial^2 \Phi}{\partial y^2} + D_{16}^* \frac{\partial^2 w}{\partial x^2} \right.$$

$$\left. + 2D_{66}^* \frac{\partial^2 w}{\partial x \partial y} + D_{26}^* \frac{\partial^2 w}{\partial y^2} \right)$$

From Equations (2.19), (2.20), and (2.72) the following shear results are obtained:

$$Q_x = -\left[B_{21}^* \frac{\partial^3 \Phi}{\partial x^3} + (B_{26}^* - B_{61}^*) \frac{\partial^3 \Phi}{\partial x^2 \partial y} + (B_{11}^* - B_{66}^*) \frac{\partial^3 \Phi}{\partial x \partial y^2} \right.$$

$$+ B_{16}^* \frac{\partial^3 \Phi}{\partial y^3} + D_{11}^* \frac{\partial^3 w}{\partial x^3} + 3D_{16}^* \frac{\partial^3 w}{\partial x^2 \partial y}$$

$$\left. + (D_{12}^* + 2D_{66}^*) \frac{\partial^3 w}{\partial x \partial y^2} + D_{26}^* \frac{\partial^3 w}{\partial y^3} \right]$$

(2.73)

$$Q_y = -\left[B_{26}^* \frac{\partial^3 \Phi}{\partial x^3} + (B_{22}^* - B_{66}^*) \frac{\partial^3 \Phi}{\partial x^2 \partial y} + (B_{16}^* - B_{62}^*) \frac{\partial^3 \Phi}{\partial x \partial y^2} \right.$$

$$+ B_{12}^* \frac{\partial^3 \Phi}{\partial y^3} + D_{16}^* \frac{\partial^3 w}{\partial x^3} + (D_{12}^* + 2D_{66}^*) \frac{\partial^3 w}{\partial x^2 \partial y}$$

$$\left. + 3D_{26}^* \frac{\partial^3 w}{\partial x \partial y^2} + D_{22}^* \frac{\partial^3 w}{\partial y^3} \right]$$

Integration of the first two constitutive Equations in (2.64) with respect to x and y, respectively, yield the following midplane displacements:

$$u^\circ = A_{12}^* \frac{\partial \Phi}{\partial x} + A_{11}^* \int \frac{\partial^2 \Phi}{\partial y^2} \, dx - A_{16}^* \frac{\partial \Phi}{\partial y} - B_{11}^* \frac{\partial w}{\partial x}$$

$$- B_{12}^* \int \frac{\partial^2 w}{\partial y^2} \, dx - 2B_{16}^* \frac{\partial w}{\partial y} + f(y)$$

$$v^\circ = - A_{26}^* \frac{\partial \Phi}{\partial x} + A_{22}^* \int \frac{\partial^2 \Phi}{\partial x^2} \, dy + A_{12}^* \frac{\partial \Phi}{\partial y}$$

$$- 2B_{26}^* \frac{\partial w}{\partial x} - B_{21}^* \int \frac{\partial^2 w}{\partial x^2} \, dy - B_{22}^* \frac{\partial w}{\partial y} + g(x)$$

(2.74)

where $f(y)$ and $g(x)$ are functions of integration and represent rigid body displacements. With the displacements determined, the stresses in each layer can be calculated from Equations (2.54–2.58).

2.7 STABILITY OF LAMINATED PLATES

In the two previous sections it was assumed that no large external inplane forces were applied to the plate. However, plate buckling occurs only under large inplane loads. Thus a stability analysis must include the effect of inplane forces on plate bending.

If Equation (2.22) is used in its present form, then for unsymmetrical laminates ($B_{ij} \neq 0$) the problem becomes nonlinear. Since our concept of a critical buckling load is based on a linear analysis, an altered form of (2.22) must be obtained. This can be done in a manner directly analogous to procedures in classical shell theory [8].

Consider the displacement field

$$u^\circ = u^{\circ i} + \lambda u^\circ, \quad v^\circ = v^{\circ i} + \lambda v^\circ, \quad w = w^i + \lambda w \qquad (2.75)$$

where the superscript i denotes the prebuckling displacements and λ is an infinitesimally small quantity. Thus a critical load is sought which causes an infinitesimally small shift in the equilibrium position. In classical stability theory this is referred to as the "adjacent equilibrium method."

Using (2.75) in conjunction with the constitutive relations (2.60) leads to the following matrix equations:

$$N = A\epsilon^{0i} + B\varkappa^i + \lambda(A\epsilon + B\varkappa) = N^i + \lambda N$$

$$(2.76)$$

$$M = B\epsilon^{0i} + D\varkappa^i + \lambda(B\epsilon + D\varkappa) = M^i + \lambda M$$

Putting (2.75) and (2.76) into Equations (2.22), collecting terms of like powers in λ, and neglecting second order terms in λ leads to the postbuckling equation

$$\frac{\partial^2 M_x}{\partial x^2} + 2 \frac{\partial^2 M_{xy}}{\partial y^2} + \frac{\partial^2 M_y}{\partial y^2} + N_x^i \frac{\partial^2 w}{\partial x^2} + N_x \frac{\partial^2 w^i}{\partial x^2} + 2N_{xy}^i \frac{\partial^2 w}{\partial x \partial y}$$

$$+ 2N_{xy} \frac{\partial^2 w^i}{\partial x \partial y} + N_y^i \frac{\partial^2 w}{\partial y^2} + N_y \frac{\partial^2 w^i}{\partial y^2} = \varrho \frac{\partial^2 w}{\partial t^2} - q$$

$$(2.77)$$

Since w^i, N_y^i, and N_{xy}^i are obtained from the solutions for the initial equilibrium position, Equation (2.77) is linear. However difficulty is still encountered as the nonlinear version of (2.22) is used to determine the initial configuration. A common simplification can be introduced by using linear theory to determine solutions corresponding to the initial equilibrium position. This assumption is the distinguishing feature of a linear stability analysis. Since the initial configuration is determined from linear theory, the terms in Equation (2.77) containing initial curvatures can be neglected. Thus Equation (2.77) becomes

$$\frac{\partial^2 M_x}{\partial x^2} + 2 \frac{\partial^2 M_{xy}}{\partial x \partial y} + \frac{\partial^2 M_y}{\partial y^2} + N_x^i \frac{\partial^2 w}{\partial x^2} + 2N_{xy}^i \frac{\partial^2 w}{\partial x \partial y}$$

$$+ N_y^i \frac{\partial^2 w}{\partial y^2} = \varrho \frac{\partial^2 w}{\partial t^2} - q$$

$$(2.78)$$

For the displacement formulation of the stability problem, the governing equations include (2.32) and (2.33). Taking (2.78) into account, Equation (2.34) becomes

$$D_{11} \frac{\partial^4 w}{\partial x^4} + 4D_{16} \frac{\partial^4 w}{\partial x^3 \partial y} + 2(D_{12} + 2D_{66}) \frac{\partial^4 w}{\partial x^2 \partial y^2} + 4D_{26} \frac{\partial^4 w}{\partial x \partial y^3}$$

$$+ D_{22} \frac{\partial^2 w}{\partial y^4} - B_{11} \frac{\partial^3 u^0}{\partial x^3} - 3B_{16} \frac{\partial^3 u^0}{\partial x^2 \partial y} - (B_{12} + 2B_{66}) \frac{\partial^3 u^0}{\partial x \partial y^2}$$

$$(2.79)$$

$$- B_{26} \frac{\partial^3 u^0}{\partial y^3} - B_{16} \frac{\partial^3 v^0}{\partial x^3} - (B_{12} + 2B_{66}) \frac{\partial^3 v^0}{\partial x^2 \partial y} - 3B_{26} \frac{\partial^3 v^0}{\partial x \partial y^2}$$

$$- B_{22} \frac{\partial^3 v^0}{\partial y^3} + \varrho \frac{\partial^2 w}{\partial t^2} = N_x^i \frac{\partial^2 w}{\partial x^2} + 2N_{xy}^i \frac{\partial^2 w}{\partial x \partial y} + N_y^i \frac{\partial^2 w}{\partial y^2} + q$$

For the stress function formulation the governing equations include (2.67) and (2.65) which becomes, after taking (2.78) into account

$$D_{11}^* \frac{\partial^4 w}{\partial x^4} + 4D_{16}^* \frac{\partial^4 w}{\partial x^3 \partial y} + 2(D_{12}^* + 2D_{66}^*) \frac{\partial^4 w}{\partial x^2 \partial y^2} + 4D_{26}^* \frac{\partial^4 w}{\partial x \partial y^3}$$

$$+ D_{22}^* \frac{\partial^4 w}{\partial y^4} + B_{21}^* \frac{\partial^4 \Phi}{\partial x^4} + (2B_{26}^* - B_{61}^*) \frac{\partial^4 \Phi}{\partial x^3 \partial y} + (B_{11}^* + B_{22}^* - 2B_{66}^*) \frac{\partial^4 \Phi}{\partial x^2 \partial y^2}$$

$$+ (2B_{16}^* - B_{62}^*) \frac{\partial^4 \Phi}{\partial x \partial y^3} + B_{12}^* \frac{\partial^4 \Phi}{\partial y^4} + \varrho \frac{\partial^2 w}{\partial t^2} = N_x^i \frac{\partial^2 w}{\partial x^2} + 2N_{xy}^i \frac{\partial^2 w}{\partial x \partial y}$$

$$+ N_y^i \frac{\partial^2 w}{\partial y^2} + q \tag{2.80}$$

For symmetric laminates ($B_{ij} = 0$) the inplane problem and bending problem uncouple with the result

$$u^0 = v^0 = N_x = N_y = N_{xy} = 0 \tag{2.81}$$

Thus it is not necessary to distinguish between the prebuckling and buckling equilibrium positions. For this case equations (2.79) and (2.80) become

$$D_{11} \frac{\partial^4 w}{\partial x^4} + 4D_{16} \frac{\partial^4 w}{\partial x^3 \partial y} + 2(D_{12} + 2D_{66}) \frac{\partial^4 w}{\partial x^2 \partial y^2}$$

$$+ 4D_{26} \frac{\partial^4 w}{\partial x \partial y^3} + D_{22} \frac{\partial^4 w}{\partial y^4} + \varrho \frac{\partial^2 w}{\partial t^2} = q \tag{2.82}$$

$$+ N_x \frac{\partial^2 w}{\partial x^2} + 2N_{xy} \frac{\partial^2 w}{\partial x \partial y} + N_y \frac{\partial^2 w}{\partial y^2}$$

For specially orthotropic laminates ($D_{16} = D_{26} = 0$), Equation (2.82) becomes

$$D_{11} \frac{\partial^4 w}{\partial x^4} + 2(D_{12} + 2D_{66}) \frac{\partial^4 w}{\partial x^2 \partial y^2} + D_{22} \frac{\partial^4 w}{\partial y^4} + \varrho \frac{\partial^2 w}{\partial t^2}$$

$$= q + N_x \frac{\partial^2 w}{\partial x^2} + 2N_{xy} \frac{\partial^2 w}{\partial x \partial y} + N_y \frac{\partial^2 w}{\partial y^2} \tag{2.83}$$

For isotropic layers (2.83) becomes the well-known equation

$$D\nabla^4 w + \varrho \frac{\partial^2 w}{\partial t^2} = q + N_x \frac{\partial^2 w}{\partial x^2} + 2N_{xy} \frac{\partial^2 w}{\partial x \partial y} + N_y \frac{\partial^2 w}{\partial y^2} \tag{2.84}$$

2.8 BOUNDARY CONDITIONS

The proper boundary conditions are those which are sufficient to guarantee unique solutions to the governing equations. By applying energy principles in conjunction with calculus of variations (see Chapter 3), we find that the necessary boundary conditions are those of classical homogeneous plate theory plus those of an inplane elasticity problem. Thus, one member of each pair of the following four quantities must be prescribed along the boundary:

$$u_n^0; N_n \quad u_s^0; N_{ns} \quad \frac{\partial w}{\partial n}; M_n \quad w; \frac{\partial M_{ns}}{\partial s} + Q_n \tag{2.85}$$

where n and s denote coordinates normal and tangential to the plate edge respectively. The quantity $\partial M_{ns}/\partial s + Q_n$ is the well known Kirchhoff condition. For stability problems (or problems where inplane force effects are considered), this condition becomes

$$\frac{\partial M_{ns}}{\partial s} + Q_n + N_n^i \frac{\partial w}{\partial n} + N_{ns}^i \frac{\partial w}{\partial s} \tag{2.86}$$

The following are some of the more useful boundary conditions:

(1) Simply-Supported

$$N_n = N_{ns} = w = M_n = 0 \tag{2.87}$$

(2) Hinged-Free in the Normal Direction

$$N_n = u_s^0 = w = M_n = 0 \qquad (2.88)$$

(3) Hinged-Free in the Tangential Direction

$$u_n^0 = N_{ns} = w = M_n = 0 \qquad (2.89)$$

(4) Clamped

$$u_n^0 = u_s^0 = w = \frac{\partial w}{\partial n} = 0 \qquad (2.90)$$

(5) Free

$$N_n = N_{ns} = M_n = \frac{\partial M_{ns}}{\partial s} + Q_n = 0 \qquad (2.91)$$

REFERENCES

1. Timoshenko, S. and S. Woinowsky-Krieger. *Theory of Plates and Shells*, McGraw-Hill (1959).
2. Ambartsumyan, S. A. "Theory of Anisotropic Shells," *NASA Report TTF-118* (1964).
3. Tsai, S. W. and N. J. Pagano. "Invariant Properties of Composite Materials," *Composite Materials Workshop*, S. W. Tsia, J. C. Halpin and N. J. Pagano, eds. Lancaster, PA:Technomic Publishing Company (1968).
4. Lekhnitskii, S. G. *Theory of Elasticity of an Anisotropic Body*, translated from the Russian by P. Fern. J. Brandstatter, ed. Holden-Day (1963).
5. Lekhnitskii, S. G. *Anisotropic Plates*, translated from the Second Russian Edition by S. W. Tsai and T. Cheron. Gordon and Breach (1968).
6. Reissner, E. and Y. Stavsky. "Bending and Stretching of Certain Types of Heterogeneous Aeolotropic Elastic Plates," *Journal of Applied Mechanics*, 28:402–408 (1961).
7. Dong, S. B., R. B. Matthiesen, K. S. Pister and R. L. Taylor. "Analysis of Structural Laminates," *Air Force Report ARL-76* (1961).
8. Baker, E. H. A. P. Cappelli and L. Kovalevsky. "Shell Analysis Manual," *NASA Report CR-912* (1968).

CHAPTER 3

Energy Formulation of Governing Equations

3.1 INTRODUCTION

IN THIS CHAPTER we consider the application of energy principles to the analysis of anisotropic laminated plates. These principles will be used in conjunction with the calculus of variations to obtain the governing equations and natural boundary conditions of an anisotropic laminated plate.

Although this approach leads to the same governing equations and boundary conditions as derived in Chapter 2 from the basic equations of classical theory of elasticity, it also provides the basis for the development of approximate solution methods. In particular, the Ritz and Galerkin methods are based on energy principles. These two methods will be used in subsequent chapters to obtain an approximate solution to a number of complex laminated plate problems. It should also be noted that energy principles are the basis for finite element formulations.

3.2 STRAIN ENERGY OF A LAMINATED PLATE

The strain energy of an elastic body in terms of an x,y,z coordinate system is given by the relationship [1]

$$U = \frac{1}{2} \int \int \int (\sigma_x \epsilon_x + \sigma_y \epsilon_y + \sigma_z \epsilon_z + \sigma_{xz} \epsilon_{xz} + \sigma_{yz} \epsilon_{yz} + \sigma_{xy} \epsilon_{xy}) dx\, dy\, dz \tag{3.1}$$

where the triple integration is performed over the volume of the body. Taking into account the basic assumptions of laminated plate theory as discussed in Chapter 2, i.e., $\epsilon_z = \epsilon_{xz} = \epsilon_{yz} = 0$, along with the ply stress-strain relations, Equation (2.24), we find that Equation (3.1) becomes

$$U = \frac{1}{2} \int \int \int (Q_{11}^{(k)} \epsilon_x^2 + 2Q_{12}^{(k)} \epsilon_x \epsilon_y + 2Q_{16}^{(k)} \epsilon_x \epsilon_{xy}$$
$$+ 2Q_{26}^{(k)} \epsilon_y \epsilon_{xy} + Q_{22}^{(k)} \epsilon_y^2 + Q_{66}^{(k)} \epsilon_{xy}^2) dx\, dy\, dz \tag{3.2}$$

41

This relationship can be expressed in terms of the laminate displacements by substituting the strain-displacements relations as given by Equations (2.5)–(2.7). These relations are repeated here for clarity:

$$\epsilon_x = \frac{\partial u^0}{\partial x} - z\frac{\partial^2 w}{\partial x^2}$$

$$\epsilon_y = \frac{\partial v^0}{\partial y} - z\frac{\partial^2 w}{\partial y^2} \tag{3.3}$$

$$\epsilon_{xy} = \frac{\partial u^0}{\partial y} + \frac{\partial v^0}{\partial x} - 2z\frac{\partial^2 w}{\partial x \partial y}$$

Substituting Equation (3.3) into Equation (3.2) and integrating with respect to z, we obtain the area integral

$$U = \frac{1}{2} \int \int \Bigg\{ A_{11}\left(\frac{\partial u^0}{\partial x}\right)^2 + 2A_{12}\frac{\partial u^0}{\partial x}\frac{\partial v^0}{\partial y} + A_{22}\left(\frac{\partial v^0}{\partial y}\right)^2$$

$$+ 2\left(A_{16}\frac{\partial u^0}{\partial x} + A_{26}\frac{\partial v^0}{\partial y}\right)\left(\frac{\partial u^0}{\partial y} + \frac{\partial v^0}{\partial x}\right) + A_{66}\left(\frac{\partial u^0}{\partial y} + \frac{\partial v^0}{\partial x}\right)^2$$

$$- B_{11}\frac{\partial u^0}{\partial x}\frac{\partial^2 w}{\partial x^2} - 2B_{12}\left(\frac{\partial v^0}{\partial y}\frac{\partial^2 w}{\partial x^2} + \frac{\partial u^0}{\partial x}\frac{\partial^2 w}{\partial y^2}\right)$$

$$- B_{22}\frac{\partial v^0}{\partial y}\frac{\partial^2 w}{\partial y^2} - 2B_{16}\left[\frac{\partial^2 w}{\partial x^2}\left(\frac{\partial u^0}{\partial y} + \frac{\partial v^0}{\partial x}\right) + 2\frac{\partial u^0}{\partial x}\frac{\partial^2 w}{\partial x \partial y}\right]$$

$$- 2B_{26}\left[\frac{\partial^2 w}{\partial y^2}\left(\frac{\partial u^0}{\partial y} + \frac{\partial v^0}{\partial x}\right) + 2\frac{\partial v^0}{\partial y}\frac{\partial^2 w}{\partial x \partial y}\right] - \tag{3.4}$$

$$- 4B_{66}\frac{\partial^2 w}{\partial x \partial y}\left(\frac{\partial u^0}{\partial y} + \frac{\partial v^0}{\partial x}\right) + D_{11}\left(\frac{\partial^2 w}{\partial x^2}\right)^2 + 2D_{12}\frac{\partial^2 w}{\partial x^2}\frac{\partial^2 w}{\partial y^2}$$

$$+ D_{22}\left(\frac{\partial^2 w}{\partial y^2}\right)^2 + 4\left(D_{16}\frac{\partial^2 w}{\partial x^2} + D_{26}\frac{\partial^2 w}{\partial y^2}\right)\left(\frac{\partial^2 w}{\partial x \partial y}\right)$$

$$+ 4D_{66}\left(\frac{\partial^2 w}{\partial x \partial y}\right)^2 \Bigg\} \, dx\, dy$$

where A_{ij}, B_{ij}, and D_{ij} are as previously defined by equation (2.27). The strain energy expression given by Equation (3.4) contains coupling between the inplane displacements u^0, v^0, and the transverse displacement w, due to the presence of products of these terms. As discussed in Chapter 2, this bending-stretching coupling is due to the B_{ij} stiffness terms. For symmetric laminates the B_{ij} terms are identically zero, and Equation (3.4) uncouples and reduces to the following:

$$
\begin{aligned}
U = \frac{1}{2} \int \int \Bigg[& A_{11} \left(\frac{\partial u^0}{\partial x} \right)^2 + 2A_{12} \frac{\partial u^0}{\partial x} \frac{\partial v^0}{\partial y} + A_{22} \left(\frac{\partial v^0}{\partial y} \right)^2 \\
& + 2 \left(A_{16} \frac{\partial u^0}{\partial x} + A_{26} \frac{\partial v^0}{\partial y} \right) \left(\frac{\partial u^0}{\partial y} + \frac{\partial v^0}{\partial x} \right) + A_{66} \left(\frac{\partial u^0}{\partial y} \right. \\
& \left. + \frac{\partial v^0}{\partial x} \right)^2 \Bigg] dx\,dy + \int \int \Bigg[D_{11} \left(\frac{\partial^2 w}{\partial x^2} \right)^2 + 2D_{12} \frac{\partial^2 w}{\partial x^2} \frac{\partial^2 w}{\partial y^2} + D_{22} \left(\frac{\partial^2 w}{\partial y^2} \right)^2 \\
& + 4 \left(D_{16} \frac{\partial^2 w}{\partial x^2} + D_{26} \frac{\partial^2 w}{\partial y^2} \right) \frac{\partial^2 w}{\partial x \partial y} + 4D_{66} \left(\frac{\partial^2 w}{\partial x \partial y} \right)^2 \Bigg] dx\,dy
\end{aligned}
$$

(3.5)

In this uncoupled form the first term on the right-hand side of Equation (3.5) contains only the inplane displacements u^0 and v^0, while the second term contains only the transverse displacement w. Thus, for pure bending problems the first expression can be considered an arbitrary constant and the strain energy for transverse bending of a laminated plate can be written in the form

$$
\begin{aligned}
U = \frac{1}{2} \int \int \Bigg[& D_{11} \left(\frac{\partial^2 w}{\partial x^2} \right)^2 + 2D_{12} \frac{\partial^2 w}{\partial x^2} \frac{\partial^2 w}{\partial y^2} + D_{22} \left(\frac{\partial^2 w}{\partial y^2} \right) + 4 \left(D_{16} \frac{\partial^2 w}{\partial x^2} \right. \\
& \left. + D_{26} \frac{\partial^2 w}{\partial y^2} \right) \frac{\partial^2 w}{\partial x \partial y} + 4D_{66} \left(\frac{\partial^2 w}{\partial x \partial y} \right)^2 \Bigg] dx\,dy + C
\end{aligned}
$$

(3.6)

where C is an arbitrary constant. This relationship is identical to the expression found for the bending strain energy of a homogeneous anisotropic plate [2].

If the plate is specially orthotropic, i.e., if $D_{16} = D_{26} = 0$, then Equation (3.6) takes on the simpler form

$$
\begin{aligned}
U = \frac{1}{2} \int \int \Bigg[& D_{11} \left(\frac{\partial^2 w}{\partial x^2} \right)^2 + 2D_{12} \frac{\partial^2 w}{\partial x^2} \frac{\partial^2 w}{\partial y^2} + D_{22} \left(\frac{\partial^2 w}{\partial y^2} \right)^2 \\
& + 4D_{66} \left(\frac{\partial^2 w}{\partial x \partial y} \right)^2 \Bigg] dx\,dy + C
\end{aligned}
$$

(3.7)

For the case of an isotropic material, or a symmetric laminate constructed of layers of isotropic materials, we have

$$D_{11} = D_{22} = D$$

$$D_{12} = \nu D \qquad (3.8)$$

$$D_{66} = \frac{(1 - \nu)}{2} D$$

and Equation (3.7) reduces to the form

$$U = \frac{1}{2} \int \int D \left\{ \left(\frac{\partial^2 w}{\partial x^2} + \frac{\partial^2 w}{\partial y^2} \right) + 2(1 - \nu) \left[\left(\frac{\partial^2 w}{\partial x \partial y} \right)^2 - \frac{\partial^2 w}{\partial x^2} \frac{\partial^2 w}{\partial y^2} \right] \right\} dx\, dy$$
$$(3.9)$$

3.3 KINETIC ENERGY OF A LAMINATED PLATE

The kinetic energy of an elastic body in terms of an x,y,z coordinate system is of the form [1]

$$T = \frac{1}{2} \int \int \int \varrho_0 \left[\left(\frac{\partial u}{\partial t} \right)^2 + \left(\frac{\partial v}{\partial t} \right)^2 + \left(\frac{\partial w}{\partial t} \right)^2 \right] dx\, dy\, dz \qquad (3.10)$$

where, as previously, ϱ_0 is the density of the material and the triple integration is performed over the volume of the body. Combining Equations (2.2)–(2.4), we obtain the displacement field

$$u = u^0 - z \frac{\partial w}{\partial x}$$

$$(3.11)$$

$$v = v^0 - z \frac{\partial w}{\partial y}$$

where w is independent of z. Substituting these displacement relations into Equation (3.10), we find the kinetic energy of a laminated plate is given by

$$T = \frac{1}{2} \int \int \int \varrho_0^{(k)} \left[\left(\frac{\partial u^0}{\partial t} - z \frac{\partial^2 w}{\partial x \partial t} \right)^2 + \left(\frac{\partial v^0}{\partial t} - z \frac{\partial^2 w}{\partial y \partial t} \right)^2 \right.$$

$$\omega^2 = C_{00} \left(m \right)$$

$$+ \left(\frac{\partial w}{\partial t} \right)^2 \right] dx \, dy \, dz \quad \text{Amplitude} \tag{3.12}$$

where $\varrho_0^{(k)}$ denotes the density of the kth layer. Integrating Equation (3.12) with respect to z and neglecting time derivatives of plate slopes in accordance with assumption 11 of the previous chapter, we arrive at the expression

$$T = \frac{1}{2} \iint \left[\varrho \left(\frac{\partial u^0}{\partial t} \right)^2 + \left(\frac{\partial v^0}{\partial t} \right)^2 + \left(\frac{\partial w}{\partial t} \right)^2 \right] dx \, dy \tag{3.13}$$

where ϱ is the integral of the density through-the-thickness of the plate as defined by Equation (2.12).

3.4 POTENTIAL ENERGY OF EXTERNAL LOADS

We now consider potential energy due to transverse loads and inplane loads. For transverse bending, we are concerned with loads generated by applying normal tractions to the top and bottom surfaces of the plate which lead to the potential energy expression

$$W = - \iint [\sigma_z(h/2) - \sigma_z(-h/2)]w \, dx \, dy \tag{3.14}$$

Taking into account Equation (2.15), we may write the potential energy given by Equation (3.14) in the form

$$W = - \iint q \, w \, dx \, dy \tag{3.15}$$

The potential energy, V, of inplane loads due to a deflection w is [3]

$$V = \iint (N_x^i \epsilon_x' + N_y^i \epsilon_y' + N_{xy}^i \epsilon_{xy}') dx \, dy \tag{3.16}$$

where N_x^i, N_y^i, N_{xy}^i are initial inplane force resultants applied to the plate in a pre-buckled state and ϵ_x', ϵ_y', ϵ_{xy}' are the midplane strains due to the deflection w. These strains are usually associated with large deflection analysis and are arrived at by considering Equations (1.4), (1.5), and (1.9) from the Green strain tensor and retaining only the nonlinear terms involving w. In the context of linear theory, these strains are applied for the purpose of determining critical buckling loads. These strains are of the following form:

$$\epsilon'_x = \frac{1}{2}\left(\frac{\partial w}{\partial x}\right)$$

$$\epsilon'_y = \frac{1}{2}\left(\frac{\partial w}{\partial y}\right) \tag{3.17}$$

$$\epsilon'_{xy} = \frac{\partial w}{\partial x}\frac{\partial w}{\partial y}$$

and Equation (3.16) takes the form

$$V = \frac{1}{2}\iint\left[N^i_x\left(\frac{\partial w}{\partial x}\right)^2 + N^i_y\left(\frac{\partial w}{\partial y}\right)^2 + 2N^i_{xy}\frac{\partial w}{\partial x}\frac{\partial w}{\partial y}\right]dx\,dy \tag{3.18}$$

3.5 GOVERNING EQUATIONS AND NATURAL BOUNDARY CONDITIONS

From Hamilton's principle [1] the governing equations of motion and the proper boundary conditions are determined from the variational equations over the arbitrary time interval $t_0 \leq t \leq t_1$

$$\int_{t_0}^{t_1} (\delta U + \delta V + \delta W - \delta T)dt = 0 \tag{3.19}$$

The first variation of the strain energy is as follows:

$$\delta U = \int_0^b \int_0^a \left\{ \left[A_{11}\frac{\partial u^o}{\partial x} + A_{12}\frac{\partial v^o}{\partial y} + A_{16}\left(\frac{\partial u^o}{\partial y} + \frac{\partial v^o}{\partial x}\right)\right.\right.$$

$$\left. - B_{11}\frac{\partial^2 w}{\partial x^2} - B_{12}\frac{\partial^2 w}{\partial y^2} - 2B_{16}\frac{\partial^2 w}{\partial x\partial y}\right]\frac{\partial}{\partial x}(\delta u^o)$$

$$+ \left[A_{12}\frac{\partial u^o}{\partial x} + A_{22}\frac{\partial v^o}{\partial y} + A_{26}\left(\frac{\partial u^o}{\partial y} + \frac{\partial v^o}{\partial x}\right)\right. \tag{3.20}$$

$$
\left. - B_{12} \frac{\partial^2 w}{\partial x^2} - B_{22} \frac{\partial^2 w}{\partial y^2} - 2B_{26} \frac{\partial^2 w}{\partial x \partial y} \right] \frac{\partial}{\partial y} (\delta v^o)
$$

$$
+ \left[A_{16} \frac{\partial u^o}{\partial x} + A_{26} \frac{\partial v^o}{\partial y} + A_{66} \left(\frac{\partial u^o}{\partial y} + \frac{\partial v^o}{\partial x} \right) - B_{16} \frac{\partial^2 w}{\partial x^2} \right.
$$

$$
\left. - B_{26} \frac{\partial^2 w}{\partial y^2} - 2B_{66} \frac{\partial^2 w}{\partial x \partial y} \right] \left[\frac{\partial}{\partial y} (\delta u^o) + \frac{\partial}{\partial x} (\delta v^o) \right]
$$

$$
- \left[B_{11} \frac{\partial u^o}{\partial x} + B_{12} \frac{\partial v^o}{\partial y} + B_{16} \left(\frac{\partial u^o}{\partial y} + \frac{\partial v^o}{\partial x} \right) - D_{11} \frac{\partial^2 w}{\partial x^2} \right.
$$

$$
\left. - D_{12} \frac{\partial^2 w}{\partial y^2} - 2D_{16} \frac{\partial^2 w}{\partial x \partial y} \right] \frac{\partial^2}{\partial x^2} (\delta w) - \left[B_{12} \frac{\partial u^o}{\partial x} \right.
$$

$$
+ B_{22} \frac{\partial v^o}{\partial y} + B_{26} \left(\frac{\partial u^o}{\partial y} + \frac{\partial v^o}{\partial x} \right) - D_{12} \frac{\partial^2 w}{\partial x^2} - D_{22} \frac{\partial^2 w}{\partial y^2}
$$

$$
\left. - 2D_{26} \frac{\partial^2 w}{\partial x \partial y} \right] \frac{\partial^2}{\partial y^2} (\delta w) - 2 \left[B_{16} \frac{\partial u^o}{\partial x} + B_{26} \frac{\partial v^o}{\partial y} \right.
$$

$$
+ B_{66} \left(\frac{\partial u^o}{\partial y} + \frac{\partial v^o}{\partial x} \right) - D_{16} \frac{\partial^2 w}{\partial x^2} - D_{26} \frac{\partial^2 w}{\partial y^2}
$$

$$
\left. - 2D_{66} \frac{\partial^2 w}{\partial x \partial y} \right] \frac{\partial^2}{\partial x \partial y} (\delta w) \right\} dx \, dy
$$

Comparing terms in this equation with the laminate constitutive relations, Equation (2.26), we can see that

$$
\delta U = \int_0^b \int_0^a \left\{ N_x \frac{\partial}{\partial x} (\delta u^o) + N_y \frac{\partial}{\partial y} (\delta v^o) + N_{xy} \left[\frac{\partial}{\partial y} (\delta u^o) \right. \right.
$$

$$
\left. \left. + \frac{\partial}{\partial x} (\delta v^o) \right] - M_x \frac{\partial^2}{\partial x^2} (\delta w) - M_y \frac{\partial^2}{\partial y^2} (\delta w) \right. \tag{3.21}
$$

$$- 2M_{xy} \frac{\partial^2}{\partial x \partial y} \left. (\delta w) \right\} \, dx \, dy$$

Green's theorem [4] can be used in conjunction with integration by parts to transform Equation (3.21) into the form

$$\delta U = - \int \int \left[\left(\frac{\partial N_x}{\partial x} + \frac{\partial N_{xy}}{\partial y} \right) \delta u^0 \right.$$

$$+ \left(\frac{\partial N_{xy}}{\partial x} + \frac{\partial N_y}{\partial y} \right) \delta v^0 + \left(\frac{\partial^2 M_x}{\partial x^2} \right.$$

$$\left. + 2 \frac{\partial^2 M_{xy}}{\partial x \partial y} + \frac{\partial^2 M_y}{\partial y^2} \right) \delta w \left. \right] dx \, dy - \int_{S_y} \left[N_{xy} \delta u^0 \right.$$

$$+ N_y \delta v^0 + M_y \frac{\partial}{\partial y} (\delta w) - \left(2 \frac{\partial M_{xy}}{\partial x} + \frac{\partial M_y}{\partial y} \right) \delta w \left. \right] dx$$

$$+ \int_{S_x} \left[N_x \delta u^0 + N_{xy} \delta v^0 + M_x \frac{\partial}{\partial x} (\delta w) \right.$$

$$\left. - \left(2 \frac{\partial M_{xy}}{\partial y} + \frac{\partial M_x}{\partial x} \right) \delta w \right] dy$$

(3.22)

where S_x is defined along the edges $x =$ constant and S_y along the edges $y =$ constant.

Performing similar operations with Equations (3.15) and (3.18), we obtain the following:

$$\delta W = - \int \int q \, \delta w \, dx \, dy \qquad (3.23)$$

$$\delta V = - \int \int \left(N_x^i \frac{\partial^2 w}{\partial x^2} + 2N_{xy}^i \frac{\partial^2 w}{\partial x \partial y} + N_y^i \frac{\partial^2 w}{\partial y^2} \right) \delta w \, dx \, dy$$

$$- \int_{S_y} \left(N_{xy}^i \, \frac{\partial w}{\partial x} + N_y^i \, \frac{\partial w}{\partial y} \right) \delta w \, dx$$

$$+ \int_{S_x} \left(N_x^i \, \frac{\partial w}{\partial x} + N_{xy}^i \, \frac{\partial w}{\partial y} \right) \delta w \, dy \tag{3.24}$$

It should be noted that terms involving derivatives of the initial force resultants are neglected in the derivation of equation (3.24).

The variation of Equation (3.12) yields

$$\delta T = \int \int \varrho \left[\frac{\partial u^0}{\partial t} \, \frac{\partial}{\partial t} \, (\delta u^0) + \frac{\partial v^0}{\partial t} \, \frac{\partial}{\partial t} \, (\delta v^0) + \frac{\partial w}{\partial t} \, \frac{\partial}{\partial t} \, (\delta w) \right] dx \, dy \tag{3.25}$$

Integrating Equation (3.25) by parts, we obtain the relationship

$$\delta T = - \int \int \varrho \left[\left(\frac{\partial^2 u^0}{\partial t^2} \, \delta u^0 + \frac{\partial^2 v^0}{\partial t^2} \, \delta v^0 + \frac{\partial^2 w}{\partial t^2} \right) \delta w - \frac{\partial}{\partial t} \right.$$

$$\left. \left(\frac{\partial u^0}{\partial t} \, \delta u^0 + \frac{\partial v^0}{\partial t} \, \delta v^0 + \frac{\partial w}{\partial t} \, \delta w \right) \right] dx \, dy \tag{3.26}$$

If we integrate this relationship with respect to time over the interval $t_0 \leq t \leq t_1$ and assume

$$\delta u^0(t_0) = \delta u^0(t_1) = \delta v^0(t_0) = \delta v^0(t_1)$$

$$= \delta w(t_0) = \delta w(t_1) = 0 \tag{3.27}$$

then the second expression in the integrand of Equation (3.26) vanishes. Thus, in view of Equation (3.27), we can write Equation (3.26) in the form

$$\delta T = - \int \int \varrho \left(\frac{\partial^2 u^0}{\partial t^2} \, \delta u^0 + \frac{\partial^2 v^0}{\partial t^2} \, \delta v^0 + \frac{\partial^2 w}{\partial t^2} \, \delta w \right) dx \, dy \tag{3.28}$$

Substituting Equations (3.22), (3.23), (3.24), and (3.28) into Equation (3.19), we obtain the following results:

$$
\int_{t_1}^{t_0} \left\{ \iint \left[\left(\frac{\partial N_x}{\partial x} + \frac{\partial N_{xy}}{\partial y} - \varrho \frac{\partial^2 u^0}{\partial t^2} \right) \delta u^0 + \left(\frac{\partial N_{xy}}{\partial x} + \frac{\partial N_y}{\partial y} \right. \right. \right.
$$

$$
\left. - \varrho \frac{\partial^2 v^0}{\partial t^2} \right) \delta v^0 + \left(\frac{\partial^2 M_x}{\partial x^2} + 2 \frac{\partial^2 M_{xy}}{\partial x \partial y} + \frac{\partial^2 M_y}{\partial y^2} + N_x^i \frac{\partial^2 w}{\partial x^2} \right.
$$

$$
\left. + 2 N_{xy}^i \frac{\partial^2 w}{\partial x \partial y} + N_y^i \frac{\partial^2 w}{\partial y^2} + q - \varrho \frac{\partial^2 w}{\partial t^2} \right) \delta w \right] dx\, dy
$$

$$
- \int_{S_x} \left[N_x \delta u^0 + N_{xy} \delta v^0 + M_x \frac{\partial}{\partial x} (\delta w) - \left(\frac{\partial M_x}{\partial x} + 2 \frac{\partial M_{xy}}{\partial y} \right. \right.
$$ (3.29)

$$
\left. \left. + N_x^i \frac{\partial w}{\partial x} + N_{xy}^i \frac{\partial w}{\partial y} \right) \delta w \right] dy
$$

$$
+ \int_{S_y} \left[N_{xy} \delta u^0 + N_y \delta v^0 + M_y \frac{\partial}{\partial y} (\delta w) \right.
$$

$$
\left. \left. - \left(2 \frac{\partial M_{xy}}{\partial x} + \frac{\partial M_y}{\partial y} + N_{xy}^i \frac{\partial w}{\partial x} + N_y^i \frac{\partial w}{\partial y} \right) \delta w \right] dx \right\} dt = 0
$$

The surface integral in Equation (3.29) will vanish if the following equations of motion are satisfied:

$$
\frac{\partial N_x}{\partial x} + \frac{\partial N_{xy}}{\partial y} = \varrho \frac{\partial^2 u^0}{\partial t^2}
$$ (3.30)

$$
\frac{\partial N_{xy}}{\partial x} + \frac{\partial N_y}{\partial y} = \varrho \frac{\partial^2 v^0}{\partial t^2}
$$ (3.31)

$$
\frac{\partial^2 M_x}{\partial x^2} + 2 \frac{\partial^2 M_{xy}}{\partial x \partial y} + \frac{\partial^2 M_y}{\partial y^2} + N_x^i \frac{\partial^2 w}{\partial x^2} + 2 N_{xy}^i \frac{\partial^2 w}{\partial x \partial y}
$$

$$+ N_y^i \; \frac{\partial^2 w}{\partial y^2} + q = \varrho \; \frac{\partial^2 w}{\partial t^2} \tag{3.32}$$

These relations are identical with Equations (2.11), (2.13), and (2.22) as derived in the previous chapter. The line integrals in Equation (3.29) define the natural boundary conditions, which are

$$N_x \delta u^0 = 0 \text{ on } S_x$$

$$N_{xy} \delta v^0 = 0 \text{ on } S_x \tag{3.33}$$

$$M_x \frac{\partial}{\partial x} (\delta w) = 0 \text{ on } S_x$$

$$\left(\frac{\partial M_x}{\partial x} + 2 \frac{\partial M_{xy}}{\partial y} + N_x^i \frac{\partial w}{\partial x} + N_{xy}^i \frac{\partial w}{\partial y} \right) \delta w = 0 \text{ on } S_x$$

$$N_{xy} \delta u^0 = 0 \text{ on } S_y$$

$$N_y \delta v^0 = 0 \text{ on } S_y \tag{3.34}$$

$$M_y \frac{\partial}{\partial y} (\delta w) = 0 \text{ on } S_y$$

$$\left(2 \frac{\partial^2 M_{xy}}{\partial x} + \frac{\partial M_y}{\partial y} + N_{xy}^i \frac{\partial w}{\partial x} + N_y^i \frac{\partial w}{\partial y} \right) \delta w = 0 \text{ on } S_y$$

Equations (3.33) and (3.34) are consistent with the boundary conditions given by Equations (2.85) and (2.86) in the previous chapter.

3.6 THE RITZ METHOD

The Ritz method [5] provides a convenient method for obtaining approximate solutions to boundary value problems. This approach is equally applicable to bending, buckling, and free vibration problems. Each of these problems is governed by an energy condition which can be written in the following form:

$$\Pi(u^0, v^0 w) = \text{stationary value} \tag{3.35}$$

where

$$\Pi = U + W \qquad \text{(transverse bending)} \qquad (3.36)$$

$$\Pi = U + W + V \qquad \text{(buckling)} \qquad (3.37)$$

$$\Pi = U + W + V - T \qquad \text{(free vibration)} \qquad (3.38)$$

For free vibration time can be removed from Equation (3.38) by considering displacement fields of the form

$$u^0(t) = u^0 e^{i\omega t}$$

$$v^0(t) = v^0 e^{i\omega t} \qquad (3.39)$$

$$w(t) = w e^{i\omega t}$$

Under these assumed displacements, the potential energy can be used in the form

$$T = \frac{1}{2} \int \int \varrho \omega^2 [(u^0)^2 + (v^0)^2 + w^2] dx \, dy \qquad (3.40)$$

This allows the free vibration problem to be solved as a static problem by considering the kinetic energy to be simply additional energy.

In the Ritz method a solution is sought in the form

$$u^0 = \sum_{m=1}^{M_1} \sum_{n=1}^{N_1} A_{mn} U_{mn}(x,y)$$

$$v^0 = \sum_{m=1}^{M_2} \sum_{n=1}^{N_2} B_{mn} V_{mn}(x,y) \qquad (3.41)$$

$$w = \sum_{m=1}^{M_3} \sum_{n=1}^{N_3} C_{mn} W_{mn}(x,y)$$

where A_{mn}, B_{mn}, and C_{mn} are undetermined coefficients. The functionns U_{mn}, V_{mn}, and W_{mn} are known and usually chosen in the variables separable form $X_m(x)$ $Y_n(y)$. The geometric boundary conditions must be satisfied by these functions.

In addition they should be continuous through at least the same order derivative as required in the corresponding differential equations. Substituting Equations (3.41) into the energy condition (3.35) leads to a minimization problem relative to the undetermined coefficients. In particular, Π is a function of A_{mn}, B_{mn}, and C_{mn} only and the conditions given by Equation (3.35) reduce to the equations:

$$\frac{\partial \Pi}{\partial A_{mn}} = 0 \qquad \begin{cases} m = 1,2, \ldots, M_1 \\ n = 1,2, \ldots, N_1 \end{cases}$$

$$\frac{\partial \Pi}{\partial B_{mn}} = 0 \qquad \begin{cases} m = 1,2, \ldots, M_2 \\ n = 1,2, \ldots, N_2 \end{cases} \qquad (3.42)$$

$$\frac{\partial \Pi}{\partial C_{mn}} = 0 \qquad \begin{cases} m = 1,2, \ldots, M_3 \\ n = 1,2, \ldots, N_3 \end{cases}$$

For the formulation presented here, Π is always quadratic in the undetermined coefficients. Thus, the conditions (3.42) lead to a $\Sigma^3_{i=1} M_i \times N_i$ set of linear simultaneous equations. For buckling and free vibration problems, (3.42) leads to a classic eigenvalue problem, that is, buckling loads and free vibration frequencies are chosen such that the determinant of the coefficients of A_{mn}, B_{mn}, and C_{mn} vanish.

3.7 THE GALERKIN METHOD

In the Galerkin method [5], as in the Ritz method, an approximate solution in the form of Equation (3.41) is sought in conjunction with the variational equation

$$\delta U + \delta W + \delta V - \delta T = 0 \qquad (3.43)$$

In this relationship the static equivalent form of T as given by Equation (3.40) is utilized. Substituting the constitutive relations (2.26) into Equation (3.32), we can write Equation (3.43) in the form

$$\iint \left\{ [(L_1 + \varrho\omega^2)u^0 + L_2 v^0 + L_3 w^0]\delta u^0 + [L_2 u^0 + (L_4 + \varrho\omega^2)v^0 \right.$$

$$+ L_5 w^0]\delta v^0 + [L_3 u^0 + L_5 v^0 + (L_6 - \varrho\omega^2)w - N_x^i \frac{\partial^2 w}{\partial x} - 2N_{xy}^i \frac{\partial^2 w}{\partial x \partial y}$$

$$\left. - N_y^i \frac{\partial^2 w}{\partial y^2} - q]\delta w \right\} dx\, dy - \int_{S_x} \left[N_x \delta u^0 + N_{xy}\delta v^0 + M_x \frac{\partial}{\partial x}(\delta w) \right.$$

$$-\left(\frac{\partial M_x}{\partial x} + 2 \frac{\partial M_{xy}}{\partial y} + N_x^i \frac{\partial w}{\partial x} + N_{xy}^i \frac{\partial w}{\partial y} \right) \delta w \right] dy + \int_{S_y} \left[N_{xy} \delta u^0 \right.$$

$$+ N_y \delta v^0 + M_y \frac{\partial}{\partial y}(\delta w) - \left(2 \frac{\partial M_{xy}}{\partial x} + \frac{\partial M_y}{\partial y} + N_{xy}^i \frac{\partial w}{\partial x} \right.$$

$$\left. + N_y^i \frac{\partial w}{\partial y} \right) \delta w \right] dy = 0 \tag{3.44}$$

where L_i are linear operators defined as follows:

$$L_1 = A_{11} \frac{\partial^2}{\partial x^2}(\ ) + 2A_{16} \frac{\partial^2}{\partial x \partial y}(\ ) + A_{66} \frac{\partial^2}{\partial y^2}(\ )$$

$$L_2 = A_{16} \frac{\partial^2}{\partial x^2}(\ ) + (A_{12} + A_{66}) \frac{\partial^2}{\partial x \partial y}(\ ) + A_{26} \frac{\partial^2}{\partial y^2}(\ )$$

$$L_3 = -B_{11} \frac{\partial^3}{\partial x^3}(\ ) - 3B_{16} \frac{\partial^3}{\partial x^2 \partial y}(\ ) - (B_{12} + 2B_{66}) \frac{\partial^3}{\partial x \partial y^2}$$

$$- B_{26} \frac{\partial^3}{\partial y^3}(\ )$$

$$L_4 = A_{66} \frac{\partial}{\partial x^2}(\ ) + 2A_{26} \frac{\partial}{\partial x \partial y}(\ ) + A_{22} \frac{\partial^2}{\partial y^2}(\ )$$

$$L_5 = -B_{16} \frac{\partial^3}{\partial x^3}(\ ) - (B_{12} + 2B_{66}) \frac{\partial^3}{\partial x^2 \partial y}(\ )$$

$$- 3B_{26} \frac{\partial^3}{\partial x \partial y^2}(\ ) - B_{22} \frac{\partial^3}{\partial y^3}(\ )$$

$$L_6 = D_{11} \frac{\partial^4}{\partial x^4}(\ ) + 4D_{16} \frac{\partial^4}{\partial x^3 \partial y}(\ ) + 2(D_{12} +$$

$$+ 2D_{66}) \frac{\partial^4}{\partial x^2 \partial y^2}(\ ) + 4D_{26} \frac{\partial^4}{\partial x \partial y^3}(\ ) + D_{22} \frac{\partial}{\partial y^4}(\ )$$

If we take the variation of Equation (3.44) with respect to the undetermined coefficients A_{mn}, B_{mn}, and C_{mn} in Equations (3.41), we obtain the result

$$\delta u^0 = \sum_{m=1}^{M_1} \sum_{n=1}^{N_1} U_{mn}(x,y)\, \delta A_{mn}$$

$$\delta v^0 = \sum_{m=1}^{M_2} \sum_{n=1}^{N_2} V_{mn}(x,y)\, \delta B_{mn}$$

$$\delta w = \sum_{m=1}^{M_3} \sum_{n=1}^{N_3} W_{mn}(x,y)\, \delta C_{mn} \tag{3.45}$$

$$\frac{\partial}{\partial x}(\delta w) = \sum_{m=1}^{M_3} \sum_{n=1}^{N_3} \frac{\partial W_{mn}}{\partial x}(x,y)\, \delta C_{mn}$$

$$\frac{\partial}{\partial y}(\delta w) = \sum_{m=1}^{M_3} \sum_{n=1}^{N_3} \frac{\partial W_{mn}}{\partial y}(x,y)\, \delta C_{mn}$$

Substituting these relationships into Equation (3.44), and recognizing that the resulting equation can only be zero if the coefficients of δA_{mn}, δB_{mn}, and δC_{mn} vanish identically, we obtain the following Galerkin equations:

$$\iint [(L_1 + \varrho\omega^2)u^0 + L_2 v^0 + L_3 w]U_{mn}(x,y)dx\,dy$$

$$-\int_{S_x} N_x U_{mn}(S_x,y)dy + \int_{S_y} N_{xy} U_{mn}(x,S_y)dx = 0 \tag{3.46}$$

$$\begin{cases} m = 1,2, \ldots ,M_1 \\ n = 1,2, \ldots ,N_1 \end{cases}$$

$$\iint [L_2 u^0 + (L_4 + \varrho\omega^2)v^0 + L_5 w]V_{mn}(x,y)dx\,dy \tag{3.47}$$

$$-\int_{S_x} N_{xy} V_{mn}(S_x,y)dy + \int_{S_y} N_y V_{mn}(x,S_y)dx = 0 \qquad (3.47)$$

$$\begin{cases} m = 1,2, \ldots,M_2 \\ n = 1,2, \ldots,N_2 \end{cases}$$

$$\iint [L_3 u^0 + L_5 v^{0} + (L_6 - N_x^i \frac{\partial^2}{\partial x^2} - 2N_{xy}^i \frac{\partial^2}{\partial x \partial y} - N_y^i \frac{\partial^2}{\partial y^2}$$

$$- q - \varrho\omega^2)W_{mn}(x,y)dx\,dy - \int_{S_x}\left[M_x \frac{\partial W_{mn}}{\partial x}(S_x,y) - \left(\frac{\partial M_x}{\partial x} \right.\right.$$

$$+ 2\frac{\partial M_{xy}}{\partial y} + N_x^i\frac{\partial}{\partial x} + N_{xy}^i\frac{\partial}{\partial y} \left.\right) W_{mn}(S_x,y) \left.\right] dy + \int_{S_y}\left[M_y \frac{\partial W_{mn}}{\partial y}(x,S_y) \right.$$

$$- \left(2\frac{\partial M_{xy}}{\partial x} + \frac{\partial M_y}{\partial y} + N_{xy}^i\frac{\partial}{\partial x} + N_y^i\frac{\partial}{\partial y} \right) W_{mn}(x,S_y) \left.\right] dx = 0$$

$$\begin{cases} m = 1,2, \ldots,M_3 \\ n = 1,2, \ldots,N_3 \end{cases}$$

(3.48)

Substitution of Equations (3.45) into Equations (3.46), (3.47), and (3.48) leads to a set of $\Sigma_{i=1}^3 M_i \times N_i$ algebraic equations.

If all of the boundary conditions in a given problem are geometric, then the boundary integrals in Equations (3.46), (3.47) and (3.48) vanish. For cases involving nongeometric boundary conditions, the appropriate line integrals in these equations are required.

It should be noted that if the Ritz and Galerkin methods are both applied to the same problem, and the same displacement functions, which satisfy the geometric boundary conditions, are chosen, then both approaches will lead to the same result. In particular the set of linear algebraic equations obtained from both the Ritz and Galerkin methods will be identical.

3.8 CONVERGENCE OF THE RITZ AND GALERKIN METHODS

In general, both the Ritz and Galerkin methods lead to approximate solutions in that the equilibrium equations, or equations of motion for dynamic problems, are only approximately satisfied. If the functions chosen for $U_{mn}(x,y)$, $V_{mn}(x,y)$, and $W_{mn}(x,y)$ each form a complete set over the plate domain, in addition to satisfying the geometric boundary conditions, then an exact solution can be obtained in the limit [5]. The rate of convergence is determined by how suitable the assumed displacements are for representing the exact solution. If a poor representation of the exact solution is chosen, convergence will be very slow. For problems involving natural boundary conditions, more rapid convergence is often obtained if all of the boundary conditions are satisfied by the chosen displacement functions. It should also be noted that if one chooses displacement functions which coincide with the exact solution to a boundary value problem, then both the Ritz and Galerkin procedures will lead to an exact solution.

The key to obtaining convergence to an exact solution for both the Ritz and Galerkin methods is the choice of a complete set of functions to represent the displacements. This may be difficult to do. For example, consider the Galerkin method in conjunction with the bending of a symmetric laminate under transverse loading. In this case the governing equation is obtained from Equation (3.47) with the result

$$
\iint \left[D_{11} \frac{\partial^4 w}{\partial x^4} + 4D_{16} \frac{\partial^4 w}{\partial x^3 \partial y} + 2(D_{16} + 2D_{66}) \frac{\partial^4 w}{\partial x^2 \partial y^2} + 4D_{26} \frac{\partial^4 w}{\partial x \partial y^3} \right.
$$

$$
\left. + D_{22} \frac{\partial^4 w}{\partial y^4} + q \right] W_{mn}(x,y)dx\,dy - \int_{S_x} \left[M_x \frac{\partial W_{mn}}{\partial x} (S_x,y) \right.
$$

$$
- \left(\frac{\partial M_x}{\partial x} + 2\frac{\partial M_{xy}}{\partial y} \right) W_{mn}(S_x,y) \right] dy + \int_{S_y} \left[M_y \frac{\partial W_{mn}}{\partial y} (x,S_y) \right.
$$

$$
- \left(2\frac{\partial M_{xy}}{\partial x} + \frac{\partial M_y}{\partial y} \right) W_{mn}(x,S_y) \right] dx = 0
$$

(3.49)

In addition to $W_{mn}(x,y)$ exactly satisfying any geometric boundary conditions, convergence to the exact solution is assured [5] only if for every $\epsilon < 0$, there exists a set of C_{mn} such that

$$
\left| w - \sum_{m=1}^{M} \sum_{n=1}^{N} C_{mn} W_{mn} \right| < \epsilon
$$

$$
\left| \frac{\partial w}{\partial x} - \sum_{m=1}^{M} \sum_{n=1}^{N} C_{mn} \frac{\partial W_{mn}}{\partial x} \right| < \epsilon
$$

(3.50)

$$
\left| \frac{\partial w}{\partial y} - \sum_{m=1}^{M} \sum_{n=1}^{N} C_{mn} \frac{\partial W_{mn}}{\partial y} \right| < \epsilon
$$

$$\left| \frac{\partial^2 w}{\partial x \partial y} - \sum_{m=1}^{M} \sum_{n=1}^{N} C_{mn} \frac{\partial^2 W_{mn}}{\partial x \partial y} \right| < \epsilon$$

$$\left| \frac{\partial^2 w}{\partial x^2} - \sum_{m=1}^{M} \sum_{n=1}^{N} C_{mn} \frac{\partial^2 W_{mn}}{\partial x^2} \right| < \epsilon$$

$$\left| \frac{\partial^2 w}{\partial y^2} - \sum_{m=1}^{M} \sum_{n=1}^{N} C_{mn} \frac{\partial^2 W_{mn}}{\partial y^2} \right| < \epsilon$$

These six conditions must be satisfied in order to assure W_{mn} is a complete set of functions. In certain classes of problems this will not be possible because of the fact that the exact solution cannot be represented by a displacement function in a variables separable form. A classic example is the transverse bending of a simply-supported anisotropic plate [6].

In many cases reasonably rapid convergence can be obtained with the Ritz and Galerkin methods for maximum plate deflections, critical buckling loads, and minimum vibration frequencies. Bending moments, which involve derivatives of the displacements, may converge very slowly, or not at all. This will be discussed in more detail in Chapter 6.

Despite the difficulties discussed here, the Ritz and Galerkin methods often provide useful tools for obtaining solutions to complex boundary value problems. These techniques will be used frequently in the following chapters.

REFERENCES

1. Langhaar, H. L. *Energy Methods in Applied Mechanics*. John Wiley and Sons (1962).
2. Lekhnitskii, S. G. *Anisotropic Plates*. Translated from Russian by S. W. Tsai and T. Cheron, Gordon and Breach (1968).
3. Timoshenko, S. and S. Woinowsky-Krieger. *Theory of Plates and Shells*. McGraw-Hill (1959).
4. Taylor, A. E. *Advanced Calculus*. Ginn and Company (1955).
5. Kantorovich, L. V. and V. I. Krylov. *Approximate Methods in Higher Order Analysis*. Translated by C. D. Benster. Interscience (1958).
6. Wang, James Ting-Shun. "On the Solution of Plates of Composite Materials," *Journal of Composite Materials*, 3:590–592 (1969).

HAPTER 4

One-Dimensional Analysis of Laminated Plates

4.1 INTRODUCTION

IN THIS CHAPTER we consider two important simplifications of the classical two-dimensional operations of laminated plates which result in one-dimensional theories. In the first simplification we consider the plate to have a very high length-to-width ratio such that the plate deformation may be considered to be independent of the length coordinate. Such behavior is referred to as *cylindrical bending* [1]. The second one-dimensional analysis involves the development of a laminated beam theory. Both of these theories are derived directly from the two-dimensional laminated plate equations developed in Chapter 2. Solutions in conjunction with these theories are also presented for the purpose of illustrating special behavior unique to anisotropic laminated plates. A number of more complex solutions involving the two-dimensional plate theory are discussed in the next three chapters.

4.2 CYLINDRICAL BENDING

Consider a laminate composed of an arbitrary number of layers having infinite length in the y direction, and uniformly supported along the edges $x = 0, a$ (see Figure 4.1). If the transverse surface loads are of the form $q = q(x)$ and inplane loads of the form $N_x^i = $ constant, $N_y^i = N_{xy}^i = 0$, then the deflected surface is cylindrical, i.e.,

$$u^0 = u^0(x,t), \ v^0 = v^0(x,t), \ w = w(x,t) \tag{4.1}$$

As in the derivation of Equation (3.40), time can be removed from Equation (4.1) for the case of free vibration by considering displacements of the form

$$u^0 = u^0(x) \ e^{i\omega t}, \ v^0 = v^0(x) \ e^{i\omega t}, \ w = w(x) \ e^{i\omega t} \tag{4.2}$$

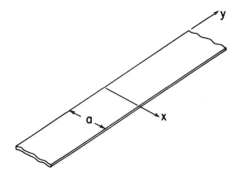

Figure 4.1. *Infinite strip.*

Substituting Equation (4.2) into the equations of motion (2.32), (2.33), and (2.79), we obtain the following one-dimensional equations:

$$A_{11} \frac{d^2 u^0}{dx^2} + A_{16} \frac{d^2 v^0}{dx^2} - B_{11} \frac{d^3 w}{dx^3} - \varrho \omega^2 u^0 = 0 \qquad (4.3)$$

$$A_{16} \frac{d^2 u^0}{dx^2} + A_{66} \frac{d^2 v^0}{dx^2} - B_{16} \frac{d^3 w}{dx^3} - \varrho \omega^2 v^0 = 0 \qquad (4.4)$$

$$D_{11} \frac{d^4 w}{dx^4} - B_{11} \frac{d^3 u^0}{dx^3} - B_{16} \frac{d^3 v^0}{dx^3} - N_x^i \frac{d^2 w}{dx^2} - q - \varrho \omega^2 w = 0 \quad (4.5)$$

For static bending under transverse loading, these equations can be uncoupled by obtaining the following results from Equations (4.3) and (4.4):

$$\frac{d^2 u^0}{dx^2} = \frac{B}{A} \frac{d^3 w}{dx^3}$$

$$(4.6)$$

$$\frac{d^2 v^0}{dx^2} = \frac{C}{A} \frac{d^3 w}{dx^3}$$

where

$$A = A_{11} A_{66} - A_{16}^2, \; B = A_{66} B_{11} - A_{16} B_{16},$$

$$C = A_{11} B_{16} - A_{16} B_{11}$$

Differentiating Equations (4.6) and substituting the results into Equation (4.4), we obtain a differential equation in w.

$$\frac{d^4w}{dx^4} = \frac{A}{D} q \tag{4.7}$$

where

$$D = D_{11}A - B_{11}B - B_{16}C$$

Equation (4.7) can be integrated to obtain w and the result substituted into Equation (4.6), which yields the following relationships:

$$\frac{d^2u^0}{dx^2} = \frac{B}{D} \int q(x) \ dx$$

$$\frac{d^2v^0}{dx^2} = \frac{C}{D} \int q(x) \ dx \tag{4.8}$$

From the constitutive relations (2.26) we obtain the following relationships for the force and moment resultants:

$$N_x = A_{11}\frac{du^0}{dx} + A_{16}\frac{dv^0}{dx} - B_{11}\frac{d^2w}{dx^2}$$

$$N_y = A_{12}\frac{du^0}{dx} + A_{26}\frac{dv^0}{dx} - B_{12}\frac{d^2w}{dx^2}$$

$$N_{xy} = A_{16}\frac{du^0}{dx} + A_{66}\frac{dv^0}{dx} - B_{16}\frac{d^2w}{dx^2} \tag{4.9}$$

$$M_x = B_{11}\frac{du^0}{dx} + B_{16}\frac{dv^0}{dx} - D_{11}\frac{d^2w}{dx^2}$$

$$M_y = B_{12}\frac{du^0}{dx} + B_{26}\frac{dv^0}{dx} - D_{12}\frac{d^2w}{dx^2}$$

$$M_{xy} = B_{16}\frac{du^0}{dx} + B_{66}\frac{dv^0}{dx} - D_{16}\frac{d^2w}{dx^2}$$

If $q(x) = q_0 = $ constant, direct integration of Equations (4.7) and (4.8) yields

$$u^0 = \frac{B}{D}(q_0 \frac{x^3}{6} + c_1 \frac{x^2}{2} + c_2 x + c_3)$$

$$v^0 = \frac{C}{D}(q_0 \frac{x^3}{6} + c_1 \frac{x^2}{2} + c_4 x + c_5) \tag{4.10}$$

$$w = \frac{A}{D}(q_0 \frac{x^4}{24} + c_1 \frac{x^3}{6} + c_6 \frac{x^2}{2} + c_7 x + c_8)$$

We now consider two sets of boundary conditions.

Case 1: Simple-Supports

at $x = 0$ and $x = a$

$$N_x = N_{xy} = M_x = 0 \tag{4.11}$$

$$w = 0 \tag{4.12}$$

From Equation (4.9) we find that the vanishing of the force and moment resultants at the plate edges, as required by Equation (4.11), can be accomplished only if
at $x = 0$ and a

$$\frac{du^0}{dx} = \frac{dv^0}{dx} = \frac{d^2w}{dx^2} = 0 \tag{4.13}$$

Thus, the conditions given by (4.11) translate into (4.13). In order to prevent rigid body displacements, we assume the supports are fixed at the origin, i.e.,
at $x = 0$

$$u^0 = v^0 = 0 \tag{4.14}$$

Combining Equations (4.10), (4.12), (4.13), and (4.14), we obtain the following displacement functions:

$$u^0 = \frac{Bq_0x^2}{12D}(2x - 3a)$$

$$v^0 = \frac{Cq_0x^2}{12D}(2x - 3a) \tag{4.15}$$

$$w = \frac{Aq_0x}{24D}(x^3 - 2ax^2 + a^3)$$

Substituting Equations (4.15) into Equation (4.9), we obtain the following force and moment resultants:

$$N_x = N_{xy} = 0$$

$$N_y = \frac{q_0x}{2D}(A_{12}B + A_{26}C - B_{12}A)(x - a)$$

$$M_x = -q_0\frac{x}{2}(x - a) \tag{4.16}$$

$$M_y = \frac{q_0x}{2D}(B_{12}B + B_{26}C - D_{12}A)(x - a)$$

$$M_{xy} = \frac{q_0x}{2D}(B_{16}B + B_{66}C - D_{16}A)(x - a)$$

Case 2: Clamped-Supports

at $x = 0$ and a

$$u^0 = v^0 = \frac{dw}{dx} = w = 0 \tag{4.17}$$

Using Equation (4.10) in conjunction with the boundary condition (4.17), we obtain the following displacements:

$$u^0 = \frac{Bq_0x}{12D}(2x^2 - 3ax + a^2)$$

$$v^0 = \frac{Cq_0x}{12D}(2x^2 - 3ax + a^2)$$ (4.18)

$$w = \frac{Aq_0x^2}{24D}(x^2 - 2ax + a^2)$$

Substituting these relationships into Equation (4.9), we obtain the following force and moment resultants:

$$N_x = N_{xy} = 0$$

$$N_y = \frac{q_0}{12D}(A_{12}B + B_{26}C - D_{12}A)(6x^2 - 6ax + a^2)$$

$$M_x = -\frac{q_0}{12}(6x^2 - 6ax + a^2)$$ (4.19)

$$M_y = \frac{q_0}{12D}(B_{12}B + B_{26}C - D_{12}A)(6x^2 - 6ax + a^2)$$

$$M_{xy} = \frac{q_0}{12D}(B_{16}B + B_{26}C - D_{16}A)(6x^2 - 6ax + a^2)$$

In each of the above cases the maximum plate deflection can be written in the form

$$w_{max} = w'(1 + E)$$ (4.20)

where

$$E = \frac{B_{11}B + B_{16}C}{D}$$

and w' denotes the maximum deflection obtained when the coupling coefficients,

B_{ij}, are neglected in the governing equations. In particular,

$$w' = \frac{5q_0a^4}{384D_{11}} \tag{4.21}$$

for simple-supports, and

$$w' = \frac{q_0a^4}{384D_{11}} \tag{4.22}$$

for clamped-supports. It can be shown that E is always positive. Thus, coupling tends to increase the maximum plate deflection. The magnitude of this increase depends on the individual ply properties and the number of plies in the laminate. As an example consider unsymmetric laminates of the class $[0°/90°]_n$ where the subscript n denotes the number of repeating bidirectional units. Since this class of laminates does not contain angle-ply layers, the shear coupling stiffnesses vanish, i.e., $A_{16} = B_{16} = D_{16} = 0$, and Equation (4.20) reduces to

$$w_{max} = w' \left(1 + \frac{B_{11}^2}{A_{11}D_{11} - B_{11}^2}\right) \tag{4.23}$$

where

$$B_{11} = \frac{E_T(E_L/E_T - 1)h^2}{2(1 - v^2_{LT}E_L/E_T)n} \tag{4.24}$$

Thus, in this case the effect of bending-extensional coupling depends on the ply modulus ratio E_L/E_T and the number of repeating bidirectional units, N. The effect of ply properties and stacking sequence on the behavior of unsymmetric laminates will be discussed in more detail in Chapter 7.

4.3 BUCKLING AND FREE-VIBRATION UNDER CYLINDRICAL BENDING

If we neglect inplane inertia effects in the absence of transverse loading ($q = 0$), and impose an initial uniform compressive load $N_x^i = -N_0$, Equation (4.5) becomes

$$D_{11}\frac{d^4w}{dx^4} - B_{11}\frac{d^3u^0}{dx^3} - B_{16}\frac{d^3v^0}{dx^3} + N_0\frac{d^2w}{dx^2} - \varrho\omega^2w = 0 \tag{4.25}$$

Differentiating Equations (4.6) and substituting the results into Equation (4.25), we obtain the following result:

$$\frac{d^4w}{dx^4} + \frac{A}{D}\left(N_0\frac{d^2w}{dx^2} - \varrho\omega^2 w\right) = 0 \tag{4.26}$$

For simply-supported boundary conditions, the following displacements satisfy Equations (4.6), (4.26), and the boundary conditions (4.12) and (4.13):

$$u^0 = A_m\cos\frac{m\pi x}{a}$$

$$v^0 = B_m\cos\frac{m\pi x}{a} \tag{4.27}$$

$$w = C_m\sin\frac{m\pi x}{a}$$

Substituting Equations (4.27) into Equations (4.6) and (4.26), we obtain the following homogeneous equations in matrix form:

$$\begin{bmatrix} 1 & 0 & -\dfrac{B}{A}\left(\dfrac{m\pi}{a}\right) \\[2ex] 0 & 1 & -\dfrac{C}{A}\left(\dfrac{m\pi}{a}\right) \\[2ex] 0 & 0 & F_m \end{bmatrix} \begin{bmatrix} A_m \\[2ex] B_m \\[2ex] C_m \end{bmatrix} = \begin{bmatrix} 0 \\[2ex] 0 \\[2ex] 0 \end{bmatrix} \tag{4.28}$$

where

$$F_m = \frac{m^4\pi^4}{a^4} - \frac{A}{D}\left(\varrho\omega^2 + N_0\frac{m^2\pi^2}{a^2}\right)$$

A nontrivial solution to Equation (4.28) exists only if F_m vanishes. Thus,

$$\omega_m = \frac{m\pi}{a}\sqrt{\frac{1}{\varrho}\frac{D}{A}\left(\frac{m\pi}{a}\right)^2 - N_0}, \qquad N_0 > 0 \tag{4.29}$$

If $N_0 = 0$, Equation (4.29) reduces to

$$\omega_m = \omega_m' \sqrt{1 - H} \qquad (4.30)$$

where

$$H = \frac{(B_{11}B + B_{16}C)}{D_{11}A} = \frac{DE}{D_{11}A}$$

and ω_m' is the flexural vibration frequency of a laminated strip in which B_{ij} is neglected and is of the form

$$\omega_m' = \frac{m^2\pi^2}{a^2} \sqrt{\frac{D_{11}}{\varrho}} \qquad (4.31)$$

The critical buckling load is obtained when ω_m vanishes. Thus,

$$N_{cr} = N_{cr}' (1 - H) \qquad (4.32)$$

where

$$N_{cr}' = D_{11} \frac{\pi^2}{a^2}$$

In Equation (4.32) N_{cr}' represents the critical buckling load for a laminated strip in which the B_{ij} coupling coefficients are neglected. As in the case of E, it can also be shown that $H > 0$. Thus, bending-extensional coupling reduces vibration frequencies and critical buckling loads.

It should be noted that for $N_0 < N_{cr}$, vibration frequencies are reduced by in-plane compressive loads. For the lowest vibration frequency, Equation (4.29) becomes

$$\omega_1 = \frac{\pi}{a} \sqrt{\frac{1}{\varrho} \frac{D}{A}\left(\frac{\pi}{a}\right)^2 - N_0}, \quad N_0 < \frac{D}{A}\left(\frac{\pi}{a}\right)^2 \qquad (4.33)$$

If an initial inplane tensile load $N_x^i = N_0 = $ constant, then the vibration frequencies are increased and the fundamental frequency is given by

$$\omega_1 = \frac{\pi}{a} \sqrt{\frac{1}{\varrho} \frac{D}{A}\left(\frac{\pi}{a}\right)^2 + N_0}, \quad N_0 > 0 \qquad (4.34)$$

Thus inplane compressive loads reduce the effective stiffness of the plate, while inplane tensile loads increase the effective stiffness.

Table 4.1. Cylindrical bending for simply-supported [±45°] laminate [3].

b/a	$w_{max}E_Th^3/q_0a^4$	N_0a^2/E_Th^3	$\omega a^2(\varrho/E_Th^3)^{1/2}$
1	1.018×10^{-2}	15.466	12.355
2	2.983×10^{-2}	5.206	7.168
3	4.186×10^{-2}	3.600	5.961
4	4.750×10^{-2}	3.060	5.495
5	4.981×10^{-2}	2.814	5.270
∞	5.396×10^{-2}	2.382	4.848

4.4 PLATE ASPECT RATIO AND CYLINDRICAL BENDING

Consider a rectangular laminated plate with dimensions a,b relative to the x,y axes, respectively. The aspect ratio b/a required in order to make the assumption of cylindrical bending depends on laminate construction. For unsymmetric cross-ply laminates of the class $[0°/90°]_n$, it has been shown [2] that the maximum deflection under transverse loading rapidly approaches cylindrical bending. For an aspect ratio $b/a = 3$ in conjunction with ply material having a modulus ratio $E_L/E_T = 40$, the plate center deflection was within 4 percent of the center deflection of an infinite strip.

In the case of angle-ply plates, however, where comparatively large values of the shear stiffness, D_{66}, can occur, the convergence to cylindrical bending with increasing aspect ratio is less rapid. This is illustrated in Table 4.1 for a $[±45°]$ angle-ply plate where the maximum deflection under uniform transverse load, buckling loads under uniaxial compression, and fundamental vibration frequencies are presented for various aspect ratios. These results are for simple-support boundary conditions with

$$E_L/E_T = 25, \; G_{LT}/E_T = 0.5, \; \nu_{LT} = 0.25 \qquad (4.35)$$

The rectangular plate solutions are obtained from Reference [3].

4.5 BENDING ANALYSIS OF LAMINATED BEAMS

While cylindrical bending provides a convenient tool for performing a one-dimensional analysis of laminated plates, a theory for laminated, anisotropic beams is also desirable. In addition to being a basic structural element, beam type specimens under concentrated loads are utilized in composite materials characterization.

Beam bending test methods are often based on homogeneous isotropic beam theory (e.g., see ASTM Standard D-790). For laminated materials the classical beam formulas must be modified to account for the stacking sequence of individual plies.

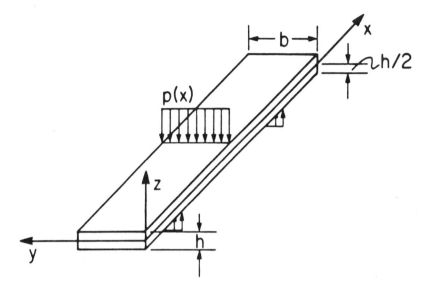

Figure 4.2. Laminated beam.

Consider the laminated beam shown in Figure 4.2. It has been shown by Hoff [4] and Pagano [5] that layered beams in which the plies are oriented symmetrically about the midplane and the orthotropic axes of material symmetry in each ply are parallel to the beam edges can be analyzed by classical beam theory if the bending stiffness EI is replaced by the equivalent stiffness $E_x^b I$ defined in the following manner:

$$E_x^b I = \sum_{k=1}^{N} E_l^k I^k \qquad (4.36)$$

where E_x^b is the effective bending modulus of the beam, E_l^k is the modulus of the kth layer relative to the beam axis, I is the moment of inertia of the kth layer relative to the midplane, and N is the number of layers in the laminate.

Equations which are applicable to a general class of symmetric laminates can be derived by considering a beam as a special case of a laminated plate [6]. Unlike the case of cylindrical bending, laminated beam theory assumes that the length is much larger than the width, i.e., $L >> b$, as illustrated in Figure 4.2. The difference between cylindrical bending and beam theory is analogous to the difference between plane strain and plane stress in classical theory of elasticity.

For bending of symmetric laminates, the constitutive relations (2.26) reduce to the form

$$\begin{bmatrix} M_x \\ M_y \\ M_{xy} \end{bmatrix} = \begin{bmatrix} D_{11} & D_{12} & D_{16} \\ D_{12} & D_{22} & D_{26} \\ D_{16} & D_{26} & D_{66} \end{bmatrix} \begin{bmatrix} \varkappa_x \\ \varkappa_y \\ \varkappa_{xy} \end{bmatrix} \tag{4.37}$$

where, as previously defined in Equation (2.7),

$$\varkappa_x = -\frac{\partial^2 w}{\partial x^2}, \; \varkappa_y = -\frac{\partial^2 w}{\partial y^2}, \; \varkappa_{xy} = -2\frac{\partial^2 w}{\partial x \partial y} \tag{4.38}$$

For present purposes it is useful to consider Equation (4.37) in the inverted form

$$\begin{bmatrix} \varkappa_x \\ \varkappa_y \\ \varkappa_{xy} \end{bmatrix} = \begin{bmatrix} D_{11}^* & D_{12}^* & D_{16}^* \\ D_{12}^* & D_{22}^* & D_{26}^* \\ D_{16}^* & D_{26}^* & D_{66}^* \end{bmatrix} \begin{bmatrix} M_x \\ M_y \\ M_{xy} \end{bmatrix} \tag{4.39}$$

where D_{ij}^* are elements of the inverse matrix of D_{ij}.

In order to derive a beam theory the following assumptions are made:

$$M_y = M_{xy} = 0 \tag{4.40}$$

Using Equations (4.38) and (4.39) in conjunction with Equation (4.40), we find

$$\varkappa_x = -\frac{\partial^2 w}{\partial x^2} = D_{11}^* M_x \tag{4.41}$$

Since beams have a high length-to-width ratio, it is assumed that

$$w = w(x) \tag{4.42}$$

Caution must be exercised in applying Equation (4.42) to laminated anisotropic materials. In particular, Equations (4.38), (4.39), and (4.40) imply that both the curvatures $\varkappa_y$ and $\varkappa_{xy}$ are functions of the bending moment M_x, that is,

$$\varkappa_y = -\frac{\partial^2 w}{\partial y^2} = D_{12}^* M_x, \; \varkappa_{xy} = -2\frac{\partial^2 w}{\partial x \partial y} = D_{16}^* M_x \tag{4.43}$$

Thus, the deflection, w, cannot be independent of y. Even in homogeneous isotropic beam theory the one-dimensional assumption is not strictly correct due to

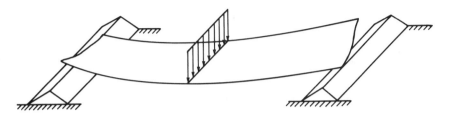

Figure 4.3. *Effect of bending-twisting coupling on the bending of anisotropic beams.*

the effect of Poisson's ratio, D_{12}^* in Equation (4.43). The effect is negligible, however, if the length-to-width ratio, R is moderately large. In the case of aniso-tropic shear coupling, as displayed by D_{16}^* in Equation (4.43), the effect can be more severe [7]. This is of particular importance for angle-ply laminates, as the length-to-width ratio is not large in the case of a laboratory type flexure spec-imen, that is, the specimen is more like a plate strip than a beam. The twisting curvature induced by the D_{16}^* term in Equation (4.43) can cause the specimen to lift off its supports at the corners [7,8]. This phenomenon is illustrated in Figure 4.3. Thus, for angle-ply laminates R must be rather large for Equation (4.42) to be valid.

Combining Equations (4.41) and (4.42), we obtain the following result:

$$\frac{d^2w}{dx^2} = - \frac{M}{E_x^b I} \tag{4.44}$$

where

$$E_x^b = \frac{12}{h^3 D_{11}^*} , \ M = bM_x, \ I = \frac{bh^3}{12}$$

and b is the width of the beam. Equation (4.44) is in the same form as classical beam theory with the homogeneous, isotropic modulus E replaced by the effec-tive bending modulus of the laminated beam, E_x^b.

For static bending in the absence of body moments and inplane force effects, the equation of motion (2.22) becomes

$$\frac{\partial^2 M_x}{\partial x^2} + 2 \frac{\partial^2 M_{xy}}{\partial x \partial y} + \frac{\partial^2 M_y}{\partial y^2} + q = 0 \tag{4.45}$$

Substituting Equation (4.41) into Equation (4.45) and taking Equation (4.42) into account, we obtain the relationship

$$\frac{d^4w}{dx^4} = D_{11}^* q \tag{4.46}$$

Multiplying by the beam width, Equation (4.46) becomes

$$\frac{d^4w}{dx^4} = \frac{p}{E_x^b I} \tag{4.47}$$

where

$$p = bq$$

Again, Equation (4.47) is analolgous to classical homogeneous, isotropic beam theory.

Under static loading in the absence of body forces, Equation (2.19) becomes

$$\frac{\partial M_x}{\partial x} + \frac{\partial M_{xy}}{\partial y} - Q_x = 0 \tag{4.48}$$

Substituting Equation (4.40) into Equation (4.48) and multiplying the result by b, we obtain the relationship

$$Q = \frac{dM}{dx} \tag{4.49}$$

where

$$Q = bQ_x$$

Equation (4.49) is exactly the same as in homogeneous, isotropic beam theory. Such a result is anticipated since this relationship is simply a statement of equilibrium between the bending moment and transverse shear resultant.

Stresses in the kth layer of the beam are given by equation (2.24). For symmetric laminates under bending loads

$$\epsilon_x = z\varkappa_x, \ \epsilon_y = z\varkappa_y, \ \epsilon_{xz} = z\varkappa_{xy} \tag{4.50}$$

and Equation (2.24) becomes

$$\begin{bmatrix} \sigma_x^{(k)} \\ \sigma_y^{(k)} \\ \sigma_{xy}^{(k)} \end{bmatrix} = z \begin{bmatrix} Q_{11}^{(k)} & Q_{12}^{(k)} & Q_{16}^{(k)} \\ Q_{12}^{(k)} & Q_{22}^{(k)} & Q_{26}^{(k)} \\ Q_{16}^{(k)} & Q_{26}^{(k)} & Q_{66}^{(k)} \end{bmatrix} \begin{bmatrix} \varkappa_x \\ \varkappa_y \\ \varkappa_{xy} \end{bmatrix} \tag{4.51}$$

Combining Equations (4.51) and (4.39), and multiplying the results by b, we obtain the following ply stresses:

$$\sigma_x^{(k)} = z f_1^{(k)} \frac{M}{I} \tag{4.52}$$

$$\sigma_y^{(k)} = z f_2^{(k)} \frac{M}{I} \tag{4.53}$$

$$\sigma_{xy}^{(k)} = z f_3^{(k)} \frac{M}{I} \tag{4.54}$$

where

$$f_1^{(k)} = (Q_{11}^{(k)} D_{11}^* + Q_{12}^{(k)} D_{12}^* + Q_{16}^{(k)} D_{16}^*) \frac{h^3}{12}$$

$$f_2^{(k)} = (Q_{12}^{(k)} D_{11}^* + Q_{22}^{(k)} D_{12}^* + Q_{26}^{(k)} D_{16}^*) \frac{h^3}{12}$$

$$f_3^{(k)} = (Q_{16}^{(k)} D_{11}^* + Q_{26}^{(k)} D_{12}^* + Q_{66}^{(k)} D_{16}^*) \frac{h^3}{12}$$

For homogeneous beams $f_1^{(k)} = $ unity, $f_2^{(k)} = f_3^{(k)} = 0$, and Equation (4.52) reduces to classical beam theory, while Equations (4.53) and (4.54) vanish. Attention should be focused on the multidirectional stresses in a laminated beam as displayed in Equations (4.52–4.54). These stresses are not correct in a zone of approximately one laminate thickness, h, away from the free edge where the state of stress is three-dimensional in nature [9,10]. Thus, the dimensions of a laminated beam must be such that b/h is large enough to minimize free-edge effects.

The interlaminar shear stress is often of interest in a laminated beam because of the relatively weak interlaminar shear strength encountered in composite materials. Determination of interlaminar shear stresses for layered plates was discussed in Chapter 2. For laminated beams the interlaminar shear stress σ_{xz} can be determined by substituting Equations (4.52) and (4.54) into the first of Equations (2.8). Integrating the results we find that

$$\sigma_{xz}^{(k)} = -\frac{1}{I} \int_{-h/2}^{z_k} f_1^{(k)} \frac{dM}{dx} z \, dz \tag{4.55}$$

Using Equation (4.39) in conjunction with Equation (4.55), we obtain

$$\sigma_{xz}^{(k)} = - \frac{Q}{I} \int_{-h/2}^{z_k} f_1^{(k)} z \, dz \qquad (4.56)$$

The integral in Equation (4.56) assures continuity of the transverse shear stress at layer interfaces. For homogeneous materials the integration can be accomplished in closed form with the result

$$\sigma_{xz} = \left[1 - 4 \left(\frac{z}{h} \right)^2 \right] \frac{h^2 Q}{8I} \qquad (4.57)$$

which is the relationship obtained from classical beam theory.

4.6 BENDING OF LAMINATED BEAMS UNDER CONCENTRATED LOADS

Consider a beam under 3-point bending, as illustrated in Figure 4.4. The bending moment in the left half of the beam is given by the relationship

$$M = - \frac{Px}{2}, \, 0 \le x \le L/2 \qquad (4.58)$$

Substituting this relationship into Equation (4.44), we obtain

$$\frac{d^2 w}{dx^2} = \frac{Px}{2 E_x^b I}, \, 0 \le x \le L/2 \qquad (4.59)$$

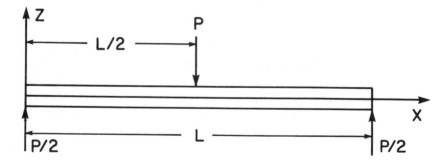

Figure 4.4. Beam under 3-point bending.

Symmetry of the loading allows a solution to be obtained by considering the left side of the beam only. The boundary conditions for simple-supports are
   at $x = 0$:

$$M = w = 0 \tag{4.60}$$

Equation (4.58) assures that the bending moment vanishes at $x = 0$. Symmetry conditions require that
   at $x = L/2$:

$$\frac{dw}{dx} = 0 \tag{4.61}$$

Integrating Equation (4.59) twice and applying the displacement conditions in Equation (4.60) and (4.61), we obtain the result

$$w = -\frac{PL^2 x}{48 E_x^b I} \left[3 - \left(\frac{2x}{L} \right)^2 \right] \tag{4.62}$$

The modulus can be expressed in terms of the center deflection, w_c, by solving Equation (4.62) for E_x^b with $x = L/2$.

$$E_x^b = \frac{PL^3}{48 w_c I} = \frac{PL^3}{4 b h^3 w_c} \tag{4.63}$$

This relationship can be utilized for determining modulus in a 3-point flexure test.

The inplane stresses can be determined by substituting Equation (4.58) into Equations (4.52–4.54). It is easily seen that the maximum stress occurs at the center of the beam ($x = L/2$). For the maximum normal stress

$$\sigma_x^{(k)} (L/2) = -3 z f_1^{(k)} \frac{PL}{bh^3} \tag{4.64}$$

The maximum stress values of $\sigma_y^{(k)}$ and $\sigma_{xy}^{(k)}$ are of the same form as Equation (4.64). Note that the maximum stress does not always occur at the outer surface of the beam. For homogeneous beams, $f_1^{(k)} = 1$, and the maximum normal tensile stress, σ_m, is on the bottom surface ($z = -h/2$). Thus, for homogeneous beams

$$\sigma_m = \frac{3PL}{2bh^2} \text{ (HOMOGENEOUS BEAM)} \tag{4.65}$$

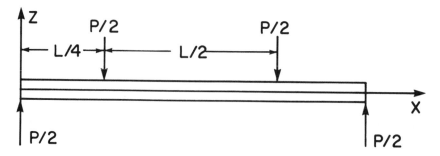

Figure 4.5. *Beam under 4-point bending.*

Now consider the beam under 4-point loading at quarter points, as illustrated in Figure 4.5. For the left half of the beam, the bending moment is given by the relationships

$$M = -\frac{Px}{2}, 0 \leq x \leq L/4 \tag{4.66}$$

$$M = -\frac{PL}{8}, L/4 \leq x \leq L/2 \tag{4.67}$$

Substituting these relationships into Equation (4.44), we obtain

$$\frac{d^2w}{dx^2} = \frac{d^2w_1}{dx^2} = \frac{Px}{2E_x^b I}, 0 \leq x \leq L/4 \tag{4.68}$$

$$\frac{d^2w}{dx^2} = \frac{d^2w_2}{dx^2} = \frac{PL}{8E_x^b I}, L/4 \leq x \leq L/2 \tag{4.69}$$

where

$$w_1 = w, 0 \leq x \leq L/4$$

$$w_2 = w, L/4 \leq x \leq L/2$$

As in the case of 3-point loading, symmetry allows a solution to be obtained by considering the left half of the beam only. The simple support boundary conditions as given by Equation (4.60) are applicable to the 4-point loading case. The moment condition is satisfied by Equation (4.68). Symmetry again requires the slope of the bending deflection to vanish at the center of the beam. In addi-

tion, continuity of deflection and deflection slope must be attained at $x = L/4$. Thus, a solution for bending deflection can be obtained by integrating Equations (4.68) and (4.69), and applying the conditions
  at $x = 0$:

$$w_1 = 0 \tag{4.70}$$

  at $x = L/4$:

$$w_1 = w_2, \frac{dw_1}{dx} = \frac{dw_2}{dx} \tag{4.71}$$

  at $x = L/2$:

$$\frac{dw_2}{dx} = 0 \tag{4.72}$$

The resulting relationships for beam deflection are

$$w_1 = \frac{-PL^2x}{192E_x^b I}\left[9 - 16\left(\frac{x}{L}\right)^2\right] \tag{4.73}$$

$$w_2 = \frac{PL^3}{768E_x^b I}\left[1 - 48\left(\frac{x}{L}\right) + 48\left(\frac{x}{L}\right)^2\right] \tag{4.74}$$

The modulus can be expressed in terms of either the quarter-point deflection or the center deflection by solving Equation (4.74) for E_x^b with $x = L/4$ and $x = L/2$, respectively. Such a procedure leads to the following results:

$$E_x^b = \frac{PL^3}{96w_q I} = \frac{PL^3}{8bh^3 w_q} \tag{4.75}$$

$$E_x^b = \frac{11PL^3}{768w_c I} = \frac{11PL^3}{64bh^3 w_c} \tag{4.76}$$

where w_q is the quarter-point deflection.

A cursory examination of Equations (4.66) and (4.67) reveals that the maximum stresses occur in the center section of the beam ($L/4 \leq x \leq L/2$). Substituting Equation (4.67) into Equation (4.52), we find that

$$\sigma_x^{(k)} = -\frac{3zf_1^{(k)}PL}{2bh^3}, \ L/4 \leq x \leq L/2 \tag{4.77}$$

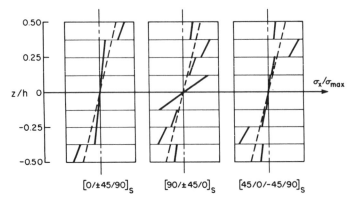

Figure 4.6. *Effect of stacking sequence on stress distribution through laminate thickness.*

As in the case of the 3-point flexure beam, the maximum stress does not necessarily occur at the outer surface. For the homogeneous case $f_1^{(k)} = 1$, and the maximum stress occurs at $z = -h/2$. Thus,

$$\sigma_m = \frac{3PL}{4bh^2} \text{ (HOMOGENEOUS BEAM)} \tag{4.78}$$

As a numerical sample of the heterogeneous nature of a multi-directional composite, we consider laminated beams containing various stacking sequences of $0°$, $\pm45°$, and $90°$ ply orientations subjected to 4-point bending. The following unidirectional graphite/epoxy ply properties taken from Reference [6] are utilized in the calculations:

$$E_L = 126 \text{ } GPa \text{ (18.3} \times 10^6 \text{ psi), } G_{LT} = 7.6 \text{ } GPa \text{ (1.1} \times 10^6 \text{ psi)} \tag{4.79}$$

$$E_T = 12.2 \text{ } GPa \text{ (1.77} \times 10^6 \text{ psi), } \nu_{LT} = 0.3$$

Using Equation (4.77) the normal stress distribution $\sigma_x^{(k)}$ is plotted through-the-thickness in Figure 4.6 for various stacking sequences. The stresses displayed in this figure are normalized by the maximum stress occurring through-the-thickness. In each stacking geometry considered, the dotted line represents classical homogeneous beam theory as determined from Equation (4.78). Severe stacking sequence effects on $\sigma_x^{(k)}$ are clearly evident in Figure 4.6. Only in the case where the $0°$ plies are on the outside of the beam does the maximum stress occur on the outer surface. This heterogeneous behavior will induce a strong stacking sequence dependence on the apparent flexural strength.

Substituting Equation (4.66) into Equation (4.57), we obtain the following interlaminar shear stress distribution for a homogeneous beam subjected to 4-point bending.

$$\sigma_{xz} = -\frac{3}{4}\left[1 - 4\left(\frac{z}{h}\right)^2\right]\frac{P}{bh}, \, 0 \leq x \leq L/4 \qquad (4.80)$$

It is easily seen that the maximum shear value, τ_{max}, occurs at the center of the beam cross-section.

$$\tau_{max} = -\frac{3P}{4bh} \, (\text{HOMOGENEOUS BEAM}) \qquad (4.81)$$

For the 4-point bend specimen at quarter points, the maximum shear stress is determined by combining Equations (4.49), (4.56), and (4.66) with the result:

$$\frac{\sigma_{xz}^{(k)}}{\tau_0} = -\frac{8}{h^2}\int_{-h/2}^{z_k} f_1^{(k)} z \, dz \qquad (4.82)$$

where τ_0 is the maximum stress in a homogeneous beam, as given by Equation (4.81).

The effect of stacking sequence on the interlaminar shear stress distribution, $\sigma_{xz}^{(k)}$, through-the-thickness of the beam is illustrated in Figures 4.7 and 4.8.

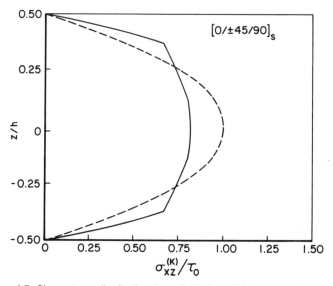

Figure 4.7. *Shear stress distribution through laminate thickness — 0° outer plies.*

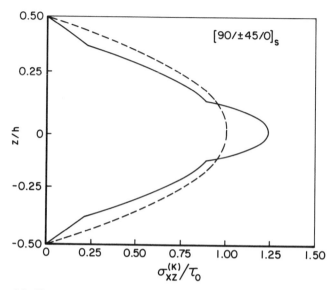

Figure 4.8. *Shear stress distribution through laminate thickness − 90° outer plies.*

These laminates are constructed of materials with the ply properties given by equation (4.79). Classical homogeneous beam theory is shown as the dotted line in these figures. Note that the maximum shear stress is highly dependent on the stacking sequence with considerable departure from classical beam theory.

Experimental data on graphite/epoxy beams with the unidirectional properties given by (4.79) have been reported in the literature [6]. This data is shown in Tables 4.2 and 4.3 for laminates with various stacking geometries of 0°, ±45°, 90° orientations. The following nomenclature is utilized for the strength data reported in Table 4.3:

S_t = Tensile strength as determined from a standard straight-sided tensile coupon.

S_f = Flexural strength as determined from laminated beam theory in conjunction with a 4-point test at quarter points.

S_f^* = Flexural strength as determined from homogeneous beam theory in conjunction with a 4-point bend test at quarter points.

Laminate flexural strength, S_f, was determined from Equation (4.77) with $z = -h/2$, i.e.,

$$S_f = \frac{3PL}{4bh^2} f_1^{(1)} \tag{4.83}$$

Table 4.2. Elastic modulus: graphite/epoxy laminates [6]: GPa (Msi).

Stacking Sequence	Theory		Experiment	
	E_x	E_x^b	E_x	E_x^b
0/±45/90	53 (7.6)	83 (12)	48 (7.0)	69 (10)
90/±45/0	53 (7.6)	24 (3.5)	46 (6.6)	19 (2.7)
+45/0/−45/90	53 (7.6)	55 (7.9)	47 (6.8)	48 (6.9)

where $f_1^{(1)}$ denotes the value of $f_1^{(k)}$ in the outer plies. Values of S_f^* were determined from Equation (4.78). A span-to-depth ratio, $L/h = 32$, was utilized in these beam experiments. In addition, $b = 19$ mm (0.75 in.) and $h = 1.9$ mm (0.075 in.).

Theoretical values of E_x in Table 4.2 were determined by inverting the constitutive relations (2.26). For symmetric laminates containing equal numbers of layers oriented at $\pm\theta$ we find that

$$E_x = \frac{A_{11}A_{22} - A_{12}^2}{A_{22}h} \tag{4.84}$$

Since the A_{ij} matrix is independent of stacking sequence, the modulus, E_x, should also be independent of stacking sequence. In a similar manner a cursory examination of Equations (2.54)–(2.56) reveals that the ply stresses for inplane loading of symmetric laminates are independent of stacking sequence. Thus, we would anticipate that the inplane tensile strengths would also be independent of stacking sequence.

Reasonably good agreement between theory and experiment was obtained in Table 4.2 for modulus measurements. As anticipated, E_x^b is strongly dependent on stacking sequence. Note that the flexural modulus data in Table 4.2 is not adjusted for bending-twisting coupling effects. For the data reported here, however, D_{16}^* is relatively small as only half of the layers contain angle-ply orientations.

Table 4.3. Strength: graphite/epoxy laminates [6]: MPa (ksi).

Stacking Sequence	S_t	S_f	S_f^*
0/±45/90	509 (73.5)	1220 (177)	823 (119)
90/±45/0	408 (58.9)	142 (20.5)	286 (41.4)
+45/0/−45/90	463 (66.9)	265 (38.3)	648 (93.7)

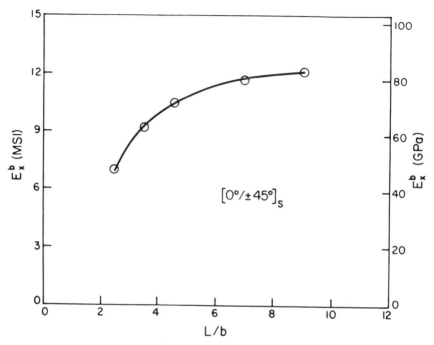

Figure 4.9. *Effect of bending-twisting coupling on experimentally measured values of bending modulus [6].*

Although the inplane tensile strengths in Table 4.3 show some dependence on stacking sequence, the flexure data is strongly dependent on stacking sequence.

The effect of bending-extensional coupling on experimental measurements of E_x^b is shown in Figure 4.9 for a $[0°/\pm45°]_s$ graphite/epoxy laminate under 4-point bending at quarter points [6]. This material has unidirectional ply properties similar to those listed in (4.79). As easily observed, the measured value of E_x^b changes with increasing values of L/b. These results suggest that for $L/b > 10$, there is little effect of shear coupling. For $5 < L/b < 10$, there will be a moderate shear coupling effect, and for $L/b < 5$, the effect will be quite severe. This laminate geometry contains twice as many angle-ply layers as unidirectional layers. Thus, the bending-twisting coupling is more severe than for the laminates considered in Tables 4.2 and 4.3.

4.7 BUCKLING AND FREE-VIBRATIONS OF LAMINATED BEAMS

Equation (2.22) is utilized for buckling and vibration analysis. For the one-dimensional case now under consideration in conjunction with the beam theory, this equation takes the form

$$\frac{\partial^2 M_x}{\partial x^2} + N_x^i \frac{\partial^2 w}{\partial x^2} = \varrho \frac{\partial^2 w}{\partial t^2} \tag{4.85}$$

Substituting Equation (4.44) into Equation (4.85) and taking the inplane load as uniform compression, $N_x^i = - N_0 = $ constant, we obtain the governing equation

$$\frac{\partial^4 w}{\partial x^4} + \frac{12}{E_x^b h^3} \left(\varrho \frac{\partial^2 w}{\partial t^2} + N_0 \frac{\partial^2 w}{\partial x^2} \right) = 0 \tag{4.86}$$

Time can be eliminated from Equation (4.86) in the usual manner by assuming

$$w(x,t) = w(x) \, e^{i\omega t} \tag{4.87}$$

Substituting this relationship into Equation (4.86), we obtain the result

$$\frac{d^4 w}{dx^4} + \frac{12}{E_x^b h^3} \left(N_0 \frac{d^2 w}{dx^2} - \varrho \omega^2 w \right) = 0 \tag{4.88}$$

Note that Equation (4.88) is exactly of the same form as Equation (4.26). Thus, a solution for the simply-supported boundary conditions $w(0) = w(L) = M(0) = M(L) = 0$, is of the form

$$w = W_m \sin \frac{m\pi x}{L}, \quad m = 1,2, \dots \tag{4.89}$$

Combining Equations (4.88) and (4.89), we obtain

$$\omega_m = \frac{m\pi}{L} \sqrt{\frac{1}{\varrho} \left(\frac{m^2 \pi^2 E_x^b h^3}{12L^2} - N_0 \right)}, \quad N_0 > 0 \tag{4.90}$$

When N_0 vanishes, Equation (4.89) reduces to

$$\omega_m = \frac{m^2 \pi^2}{2L^2} \sqrt{\frac{E_x^b h^3}{3\varrho}} \tag{4.91}$$

The critical buckling load $N_0 = N_{cr}$, which corresponds to $m = 1$, is obtained by allowing ω_m to vanish.

$$N_{cr} = \frac{\pi^2 E_x^b h^3}{12 L^2} \tag{4.92}$$

The major difference between cylindrical bending and beam theory for symmetric laminates is the bending stiffness terms. This can be determined by comparing Equations (4.91) and (4.92) with Equations (4.31) and (4.32), respectively. The bending stiffness D_{11} can be written in the form

$$D_{11} = \frac{E_x^b h^3}{12[1 - (\nu_{xy}^b)^2 E_y^b / E_x^b]} \tag{4.93}$$

Thus, the major difference is in the term $[1 - (\nu_{xy}^b)^2 E_y^b / E_x^b]$. This is analogous to the difference between plane stain and plane stress in classical theory of elasticity where cylindrical bending corresponds to plane strain and beam theory to plane stress. In many cases numerical results will be similar between cylindrical bending and beam theory. The largest difference will occur for laminates containing angle-ply layers. In such cases ν_{xy}^b can be very large.

As another example of buckling, consider the classical case of clamped ends. The boundary conditions are as follows:
at $x = 0$ and L

$$w = \frac{dw}{dx} = 0 \tag{4.94}$$

A solution to Equation (4.88), with $\omega = 0$, which satisfies the boundary conditions, Equation (4.94), is of the form

$$w = W_m (1 - \cos 2\frac{m\pi x}{L}), \ m = 1,2, \ldots \tag{4.95}$$

Substituting this relationship into Equation (4.88), we obtain

$$N_0 = \frac{E_x^b h^3 m^2 \pi^2}{3 L^2} \tag{4.96}$$

The critical load corresponds to $m = 1$. Thus,

$$N_{cr} = \frac{E_x^b h^3 \pi^2}{3 L^2} \text{(CLAMPED BOUNDARIES)} \tag{4.97}$$

A comparison between Equations (4.92) and (4.97) reveals that the clamped boundary conditions produce a buckling load which is four times higher than for simply-supported boundary conditions. It should be noted that an antisymmetric solution exists to this buckling problem. However, the symmetric mode with $m = 1$, as given by Equation (4.95), always corresponds to the critical buckling load [11].

REFERENCES

1. Whitney, J. M. "Cylindrical Bending of Unsymmetrically Laminated Plates," *Journal of Composite Materials*, 3:715–719 (October 1969).
2. Whitney, J. M. "Bending-Extensional Coupling in Laminated Plates Under Transverse Loading," *Journal of Composite Materials*, 3:20–28 (January 1969).
3. Whitney, J. M. and A. W. Leissa. "Analysis of a Simply-Supported Laminated Anisotropic Rectangular Plate," *AIAA Journal*, 7:28–33 (January 1970).
4. Hoff, N. J. "Strength of Laminates and Sandwich Structural Elements," Chapter 1, *Engineering Laminates*, edited by A. G. H. Dietz, John Wiley and Sons, New York (1949).
5. Pagano, N. J. "Analysis of the Flexure Test of Bidirectional Composites," *Journal of Composite Materials*, 1:336–342 (October 1967).
6. Whitney, J. M., C. E. Browning and A. Mair. "Analysis of the Flexure Test for Laminated Composite Materials," *Composite Materials: Testing and Design (Third Conference)*, ASTM STP 546, American Society for Testing and Materials, pp. 30–45 (1974).
7. Whitney, J. M. and R. J. Daukseys. "Flexure Experiments on Off-Axis Composites," *Journal of Composite Materials*, 4:135–137 (April 1970).
8. Halpin, J. C. *Primer on Composite Materials: Analysis*. Revised Edition, Technomic Publishing Co. (1984).
9. Pipes, R. B. and N. J. Pagano. "Interlaminar Stresses in Composite Laminates Under Uniform Axial Extension," *Journal of Composite Materials*, 4:538–548 (October 1970).
10. Pagano, N. J. and R. B. Pipes, "Influence of Stacking Sequence on Laminate Strength," *Journal of Composite Materials*, 5:50–57 (January 1971).
11. Timoshenko, S. P. and J. M. Gere. *Theory of Elastic Stability*. Second Edition, McGraw-Hill Book Company, Inc. (1961).

HAPTER 5

Specially Orthotropic Plates

5.1 INTRODUCTION

THERE ARE VARYING degrees of complexity encountered in laminated plate analysis. The least complicated are the one-dimensional formulations discussed in Chapter 4. Both symmetric and unsymmetric laminates are handled in a straightforward manner when the deformations are one-dimensional in nature. In this chapter and the next two, we consider two-dimensional problems of increasing complexity.

In the case of two-dimensional laminated plate analysis, the first degree of simplification is for the plate to be symmetric, thus uncoupling inplane and bending response. Symmetric laminates can be broken into two general categories, specially orthotropic plates and midplane symmetric laminates (anisotropic plates). In this chapter we consider bending under transverse load, buckling, and free-vibration of plates in the specially orthotropic category.

Laminates which are symmetric, $B_{ij} = 0$, and for which the bending-twisting coupling terms vanish, $D_{16} = D_{26} = 0$, are referred to as *specially orthotropic*. Since solutions are more easily attained for such plates, composite laminates are often assumed to behave as specially orthotropic plates. Most engineering laminates, however, do not qualify as specially orthotropic because of the existence of D_{16} and D_{26} bending-twisting coupling terms. In many practical applications it is a reasonable approximation to assume specially orthotropic response. The solutions presented in this chapter will be compared to more complex solutions including bending-twisting coupling terms (D_{16}, D_{26}) in Chapter 6, and with solutions including the bending-extensional coupling terms (B_{ij}) in Chapter 7. Thus, the inaccuracies associated with assuming specially orthotropic behavior will be assessed and guidelines established for using the simpler model as presented in this chapter.

5.2 BENDING OF SIMPLY-SUPPORTED RECTANGULAR PLATES

Consider the rectangular laminated plate shown in Figure 5.1 subjected to the transverse load $q = q(x,y)$. For this loading in conjunction with specially ortho-

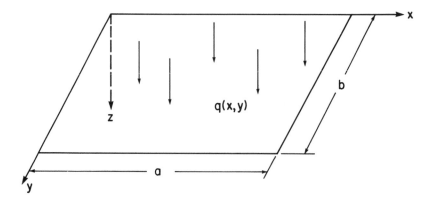

Figure 5.1. *Plate geometry.*

tropic plates, Equation (2.43) in the absence of inertia terms (static loading) reduces to

$$D_{11} \frac{\partial^4 w}{\partial x^4} + 2(D_{12} + 2D_{66}) \frac{\partial^4 w}{\partial x^2 \partial y^2} + D_{22} \frac{\partial^4 w}{\partial y^4} = q \tag{5.1}$$

For a simply-supported plate the following boundary conditions are applicable:

at $x = 0$ and a

$$w = M_x = -D_{11} \frac{\partial^2 w}{\partial x^2} - D_{12} \frac{\partial^2 w}{\partial y^2} = 0 \tag{5.2}$$

at $y = 0$ and b

$$w = M_y = -D_{12} \frac{\partial^2 w}{\partial x^2} - D_{22} \frac{\partial^2 w}{\partial y^2} = 0 \tag{5.3}$$

The moment deflection relations are obtained from the laminate constitutive relations, Equations (2.26).

We consider cases in which the transverse load can be expanded in the double Fourier sine series

$$q(x,y) = \sum_{m=1}^{\infty} \sum_{n=1}^{\infty} q_{mn} \sin \frac{m\pi x}{a} \sin \frac{n\pi y}{b} \tag{5.4}$$

where the Fourier coefficients are determined from the relationship

$$q_{mn} = \frac{4}{ab} \int_0^b \int_0^a q(x,y) \sin \frac{m\pi x}{a} \sin \frac{n\pi y}{b} \, dx \, dy \tag{5.5}$$

A solution to Equation (5.1) which satisfies the boundary conditions (5.2) and (5.3) is of the form

$$w = \sum_{m=1}^{\infty} \sum_{n=1}^{\infty} A_{mn} \sin \frac{m\pi x}{a} \sin \frac{n\pi y}{b} \tag{5.6}$$

Substituting this equation into Equation (5.1), we obtain the infinite series solution

$$w = \frac{a^4}{\pi^4} \sum_{m=1}^{\infty} \sum_{n=1}^{\infty} \frac{q_{mn}}{D_{mn}} \sin \frac{m\pi x}{a} \sin \frac{n\pi y}{b} \tag{5.7}$$

where

$$D_{mn} = D_{11}m^4 + 2(D_{12} + 2D_{66})(mnR)^2 + D_{22}(nR)^4$$

where $R = a/b$. Substituting Equation (5.7) into the laminate constitutive relations, Equations (2.26), we obtain the following expressions for the moment resultants

$$M_x = \frac{a^2}{\pi^2} \sum_{m=1}^{\infty} \sum_{n=1}^{\infty} \frac{q_{mn}}{D_{mn}} (m^2 D_{11} + n^2 R^2 D_{12}) \sin \frac{m\pi x}{a} \sin \frac{n\pi y}{b} \tag{5.8}$$

$$M_y = \frac{a^2}{\pi^2} \sum_{m=1}^{\infty} \sum_{n=1}^{\infty} \frac{q_{mn}}{D_{mn}} (m^2 D_{12} + n^2 R^2 D_{22}) \sin \frac{m\pi x}{a} \sin \frac{n\pi y}{b} \tag{5.9}$$

$$M_{xy} = -2 \frac{a^2 R}{\pi^2} D_{66} \sum_{m=1}^{\infty} \sum_{n=1}^{\infty} \frac{mn q_{mn}}{D_{mn}} \cos \frac{m\pi x}{a} \cos \frac{n\pi y}{b} \tag{5.10}$$

From Equations (2.54–2.56) we derive the following ply inplane stresses:

$$\sigma_x^{(k)} = \frac{a^2}{\pi^2} z \sum_{m=1}^{\infty} \sum_{n=1}^{\infty}$$

(5.11)

$$\frac{q_{mn}}{D_{mn}} (Q_{11}^{(k)}m^2 + Q_{12}^{(k)}n^2R^2)\sin \frac{m\pi x}{a} \sin \frac{n\pi y}{b}$$

$$\sigma_y^{(k)} = \frac{a^2}{\pi^2} z \sum_{m=1}^{\infty} \sum_{n=1}^{\infty} \frac{q_{mn}}{D_{mn}}$$

(5.12)

$$(Q_{12}^{(k)}m^2 + Q_{22}^{(k)}n^2R^2)\sin \frac{m\pi x}{a} \sin \frac{n\pi y}{b}$$

$$\tau_{xy}^{(k)} = -2 \frac{a^2R}{\pi^2} Q_{66}^{(k)} z \sum_{m=1}^{\infty} \sum_{n=1}^{\infty} \frac{mn\, q_{mn}}{D_{mn}} \cos \frac{m\pi x}{a} \cos \frac{n\pi y}{b}$$

(5.13)

The interlaminar shear stresses can be determined by substituting Equations (5.11–5.13) into the first two equilibrium Equations in (2.8) and integrating with respect to z.

$$\tau_{xz}^{(k)} = -\frac{a}{\pi} \sum_{m=1}^{\infty} \sum_{n=1}^{\infty} \int_{-h/2}^{z^{(k)}} \frac{m\, q_{mn}}{D_{mn}} [Q_{11}^{(k)}m^2 + (Q_{12}^{(k)}$$

(5.14)

$$+ 2Q_{66}^{(k)})n^2R^2] \cos \frac{m\pi x}{a} \sin \frac{n\pi y}{b} \, z \, dz$$

$$\tau_{yz}^{(k)} = \frac{a}{\pi} \sum_{m=1}^{\infty} \sum_{n=1}^{\infty} \int_{-h/2}^{z^{(k)}} \frac{n\, q_{mn}}{D_{mn}} [(Q_{12}^{(k)} + 2Q_{66}^{(k)})m^2$$

(5.15)

$$+ Q_{22}^{(k)}n^2R^2] \sin \frac{m\pi x}{a} \cos \frac{n\pi y}{b} \, z \, dz$$

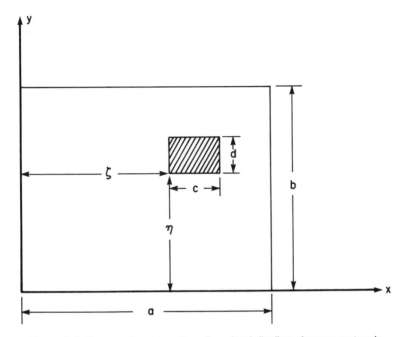

Figure 5.2. *Rectangular plate with uniform load distributed over a rectangle.*

One loading of common interest is the uniform load $q = q_0 =$ constant. For this case the integral in Equation (5.5) yields

$$q_{mn} = \frac{16 \, q_0}{\pi^2 mn} \quad (m,n \text{ odd})$$

$$q_{mn} = 0 \qquad (m,n \text{ even})$$

(5.16)

Another load of particular interest is a single load P uniformly distributed over a rectangle as shown in Figure 5.2. For this case the integral in Equation (5.5) leads to the result

$$q_{mn} = \frac{16P}{\pi^2 mn \ cd} \sin \frac{m\pi\xi}{a} \sin \frac{n\pi\eta}{b} \sin \frac{m\pi c}{2a} \sin \frac{n\pi d}{2b} \qquad (5.17)$$

where c and d are dimensions of the rectangle parallel to the x and y directions, respectively, and ξ and η are the x and y coordinates, respectively, of the center of the rectangle.

Table 5.1 Effect of stacking sequence on bending of [0/90]$_{ns}$ laminates.

	$w_{max} E_T h^3 \times 10^{-2}/a^4 q_0$		
n	$E_L/E_T = 5$	$E_L/E_T = 10$	$E_L/E_T = 15$
1	2.273	1.435	1.047
2	2.282	1.445	1.057
3	2.283	1.446	1.058
∞	2.284	1.447	1.058

By allowing c and d to approach zero, we obtain an expression for a concentrated load P located at the coordinates $x = \xi$, $y = \eta$.

$$q_{mn} = \frac{4P}{ab} \sin \frac{m\pi\xi}{a} \sin \frac{n\pi\eta}{b} \qquad (5.18)$$

As a numerical example, let us consider the class of laminates $[0°/90°]_{ns}$ with the following ply elastic properties:

$$E_L/E_T = \text{variable}, \ \nu_{LT} = 0.25, \ G_{LT}/E_T = 0.5 \qquad (5.19)$$

Using these ply properties, the maximum deflection is calculated from Equation (5.7) for various anisotropy ratios and increasing values of n. Results are shown in Table 5.1 for a uniform load $q = q_0 = $ constant. Using Equation (5.7) in conjunction with Equation (5.16), the maximum deflection becomes

$$w_{max} = \frac{16q_0 a^4}{\pi^6} \sum_{m=1,2,\ldots}^{\infty} \sum_{n=1,3,\ldots}^{\infty} \frac{(-1)^{(\frac{m+n+2}{2})}}{mn \, D_{mn}} \qquad (5.20)$$

Convergence of this series is quite rapid. For the results in Table 5.1, the series in Equation (5.20) is truncated at $m = n = 13$, which yields 49 terms. For this laminate geometry, heterogeneity is seen to have little effect on the maximum deflection. The anisotropy ratio does affect the plate stiffness, however, with reduced deflection for increasing E_L/E_T.

The solution procedure discussed here for simply-supported rectangular plates was first introduced by Navier [1].

5.3 BENDING OF RECTANGULAR PLATES WITH TWO SIMPLY-SUPPORTED EDGES

Consider a laminate which is simply-supported along the edges $y = 0, b$ and subjected to an arbitrary transverse load q. If the surface load is represented by a double Fourier series as given by Equation (5.4), then the deflection can be assumed to be of the form

$$w = \sum_{n=1}^{\infty} \phi_n(x) \sin \frac{n\pi y}{b} + \frac{a^4}{\pi^4} \sum_{m=1}^{\infty} \sum_{n=1}^{\infty} \frac{q_{mn}}{D_{mn}} \sin \frac{m\pi x}{a} \sin \frac{n\pi y}{b}$$

(5.21)

where D_{mn} is defined in Equation (5.7). This relationship satisfies the simply-supported boundary conditions, Equation (5.3). Substituting Equations (5.4) and (5.21) into Equation (5.1), we obtain the result

$$\sum_{n=1}^{\infty} \left[D_{11} \frac{d^4\phi_n}{dx^4} - 2(D_{12} + 2D_{66}) \frac{n^2\pi^2}{b^2} \frac{d^2\phi_n}{dx^2} \right.$$

(5.22)

$$\left. + D_{22} \frac{n^4\pi^4}{b^4} \phi_n \right] \sin \frac{n\pi y}{b} = 0$$

It is easily seen that $\phi_n(x)$ must satisfy the differential equation

$$D_{11} \frac{d^4\phi_n}{dx^4} - 2(D_{12} + 2D_{66}) \frac{n^2\pi^2}{b^2} \frac{d^2\phi_n}{dx^2} + D_{22} \frac{n^4\pi^4}{b^4} \phi_n = 0 \quad (5.23)$$

Solutions to this differential equation are of the form

$$\phi_n(x) = \exp \frac{n\pi\lambda x}{b}$$

(5.24)

Substituting this function into Equation (5.23), we obtain the algebraic equation

$$D_{11}\lambda^4 - 2(D_{12} + 2D_{66})\lambda^2 + D_{22} = 0$$

(5.25)

The roots of this characteristic equation are of the form

$$\lambda^2 = \frac{1}{D_{11}} [D_{12} + 2D_{66} \pm \sqrt{(D_{12} + 2D_{66})^2 - D_{11}D_{22}}]$$

(5.26)

The solution of Equation (5.23) can be written in terms of four arbitrary constants A_n, B_n, C_n, and D_n. The precise form of the solution depends on Equation (5.26).

Case I: Roots of Equation (5.25) are Real and Unequal

For this case the roots are denoted by $\pm\lambda_1$ and $\pm\lambda_2$ ($\lambda_1,\lambda_2 > 0$) and we can write the solution of Equation (5.23) in the form

$$\phi_n(x) = A_n \cosh \frac{n\pi\lambda_1 x}{b} + B_n \sinh \frac{n\pi\lambda_1 x}{b} + C_n \cosh \frac{n\pi\lambda_2 x}{b}$$

$$+ D_n \sinh \frac{n\pi\lambda_2 x}{b}$$

(5.27)

In this case the deflection is

$$w = \sum_{n=1}^{\infty} \left[A_n \cosh \frac{n\pi\lambda_1 x}{b} + B_n \sinh \frac{n\pi\lambda_1 x}{b} + C_n \cosh \frac{n\pi\lambda_2 x}{b} \right.$$

$$\left. + D_n \sinh \frac{n\pi\lambda_2 x}{b} \right] + \frac{a^4}{\pi^4} \sum_{m=1}^{\infty} \frac{q_{mn}}{D_{mn}} \sin \frac{m\pi x}{a} \sin \frac{n\pi y}{b}$$

(5.28)

Case II: Roots of Equation (5.25) are Real and Equal

For this case the roots are denoted by $\pm\lambda$ ($\lambda > 0$) and we can write the solution of Equation (5.23) in the form

$$\phi_n(x) = (A_n + B_n x) \cosh \frac{n\pi\lambda x}{b} + (C_n + D_n x) \sinh \frac{n\pi\lambda x}{b}$$

(5.29)

In this case the solution for the deflection is

$$w = \sum_{n=1}^{\infty} \left[(A_n + B_n x) \cosh \frac{n\pi\lambda x}{b} + (C_n + D_n x) \sinh \frac{n\pi\lambda x}{b} \right.$$

$$\left. + \frac{a^4}{\pi^4} \sum_{m=1}^{\infty} \frac{q_{mn}}{D_{mn}} \sin \frac{n\pi x}{a} \right] \sin \frac{n\pi y}{b}$$

(5.30)

Case III: Roots of Equation (5.25) are Complex

For this case the roots are denoted by $\lambda_1 \pm i\lambda_2$ and $-\lambda_1 \pm i\lambda_2$ ($\lambda_1,\lambda_2 > 0$) and we can write the solution of Equation (5.23) in the form

$$\phi_n(x) = (A_n \cos \frac{n\pi\lambda_2 x}{b} + B_n \sin \frac{m\pi\lambda_2 x}{b}) \cosh \frac{n\pi\lambda_1 x}{b}$$

$$+ (C_n \cos \frac{n\pi\lambda_2 x}{b} + D_n \sin \frac{m\pi\lambda_2 x}{b}) \sinh \frac{n\pi\lambda_1 x}{b} \qquad (5.31)$$

In this case the deflection becomes

$$w = \sum_{n=1}^{\infty} [(A_n \cos \frac{n\pi\lambda_2 x}{b} + B_n \sin \frac{n\pi\lambda_2 x}{b}) \cosh \frac{n\pi\lambda_1 x}{b}$$

$$+ (C_n \cos \frac{n\pi\lambda_2 x}{b} + D_n \sin \frac{n\pi\lambda_2 x}{b}) \sinh \frac{n\pi\lambda_1 x}{b} \qquad (5.32)$$

$$+ \frac{a^4}{\pi^4} \sum_{m=1}^{\infty} \frac{q_{mn}}{D_{mn}} \sin \frac{m\pi x}{a}] \sin \frac{n\pi y}{b}$$

The constants in these solutions are determined from two boundary conditions on the edges $x = 0$ and a. It should be noted that for the case of simple-supports along the edges $x = 0$ and a, the boundary conditions (5.2) in conjunction with each of the deflection functions (5.28), (5.30), and (5.32) lead to four homogeneous algebraic equations for the four constants, which yield the trivial solution $A_n = B_n = C_n = D_n = 0$. In this case the solution is identical to Equation (5.7).

A simplification to the deflections (5.28), (5.30), and (5.32) can be obtained for the case of a uniform load $q = q_0 = $ constant. In particular it is only necessary to expand the transverse load in a single Fourier series of the form

$$q(x,y) = \frac{4q_0}{\pi} \sum_{n=1,3,\ldots}^{\infty} \frac{1}{n} \sin \frac{n\pi y}{b} \qquad (5.33)$$

and the second part of Equation (5.21) is replaced by the particular solution

$$w_p = \frac{4b^4 q_0}{D_{22}\pi^5} \sum_{n=1,3,\ldots}^{\infty} \frac{1}{n^5} \sin \frac{n\pi y}{b} \qquad (5.34)$$

In this case a nontrivial solution will be obtained for the four constants A_n, B_n, C_n, and D_n in conjunction with simply-supported boundary conditions, and the resulting deflection function will differ from Equation (5.7). As an example let us consider simply-supported boundary conditions in conjunction with real and un-equal roots (Case I). Substituting the deflection function (5.28) into the simply-supported boundary conditions (5.2), and solving the resulting algebraic equations, we find that

$$A_n = \frac{-4\lambda_2^2 b^4 q_0}{D_{22}\pi^5 n^5(\lambda_2^2 - \lambda_1^2)}$$

$$B_n = \frac{-4\lambda_2^2 b^4 q_0}{D_{22}\pi^5 n^5(\lambda_2^2 - \lambda_1^2)}\left(\frac{1 - \cosh n\pi\lambda_1 R}{\sinh n\pi\lambda_1 R}\right)$$

$$C_n = \frac{4\lambda_1^2 b^4 q_0}{D_{22}\pi^5 n^5(\lambda_2^2 - \lambda_1^2)}$$ (5.35)

$$D_n = \frac{4\lambda_1^2 b^4 q_0}{D_{22}\pi^5 n^5(\lambda_2^2 - \lambda_1^2)}\left(\frac{1 - \cosh n\pi\lambda_{12} R}{\sinh n\pi\lambda_2 R}\right)$$

where $R = a/b$ and n takes on odd values (i.e. $n = 1,3,5,\ldots$).

Now consider the case of clamped edges in conjunction with real and unequal roots (Case I). The boundary conditions are
at $x = 0$ and a

$$w = \frac{\partial w}{\partial x} = 0$$ (5.36)

Substituting Equation (5.28) into Equation (5.36), and solving the resulting algebraic equations, we find that

$$A_n = \frac{4\lambda_2 b^4 q_0}{D_{22}\pi^5 n^5 H_n}\left[\lambda_1(\cosh n\pi\lambda_1 R - \cosh n\pi\lambda_2 R)(\cosh n\pi\lambda_2 R - 1)\right.$$

$$\left. - (\lambda_2\sinh n\pi\lambda_1 R - \lambda_1\sinh n\pi\lambda_2 R)\sinh n\pi\lambda_2 R\right]$$

$$B_n = \frac{4\lambda_2 b^4 q_0}{D_{22}\pi^5 n^5 H_n}\left[\lambda_2(\cosh n\pi\lambda_1 R - \cosh n\pi\lambda_2 R)(\sinh n\pi\lambda_2 R\right.$$

$$\left. - (\lambda_1\sinh n\pi\lambda_1 R - \lambda_2\sinh n\pi\lambda_2 R)(\cosh n\pi\lambda_2 R - 1)\right]$$

(5.37)

$$C_n = -\left(\frac{4b^4 q_0 + D_{22}\pi^5 n^5 H_n A_n}{D_{22}\pi^5 n^5 H_n} \right)$$

$$D_n = -\frac{\lambda_1}{\lambda_2} B_n$$

where

$$H_n = \lambda_1\lambda_2(\cosh n\pi\lambda_1 R - \cosh n\pi\lambda_2 R)^2 - (\lambda_1 \sinh n\pi\lambda_1 R$$

$$- \lambda_2 \sinh n\pi\lambda_2 R)(\lambda_2 \sinh n\pi\lambda_1 R - \lambda_1 \sinh n\pi\lambda_2 R)$$

Moments can be determined from Equations (2.49–2.51), and stresses from Equations (2.54–2.58).

The solution procedure described here for rectangular plates simply-supported along the opposite edges was first introduced by Levy (see Reference [1]). Additional solutions involving other boundary conditions along the edges $x = 0$ and a, or involving characteristic roots not of the form considered in Case I, can be found in Reference [2].

5.4 BENDING OF CLAMPED RECTANGULAR PLATES

Exact solutions to the governing equations and boundary conditions were obtained for the problems considered in the previous two sections. In practice, however, many cases are encountered in which exact solutions are not available and approximate methods must be utilized. In this section we consider the bend-

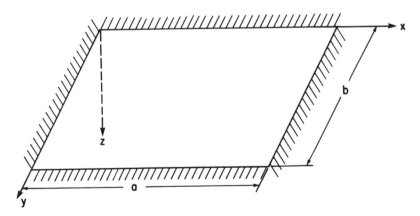

Figure 5.3. *Clamped rectangular plate.*

ing of a clamped rectangular plate subjected to a uniform transverse load $q(x,y) = q_0$ (see Figure 5.3). We seek approximate solutions in conjunction with the Ritz method as described in Section 3.6.

Combining Equations (3.35) and (3.36), we obtain the energy criterion

$$U + W = \text{stationary value} \tag{5.38}$$

For specially orthotropic plates we substitute Equations (3.7) and (3.15) into Equation (5.38) with the result

$$U = \frac{1}{2} \int_0^a \int_0^b \left[D_{11} \left(\frac{\partial^2 w}{\partial x^2} \right)^2 + 2D_{12} \frac{\partial^2 w}{\partial x^2} \frac{\partial^2 w}{\partial y^2} + D_{22} \left(\frac{\partial^2 w}{\partial y^2} \right)^2 \right.$$

$$\left. + 4D_{66} \left(\frac{\partial^2 w}{\partial x \partial y} \right)^2 \right] dx\, dy \tag{5.39}$$

$$V = - \int_0^b \int_0^a q_0 w\, dx\, dy$$

For clamped boundaries we have:
  at $x = 0, a$

$$w = \frac{\partial w}{\partial x} = 0 \tag{5.40}$$

  at $y = 0, b$

$$w = \frac{\partial w}{\partial y} = 0 \tag{5.41}$$

In order to find an approximate solution for the plate deflection, we consider the following finite series in the variables separable form:

$$w = \sum_{m=1}^{M} \sum_{n=1}^{N} A_{mn} X_m(x)\, Y_n(y) \tag{5.42}$$

The functions $X_m(x)$ and $Y_n(y)$ are chosen such that they satisfy the boundary conditions (5.40) and (5.41), respectively. Using the series (5.42), the condition

(5.38) reduces to the MxN conditions

$$\frac{\partial U}{\partial A_{mn}} = -\frac{\partial W}{\partial A_{mn}} \quad \begin{cases} m = 1,2,\ldots,M \\ n = 1,2,\ldots,N \end{cases} \tag{5.43}$$

Combining Equations (5.39), (5.42), and (5.43), we obtain the following MxN algebraic equations for the determination of the coefficients A_{mn}:

$$
\begin{aligned}
\sum_{i=1}^{M} \sum_{j=1}^{N} \Bigg\{ & D_{11} \int_0^a \frac{d^2X_i}{dx^2} \frac{d^2X_m}{dx^2}\, dx \int_0^b Y_j Y_n\, dy \\[2mm]
& + D_{12} \left[\int_0^a X_m \frac{d^2X_i}{dx^2} \cdot dx \int_0^b Y_j \frac{d^2Y_n}{dy^2}\, dy \right. \\[2mm]
& \left. + \int_0^a X_i \frac{d^2X_m}{dx^2}\, dx \int_0^b Y_n \frac{d^2Y_j}{dy^2}\, dy \right] \\[2mm]
& + D_{22} \int_0^a X_i X_m dx \int_0^b \frac{d^2Y_j}{dy^2} \frac{d^2Y_n}{dy^2} \\[2mm]
& + 4D_{66} \int_0^a \frac{dX_i}{dx} \frac{dX_m}{dx}\, dx \int_0^b \frac{dY_j}{dy} \frac{dY_n}{dy}\, dy \Bigg\} A_{ij} \\[2mm]
& = q_0 \int_0^a \left\{ X_m dx \int_0^b Y_n dy \right. \quad \begin{matrix} m = 1,2,\ldots,M \\ n = 1,2,\ldots,N \end{matrix}
\end{aligned}
\tag{5.44}
$$

We now consider two choices for the functions $X_m(x)$ and $Y_n(y)$.

As a first example, we may select the functions as polynomials, with each term satisfying the boundary conditions (5.40) and (5.41):

$$X_m(x) = (x^2 - ax)^2 \, x^{m-1} \tag{5.45}$$

$$Y_n(y) = (y^2 - by)^2 \, y^{n-1}$$

Table 5.2. First five roots to Equation (5.50).

λ_1	λ_2	λ_3	λ_4	λ_5
4.730	7.853	10.996	14.137	17.279

Substituting these functions into Equation (5.44) with $M = N = 1$ (the first approximation), we obtain

$$A_{11} = \frac{6.125 \, q_0}{7D_{11}b^4 + 4(D_{12} + 2D_{66})a^2b^2 + 7D_{22}a^4} \tag{5.46}$$

which leads to the deflection function

$$w = \frac{6.125 \, q_0 \, (x^2 - ax)^2(y^2 - by)^2}{7D_{11}b^4 + 4(D_{12} + 2D_{66})a^2b^2 + 7D_{22}a^4} \tag{5.47}$$

The maximum deflection occurs at the center, and for the first approximation as given by Equation (5.47), we have

$$w_{max} = 0.00342 \, \frac{q_0 a^4}{D_{11} + 0.571(D_{12} + 2D_{66})R^2 + D_{22}R^4} \tag{5.48}$$

where R is the plate aspect ratio a/b.

As a second choice for the functions $X_m(x)$ and $Y_n(y)$, we select the characteristic shapes found as solutions of the natural vibration of a beam with clamped ends [3]:

$$X_m(x) = \gamma_m \cos \frac{\lambda_m x}{a} - \gamma_m \cosh \frac{\lambda_m x}{a} + \sin \frac{\lambda_m x}{a} - \sinh \frac{\lambda_m x}{a}$$

$$Y_n(y) = \gamma_n \cos \frac{\lambda_n y}{b} - \gamma_n \cosh \frac{\lambda_n y}{b} + \sin \frac{\lambda_n y}{b} - \sinh \frac{\lambda_n y}{b} \tag{5.49}$$

where the λ_m and λ_n are the roots of the frequency equation

$$\cos \lambda_i \cosh \lambda_i = 1 \tag{5.50}$$

and

$$\gamma_i = \frac{\cos \lambda_i - \cosh \lambda_i}{\sin \lambda_i + \sinh \lambda_i} \tag{5.51}$$

Table 5.3. Values of λ_i determined from Equation (5.54).

λ_1	λ_2	λ_3	λ_4	λ_5
4.712	7.854	10.996	14.137	17.279

where $i = m,n$ in both Equations (5.50) and (5.51). The first five values of λ_i which satisfy Equation (5.50) have been tabulated in Reference [4] and are given in Table 5.2.

For $i > 5$ an approximate relationship for determining λ_i can be derived. In particular for large values of λ_i,

$$\cosh \lambda_i \approx \frac{e^{\lambda_i}}{2} \qquad (5.52)$$

and Equation (5.50) becomes

$$\cos \lambda_i \approx 2e^{-\lambda_i} \approx 0 \qquad (5.53)$$

Roots to this relationship can be written in the form

$$\lambda_i = (2i + 1)\frac{\pi}{2} \qquad (5.54)$$

In order to demonstrate the accuracy of this relationship, the first five values of λ_i as calculated from Equation (5.54) are listed in Table 5.3. We see that for $i > 2$, Equation (5.54) yields exact values for λ_i to five significant figures.

The functions (5.49) satisfy the boundary conditions (5.40) and (5.41). For a first approximation with the assumed set of functions, we again consider $M = N = 1$. In this case $\lambda_1 = 4.73$ and $\gamma_1 = 0.9825$. Thus

$$X(x) = 0.9825 \cos \frac{4.73x}{a} - 0.9825 \cosh \frac{4.73x}{a}$$

$$+ \sin \frac{4.73x}{a} - \sinh \frac{4.73x}{a}$$

$$\qquad (5.55)$$

$$Y(y) = 0.9825 \cos \frac{4.73y}{b} - 0.9825 \cosh \frac{4.73y}{b}$$

$$+ \sin \frac{4.73y}{b} - \sinh \frac{4.73y}{b}$$

Substituting these relationships into Equation (5.44), we obtain the single equation

$$[500.56(D_{11} + D_{22}R^4) + 302.71R^2(D_{12} + 2D_{66})]A_{11} = 0.6903\ q_0a^4 \qquad (5.56)$$

or

$$A_{11} = \frac{0.6903\ q_0a^4}{500.56[D_{11} + 0.6047(D_{12} + 2D_{66})R^2 + D_{22}R^4]} \qquad (5.57)$$

The maximum deflection can be determined by evaluating $X_1(x)$ and $Y_1(y)$ at the center of the plate with the result

$$w_{max} = 0.00348\ \frac{q_0a^4}{[D_{11} + 0.6047(D_{12} + 2D_{66})R^2 + D_{22}R^4]} \qquad (5.58)$$

The two approximate solutions appear to be almost identical. In particular if we consider an isotropic plate with $D_{11} = D_{22} = D_{12} + 2D_{66} = D$ and $R = 1$, Equation (5.48) based on the polynomial solution yields

$$w_{max} = 0.00133\ \frac{q_0a^4}{D} \qquad (5.59)$$

while Equation (5.58) yields the result

$$w_{max} = 0.00134\ \frac{q_0a^4}{D} \qquad (5.60)$$

Using a large number of terms in the series (5.49), we find the solution for an isotropic plate to be

$$w_{max} = 0.00126\ \frac{q_0}{D} \qquad (5.61)$$

The errors in the polynomial approximation (5.59) and the beam function approximation (5.60) are 5.6% and 6.3%, respectively.

The maximum moment occurs at the middle of each edge and can be calculated by differentiating w in conjunction with equation (2.72). Using the polynomial solution (5.47), we find that

$$M_{max} = -\ 0.0425\ q_0a^2 \qquad (5.62)$$

and using the beam function (5.49), we find that a one term solution yields

$$M_{max} = -\ 0.0384\ q_0 a^2 \tag{5.63}$$

Using a large number of terms in the series (5.49), we obtain the result

$$M_{max} = -\ 0.0513\ q_0 a^2 \tag{5.64}$$

and the accuracy of a one term approximation is much less acceptable for the maximum moment than for the maximum deflection.

5.5 STABILITY OF SIMPLY-SUPPORTED RECTANGULAR PLATES UNDER UNIFORM COMPRESSION

We now consider simply-supported rectangular plates compressed by uniform force resultants N_x and N_y (see Figure 5.4). In order to find the critical value of the loads N_x and N_y such that the initially flat plate is in equilibrium in a slightly deflected mode shape, we must utilize governing equations which include initial inplane force effects. For the static case with $q = 0$, Equation (2.83) reduces to

$$D_{11} \frac{\partial^4 w}{\partial x^4} + 2(D_{12} + 2D_{66}) \frac{\partial^4 w}{\partial x^2 \partial y^2} + D_{22} \frac{\partial^4 w}{\partial y^4} = N_x \frac{\partial^2 w}{\partial x^2} + N_y \frac{\partial^2 w}{\partial y^2} \tag{5.65}$$

We seek a nontrivial solution to the deflection w which satisfies Equation (5.65)

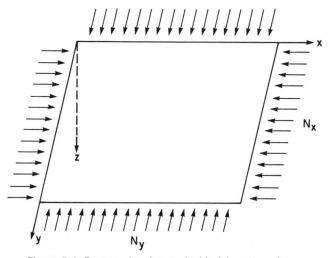

Figure 5.4. *Rectangular plate under biaxial compression.*

and the simply-supported boundary conditions
at $x = 0$ and a

$$w = M_x = -D_{11}\frac{\partial^2 w}{\partial x^2} - D_{12}\frac{\partial^2 w}{\partial y^2} = 0 \qquad (5.66)$$

at $y = 0$ and b

$$w = M_y = -D_{12}\frac{\partial^2 w}{\partial x^2} - D_{22}\frac{\partial^2 w}{\partial y^2} = 0 \qquad (5.67)$$

These boundary conditions are satisfied by a deflection of the form

$$w = A_{mn} \sin\frac{m\pi x}{a} \sin\frac{n\pi y}{b} \qquad (5.68)$$

where m and n are integers. Substituting Equation (5.68) into the governing partial differential Equation (5.65), we obtain the result

$$\pi^2 A_{mn}[D_{11}m^4 + 2(D_{12} + 2D_{66})m^2n^2R^2 + D_{22}n^4R^4]$$

$$= -A_{mn}a^2[N_x m^2 + N_y n^2 R^2] \qquad (5.69)$$

where as previously R is the plate aspect ratio a/b. Since the trivial solution $A_{mn} = 0$ is not of interest, we seek values of N_x and N_y such that

$$\pi^2[D_{11}m^4 + 2(D_{12} + 2D_{66})m^2n^2R^2 + D_{22}n^4R^4]$$

$$= -a^2[N_x m^2 + N_y n^2 R^2] \qquad (5.70)$$

We now consider uniform inplane loads of the form

$$N_x = -N_0, \; N_y = -kN_0 \qquad (5.71)$$

where $N_0 > 0$ and k denotes the ratio of N_y/N_x. The condition (5.69) now reduces to

$$N_0 = \frac{\pi^2[D_{11}m^4 + 2(D_{12} + 2D_{66})m^2n^2R^2 + D_{22}n^4R^4]}{a^2(m^2 + kn^2R^2)} \qquad (5.72)$$

This relationship gives the values of N_0 (for a given k) for any configuration of m and n. Obviously an infinite number of values of N_0 can be determined such that Equation (5.72) is satisfied. In particular, a unique value of N_0 exists for each combination of m and n.

The critical buckling load corresponds to the value of m and n which yields the lowest value of N_0. We now consider several particular cases in detail.

Case I: k = 0

In this case we have uniaxial compression in the x-direction and Equation (5.72) becomes

$$N_0 = \frac{\pi^2}{m^2 a^2} [D_{11} m^4 + 2(D_{12} + 2D_{66}) m^2 n^2 R^2 + D_{22} n^4 R^4] \qquad (5.73)$$

and the smallest value of N_0 clearly occurs for $n = 1$. The buckling load is given by the smallest value of the following relationship as m is allowed to vary:

$$N_0 = \frac{\pi^2}{m^2 a^2} [D_{11} m^4 + 2(D_{12} + 2D_{66}) m^2 n^2 R^2 + D_{22}] \qquad (5.74)$$

For given values of D_{11}, D_{12}, D_{66}, D_{22}, and R, we determine the critical buckling load from the value of m which yields the lowest value of N_0.

For example, if we consider an orthotropic plate with $D_{11}/D_{22} = 10$ and $(D_{12} + 2D_{66})/D_{22} = 1$, then Equation (5.74) becomes

$$N_0 = \frac{\pi D_{22}}{m^2 a^2} (10 m^4 + 2m^2 R^2 + R^4) \qquad (5.75)$$

The value of N_0 is a minimum when $m = 1$ if $R \leq 1.78$. In particular, for a square plate ($R = 1$) the critical buckling load is

$$N_{cr} = 13\pi^2 \frac{D_{22}}{a^2} \qquad (5.76)$$

Values of N_{cr} versus plate aspect ratio, $R = a/b$, are shown in Figure 5.5. Note that at certain aspect ratios, two possible buckled mode shapes are possible. That is, when

$$R = [m(m + 1)]^{1/2} \left(\frac{D_{11}}{D_{22}}\right)^{1/4} = 1.78 [m(m + 1)]^{1/2} \qquad (5.77)$$

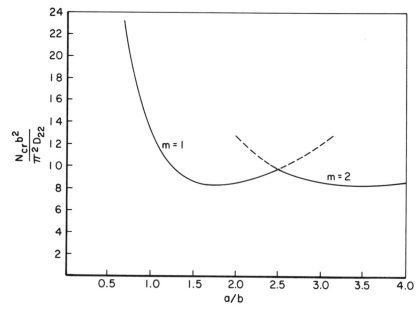

Figure 5.5. Variation of N_{cr} with aspect ratio.

the two buckled shapes

$$w = A_{ml} \sin \frac{m\pi x}{a} \sin \frac{n\pi y}{b} \tag{5.78}$$

and

$$w = A_{(m+1)1} \sin \frac{m\pi x}{a} \sin \frac{n\pi y}{b} \tag{5.79}$$

lead to identical values of N_{cr}.

Also, the absolute minimum buckling load occurs for $m = i$ such that

$$R = i^4 \left(\frac{D_{11}}{D_{22}} \right)^{1/2} = 1.78 \, i \tag{5.80}$$

and in this case

$$N_{cr} = 8.32 \, \pi^2 \frac{D_{22}}{a^2} \tag{5.81}$$

Case II: k = 1, a = b

This corresponds to the case of a square plate under biaxial compression. Equation (5.72) for this case becomes

$$N_0 = \frac{\pi^2[D_{11}m^4 + 2(D_{12} + 2D_{66})m^2n^2 + D_{22}n^4]}{a^2(m^2 + n^2)} \tag{5.82}$$

A cursory examination of Equation (5.82) reveals that N_{cr} occurs with $m = 1$ provided $D_{11} \geq D_{22}$. In this case

$$N_0 = \frac{\pi^2}{a^2} D_{22} \frac{\left[\dfrac{D_{11}}{D_{22}} + 2 \dfrac{(D_{12} + 2D_{66})}{D_{22}} n^2 + n^4\right]}{1 + n^2} \tag{5.83}$$

For the numerical values considered in Case I, the critical buckling load occurs with $n = 1$:

$$N_{cr} = 6.5 \frac{\pi^2}{a^2} D_{22} \tag{5.84}$$

For the case $D_{11}/D_{22} = 15$, $(D_{12} + 2D_{66})/D_{22} = 1$, the critical value is obtained for $n = 2$:

$$N_{cr} = 7.8 \frac{\pi^2}{a^2} D_{22} \tag{5.85}$$

Case III: k < 0

This corresponds to the case where N_y is a tensile load. For a square plate with $k = -0.5$, Equation (5.72) becomes

$$N_0 = \frac{2\pi^2[D_{11}m^4 + 2(D_{12} + 2D_{66}) + D_{22}]}{a^2(2m^2 - n^2)} \tag{5.86}$$

Since k is negative this equation indicates that the critical buckling load will be larger than for the case $N_y = 0$. For the numerical example considered in Case I, the critical buckling load occurs for $m = n = 1$ with the result

$$N_{cr} = 26 \frac{\pi^2 D_{22}}{a^2} \tag{5.87}$$

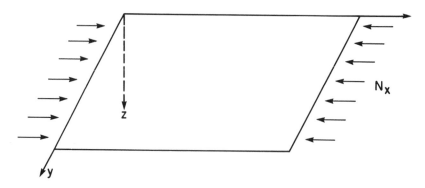

Figure 5.6. *Plate under uniaxial compression.*

Thus for a tensile value of N_y that is one-half the compression value of N_x, the critical buckling load is twice that obtained for $N_y = 0$. We can conclude from these results that the effect of a tensile value of N_y is to increase the effective bending stiffness of the plate.

5.6 STABILITY OF RECTANGULAR PLATES WITH TWO SIMPLY-SUPPORTED EDGES

In this section we consider a rectangular plate simply-supported along the edges $x = 0$, a and uniformly compressed in the x direction as shown in Figure 5.6. The boundary conditions along the edges $y = 0, b$ are arbitrary. For this case the governing equation is given by (5.65) with the inplane loads represented as follows:

$$N_x = - N_0, \quad N_y = 0 \tag{5.88}$$

A solution to Equation (5.65) can be found in the form

$$w = f(y) \sin \frac{m\pi x}{a} \tag{5.89}$$

which satisfies the simply-supported boundary conditions
at $x = 0$ and a

$$w = M_x = - D_{11} \frac{\partial^2 w}{\partial x^2} - D_{12} \frac{\partial^2 w}{\partial y^2} = 0 \tag{5.90}$$

Substituting Equation (5.89) into Equation (5.65) leads to the following differential Equation for $f(y)$:

$$D_{22} \frac{d^4 f}{dy^4} - 2 \left(\frac{m\pi}{a} \right)^2 (D_{12} + 2D_{66}) \frac{d^2 f}{dy^2}$$

$$+ D_{11} \left(\frac{m\pi}{a} \right)^4 f - N_0 \left(\frac{m\pi}{a} \right)^2 f = 0$$

(5.91)

The solution of this equation is assumed to be of the form

$$f(y) = A \cosh \frac{\lambda_1 y}{a} + B \sinh \frac{\lambda_1 y}{a} + C \cos \frac{\lambda_2 y}{a} + D \sin \frac{\lambda_2 y}{a} \qquad (5.92)$$

where

$$\lambda_1 = m\pi \sqrt{\sqrt{\left(\frac{D_{12} + 2D_{66}}{D_{22}} \right)^2 - \frac{D_{11}}{D_{22}} + \frac{N_0}{D_{22}} \left(\frac{a}{m\pi} \right)^2} + \frac{D_{12} + 2D_{66}}{D_{22}}}$$

$$\lambda_2 = m\pi \sqrt{\sqrt{\left(\frac{D_{12} + 2D_{66}}{D_{22}} \right)^2 - \frac{D_{11}}{D_{22}} + \frac{N_0}{D_{22}} \left(\frac{a}{m\pi} \right)^2} - \frac{D_{12} + 2D_{66}}{D_{22}}}$$

(5.93)

and the constants A, B, C, and D must be determined from the four conditions on the edges $y = 0,b$. In determining the constants A, B, C, and D we obtain four homogeneous equations. The critical buckling load is determined by setting the determinant of this system to zero and determining for this condition the lowest possible value of N_0.

To illustrate this procedure we consider the cases where the edges at $y = 0,b$ are clamped. The boundary conditions at $y = 0,b$ are

$$w = \frac{\partial w}{\partial y} = 0 \qquad (5.94)$$

Combining Equations (5.89) and (5.93) and substituting the results into the boundary conditions (5.94), we obtain the following algebraic equations in matrix form

$$
\begin{bmatrix}
1 & 0 & 1 & 0 \\
0 & s & 0 & t \\
\cosh \dfrac{\lambda_1}{R} & \sinh \dfrac{\lambda_1}{R} & \cos \dfrac{\lambda_2}{R} & \sin \dfrac{\lambda_2}{R} \\
\lambda_1 \sinh \dfrac{\lambda_1}{R} & \lambda_1 \cosh \dfrac{\lambda_1}{R} & -\lambda_2 \sin \dfrac{\lambda_2}{R} & \lambda_2 \cosh \dfrac{\lambda_2}{R}
\end{bmatrix}
\times
\begin{bmatrix}
A \\
B \\
C \\
D
\end{bmatrix}
=
\begin{bmatrix}
0 \\
0 \\
0 \\
0
\end{bmatrix}
\tag{5.95}
$$

where $R = a/b$.

In order to avoid a trivial solution ($A = B = C = D = 0$), the determinant of the coefficient matrix in Equation (5.95) must vanish. This condition leads to the following equation:

$$2(1 - \cos \frac{\lambda_2}{R} \cosh \frac{\lambda_1}{R}) = \left(\frac{\lambda_2^2 - \lambda_1^2}{\lambda_1 \lambda_2} \right) \sin \frac{\lambda_2}{R} \sinh \frac{\lambda_1}{R} \qquad (5.96)$$

Since λ_1 and λ_2, as given by Equation (5.93), both involve N_0, the critical buckling load corresponds to the lowest value of N_0 which satisfies Equation (5.96).

We now consider a square plate ($a = b$) with $D_{11}/D_{12} = 10$ and $(D_{12} + 2D_{66})/D_{22} = 1.67$. Then

$$\lambda_1 = m\pi \sqrt{\sqrt{-7.22 + \frac{k}{m^2}} + 1.67}$$

$$\qquad (5.97)$$

$$\lambda_2 = m\pi \sqrt{\sqrt{-7.22 + \frac{k}{m^2}} - 1.67}$$

where

$$k = \frac{N_0}{D_{22}} \left(\frac{a}{\pi} \right)^2$$

and Equation (5.96) for this case is

$$[2 - 2\cos(m\pi \sqrt{\sqrt{-7.22 + k/m^2} - 1.67})] \, x$$

$$\cosh(m\pi \sqrt{\sqrt{-7.22 + k/m^2} + 1.67}) \qquad (5.98)$$

$$= - \left[\frac{3.34}{\sqrt{-7.22 + k/m^2 - (1.67)^2}} \right] \sin(m\pi \sqrt{\sqrt{-7.22 + k/m^2} - 1.67})$$

$$x \, \sinh(m\pi \sqrt{\sqrt{-7.22 + k/m^2} + 1.67})$$

The minimum value of k occurs with $m = 1$, and may be found by iteratively solving Equation (5.98). The critical buckling load is found to be

$$N_{cr} = 19.1 \ \pi^2 \ \frac{D_{22}}{a^2} \tag{5.99}$$

Other boundary conditions on the edges $y = 0,b$ can be handled in the same manner.

5.7 STABILITY OF SIMPLY-SUPPORTED RECTANGULAR PLATES UNDER SHEAR LOAD

In this section the stability of simply-supported rectangular plates subjected to a uniform shear load, Figure 5.7, is considered. Thus, the inplane loading takes the form

$$N_x = N_y = 0, \ N_{xy} = S = \text{constant} \tag{5.100}$$

The solutions will be sought by the Galerkin method as described in Section 3.7.

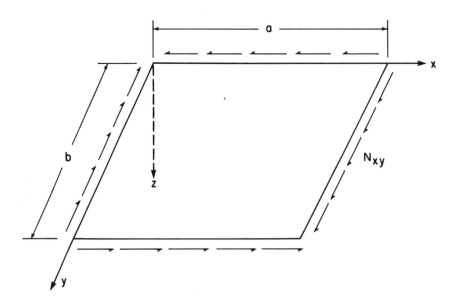

Figure 5.7. Rectangular plate under shear load.

An assumed solution of the form

$$w = \sum_{m=1}^{M} \sum_{n=1}^{N} A_{mn} \sin \frac{m\pi x}{a} \sin \frac{n\pi y}{b} \qquad (5.101)$$

satisfies the simply-supported boundary conditions prescribed by Equations (5.66) and (5.67). Since all of the boundary conditions are satisfied by Equation (5.101), the line integrals in Equation (3.47) vanish and the Galerkin equation becomes

$$\int_0^a \int_0^b \left[D_{11} \frac{\partial^4 w}{\partial x^4} + 2(D_{12} + 2D_{66}) \frac{\partial^4 w}{\partial x^2 \partial y^2} + D_{22} \frac{\partial^4 w}{\partial y^4} \right.$$
$$(5.102)$$
$$\left. - 2S \frac{\partial^2 w}{\partial x \partial y} \right] \sin \frac{m\pi x}{a} \sin \frac{n\pi y}{b} \, dx \, dy = 0 \quad \begin{cases} m = 1,2,\ldots,M \\ n = 1,2,\ldots,N \end{cases}$$

Substituting Equation (5.101) into Equation (5.102) and performing the integrations, we arrive at the following set of algebraic equations:

$$\pi^4 [D_{11} m^4 + 2(D_{12} + 2D_{66}) m^2 n^2 R^2 + D_{22} n^4 R^4] A_{mn}$$
$$- 32 \, mn \, R^3 b^2 S \sum_{i=1}^{M} \sum_{j=1}^{N} M_{ij} A_{ij} = 0 \qquad (5.103)$$

where

$$M_{ij} = \frac{ij}{(m^2 - i^2)(n^2 - j^2)} \quad \begin{cases} m \pm i \quad \text{odd} \\ n \pm j \quad \text{odd} \end{cases}$$

$$= 0 \quad \begin{cases} i = m \quad m \pm i \quad \text{even} \\ j = n \quad n \pm j \quad \text{even} \end{cases}$$

where, as previously, R is the plate aspect ratio a/b. Equation (5.103) yields a set of MxN homogeneous equations which can be divided into two sets. These sets correspond to $m + n$ odd and $m + n$ even. In each of these cases a nontrivial solution is obtained if S is chosen such that the determinant of the coefficient matrix vanishes. Equation (5.103) is in the form of a classical eigenvalue prob-

Table 5.4. Shear buckling coefficient for square simply-supported plate.

$S_{cr} = \pm k (D_{22}/a^2)$				
M = N	2	3	4	5
k	392.9	313.7	310.2	309.0

lem, and the critical buckling load, S_{cr}, corresponds to the lowest eigenvalue (smallest value of S) which can be obtained by iteration. For buckling of specially orthotropic plates under pure shear, the critical value of S does not depend on the direction of the shear stress. As a result the eigenvalues occur in positive and negative pairs, and standard iteration procedures must be modified to take this into account [5].

Convergence of the system of equations given by (5.103) is indicated in Table 5.4, where numerical results are presented for increasing values of M and N. A square plate is utilized in the example with $D_{11}/D_{22} = 10$ and $(D_{12} + 2D_{66})/D_{22} = 1.67$. Convergence of the solution in this case is much slower than in the case of an isotropic clamped plate under uniform transverse load as presented in Section 5.4. For example a four-term series ($M = N = 2$) yields a 27% error compared to a 25-term solution ($M = N = 5$). In addition, the four-term solution is nonconservative, i.e. it is greater than the more accurate 25-term solution.

Further results are given in Reference [2].

5.8 STABILITY OF CLAMPED RECTANGULAR PLATES UNDER SHEAR LOAD

We now consider a rectangular plate subjected to uniform shear, S, Figure 5.7, with all four edges clamped. Solutions will be sought by the Ritz method. Combining Equations (3.35) and (3.37) with $q = 0$ (no transverse load), we obtain the energy criterion

$$U + V = \text{stationary value} \qquad (5.104)$$

Substituting Equations (3.7) and (3.18) into Equation (5.104), we obtain the following result for the case of pure shear loading:

$$U + V = \frac{1}{2} \int_0^b \int_0^a \left[D_{11} \left(\frac{\partial^2 w}{\partial x^2} \right)^2 + 2D_{12} \frac{\partial^2 w}{\partial x^2} \frac{\partial^2 w}{\partial y^2} \right.$$
$$\left. + D_{22} \left(\frac{\partial^2 w}{\partial y^2} \right)^2 + 4D_{66} \left(\frac{\partial^2 w}{\partial x \partial y} \right)^2 + 2S \frac{\partial w}{\partial x} \frac{\partial w}{\partial y} \right] dx\, dy \qquad (5.105)$$

For clamped boundary conditions we require that
at $x = 0$ and a

$$w = \frac{\partial w}{\partial x} = 0 \tag{5.106}$$

at $y = 0$ and b

$$w = \frac{\partial w}{\partial y} = 0 \tag{5.107}$$

The solution is assumed to be of the form

$$w = \sum_{m=1}^{M} \sum_{n=1}^{N} A_{mn} X_m(x) Y_n(y) \tag{5.108}$$

where $X_m(x)$ and $Y_n(y)$ must satisfy the boundary conditions (5.106) and (5.107), respectively. Using the assumed displacement function (5.108), the condition (5.104) reduces to the MxN conditions

$$\frac{\partial(U + V)}{\partial A_{mn}} = 0 \quad \begin{array}{l} m = 1,2,\dots,M \\ n = 1,2,\dots,N \end{array} \tag{5.109}$$

Combining Equations (5.105), (5.108), and (5.109), we obtain the following MxN homogeneous algebraic equations:

$$\sum_{i=1}^{M} \sum_{j=1}^{N} \left\{ D_{11} \int_0^a \frac{d^2 X_i}{dx^2} \frac{d^2 X_m}{dx^2} \, dx \int_0^b Y_j \, Y_n \, dy \right.$$

$$+ D_{12} \left[\left[\int_0^a X_m \frac{d^2 X_i}{dx^2} \, dx \int_0^b Y_j \frac{d^2 Y_n}{dy^2} \, dy \right. \right.$$

$$\left. + \int_0^a X_i \frac{d^2 X_m}{dx^2} \, dx \int_0^b Y_n \frac{d^2 Y_j}{dy^2} \, dy \right]$$

$$+ D_{22} \int_0^a X_i \, X_m \, dx \int_0^b \frac{d^2 Y_j}{dy^2} \frac{d^2 Y_n}{dy_2} \, dy$$

$$\tag{5.110}$$

$$+ 4D_{66} \int_0^a \frac{dX_i}{dx} \frac{dX_m}{dx} dx \int_0^b \frac{dY_j}{dy} \frac{dY_n}{dy} dy$$

$$+ S \left[\int_0^a X_m \frac{dX_i}{dx} dx \int_0^b Y_j \frac{dY_n}{dy} dy \right.$$

$$\left. + \int_0^a X_i \frac{dX_m}{dx} dx \int_0^b Y_n \frac{dY_j}{dy} dy \right] \right\} A_{ij}$$

$$= 0 \quad \begin{cases} m = 1,2,\ldots,M \\ n = 1,2,\ldots,N \end{cases}$$

Note the similarity between these $M{\times}N$ equations and those derived in Section 5.4 using the Ritz method in conjunction with a bending problem.

In order to obtain a nontrivial solution to the homogeneous equations in (5.110), the determinant of the coefficient matrix must vanish. The lowest value of S satisfying this condition is the critical buckling load.

The displacement functions chosen in conjunction with Equation (5.108) are the characteristic shapes of free vibration of a beam with fixed ends. Then just as in Section 5.4 we have:

$$X_m(x) = \gamma_m \cos \frac{\lambda_m x}{a} - \gamma_m \cosh \frac{\lambda_m x}{a} + \sin \frac{\lambda_m x}{a} - \sinh \frac{\lambda_m x}{a}$$

$$(5.111)$$

$$Y_n(y) = \gamma_n \cos \frac{\lambda_n y}{b} - \gamma_n \cosh \frac{\lambda_n y}{b} + \sin \frac{\lambda_n y}{b} - \sinh \frac{\lambda_n y}{b}$$

where λ_m, λ_n, γ_m, and γ_n are determined from Equations (5.50) and (5.51).

A first approximation to the buckling load can be obtained using the terms A_{11} and A_{22} (the integrals multiplying S are zero for $M = N = 1$ and the resulting critical buckling load $\rightarrow \infty$). The resulting equations are found to be

$$[500.56 \, D_{11} + 302.71(D_{12} + 2D_{66})R^2 + 500.56 \, D_{22}R^4]A_{11}$$

$$+ 22.34 \, SRa^2A_{22} = 0$$

$$(5.112)$$

$$22.34 \, SRa^2A_{11} + [3803.5 \, D_{11} + 4241.2(D_{12} + 2D_{66})R^2$$

$$+ 3803.5 \, D_{22}R^4]A_{22} = 0$$

Table 5.5. Shear buckling coefficient for square clamped plate.

$S_{cr} = \pm k \, (D_{22}/a^2)$				
M = N	2	3	5	7
k	768.0	544.0	505.2	503.6

where, as previously, R is the plate aspect ratio a/b. In order to have a nontrivial solution to Equation (5.112), the determinant of the coefficient matrix must be zero. This condition results in our first approximation for the critical buckling load:

$$S_{cr} = \pm \{[500.56 \, D_{11} + 302.71(D_{12} + 2D_{66})R^2$$

$$+ \, 500.56 \, D_{22}R^4][3803.5 \, D_{11} + 4241.2(D_{12} + 2D_{66})R^2$$

$$+ \, 3803.5 \, D_{22}R^2]/499.08 \, Ra^2\}^{1/2} \qquad (5.113)$$

The $\pm$ sign indicates that the shear may be either positive or negative, i.e. the plate does not have a preferred direction of shear.

For an isotropic square plate, Equation (5.113) reduces to

$$S_{cr} = \pm \, 176 \, \frac{D}{a^2} \qquad (5.114)$$

The exact solution [6] is

$$S_{cr} = \pm \, 145 \, \frac{D}{a^2} \qquad (5.115)$$

and the two-term approximate solution is not very accurate.

For the case of an orthotropic plate with $D_{11}/D_{22} = 10$, $(D_{12} + 2D_{66})/D_{22} = 1.67$ and $a = b$, the two-term approximation yields

$$S_{cr} = \pm \, 768 \, \frac{D_{22}}{a^2} \qquad (5.116)$$

Solutions for the same case using a greater number of terms in the series (5.108) are presented in Table 5.5. Just as in the previous section and as found in the present case for the isotropic plate, the convergence of the solution is slow, and the two-term approximation is unacceptable.

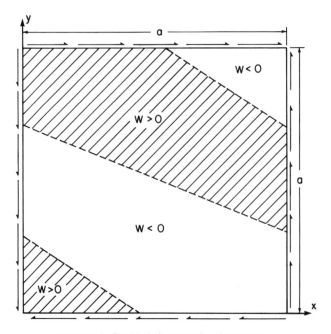

Figure 5.8. *Buckled shape under shear load.*

Figure 5.8 illustrates the character of the buckled deformation obtained using a large number of terms in the series (5.108). Note that the buckled shape possesses nodal lines (lines of zero deflection) that are not parallel to the sides of the plate. Such nodal lines do not occur for uniform compressive loadings of specially orthotropic plates. The unsymmetric nature of the solutions with respect to the lines $y = b/2$ or $x = a/2$ for shear buckling requires a larger number of terms in the approximating functions to obtain accurate solutions than is required for similar accuracy when considering compressive loadings.

5.9 STABILITY OF AN INFINITE STRIP UNDER SHEAR LOADING

In Sections 5.5 and 5.6 solutions were presented for plates with various boundary conditions under compressive loading in the x direction. These solutions converge to a constant value of the critical buckling load for long plates ($a >> b$). The solutions presented in Sections 5.7 and 5.8 for shear stability were approximate and cannot conveniently be applied to long plates. In this section we consider the shear buckling of plates infinite in length in the x direction and subjected to uniform shear loading. These exact solutions provide the limiting cases for long plates under shear.

We consider an infinite plate with the edges at $y = \pm b/2$ simply-supported, subjected to the uniformly distributed she

$$N_{xy} = S = \text{constant}$$

as shown in Figure 5.9.

The governing partial differential equation is obtained from (2.83):

$$D_{11} \frac{\partial^4 w}{\partial x^4} + 2(D_{12} + 2D_{66}) \frac{\partial^4 w}{\partial x^2 \partial y^2} + D_{22} \frac{\partial^4 w}{\partial y^4} - 2S \frac{\partial^2 w}{\partial x \partial y} = 0 \quad (5.118)$$

The boundary conditions along the edges $y = \pm b/2$ are
(1) simply-supported edges

$$w = M_y = -D_{12} \frac{\partial^2 w}{\partial x^2} - D_{22} \frac{\partial^2 w}{\partial y^2} = 0 \quad (5.119)$$

(2) clamped edges

$$w = \frac{\partial w}{\partial y} = 0 \quad (5.120)$$

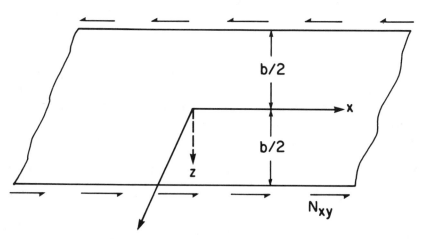

Figure 5.9. *Infinite strip under shear load.*

Using the approach suggested by Seydel [7], we consider a solution in the form

$$w = f(y) \, e^{2\xi x i / b} \tag{5.121}$$

where $i = \sqrt{-1}$. Substituting this relationship into the governing Equation (5.118), we obtain the differential equation

$$D_{11} \left(\frac{2\xi}{b} \right)^4 f - 2(D_{12} + 2D_{66}) \left(\frac{2\xi}{b} \right)^2 \frac{d^2 f}{dy^2} + D_{22} \frac{d^4 f}{dy^4}$$

$$- Si \left(\frac{2\xi}{b} \right) \frac{df}{dy} = 0 \tag{5.122}$$

The solution to Equation (5.122) can be written in the form

$$f(y) = A e^{2\lambda_1 i y / b} + B e^{2\lambda_2 i y / b} + C e^{2\lambda_3 i y / b} + D e^{2\lambda_4 i y / b} \tag{5.123}$$

where λ_1, λ_2, λ_3, and λ_4 are roots of the characteristic equation

$$D_{22}\lambda^4 + 2(D_{12} + 2D_{66})\xi^2\lambda^2 + D_{11}\xi^4 + 2 \left(\frac{b}{2} \right)^2 S\xi\lambda = 0 \tag{5.124}$$

The roots λ_1, λ_2, λ_3, and λ_4 may be real or complex.

Combining Equations (5.123) and (5.121), the solution takes the form

$$w = e^{2\xi x i / b} (A e^{\lambda_1 i y / b} + B e^{\lambda_2 i y / b} + C e^{\lambda_3 i y / b} + D e^{\lambda_4 i y / b}) \tag{5.125}$$

Combining Equations (5.125) with the boundary conditions (5.119) or (5.120), we obtain four homogeneous equations in the four unknowns A, B, C, and D. In order to obtain a nonzero solution, the determinant of the coefficient matrix must vanish. The solution is found to depend on ξ, and the critical buckling load corresponds to the value of ξ which yields the lowest value of S.

Seydel [7] presented solutions for S_{cr} and for ξ, where ξ characterizes the length between the successive buckling waves in the plate. These solutions are given for a complete range of the quantity Υ, where

$$\Upsilon = \frac{\sqrt{D_{11}D_{22}}}{(D_{12} + 2D_{66})} \tag{5.126}$$

Table 5.6. Coefficients β_1 and β_2 for simply-supported sides.

T	β_1	β_2
.0	11.71	—
.2	11.8	1.94
.5	12.2	2.07
1.0	13.17	2.49
2.0	10.8	2.28
3.0	9.95	2.16
5.0	9.25	2.13
10.0	8.7	2.08
20.0	8.4	—
40.0	8.25	—
∞	8.125	—

For $1 \le \Upsilon \le \infty$

$$S_{cr} = \frac{4\beta_1}{b^2} (D_{11}D_{22}^3)^{1/4} \tag{5.127}$$

$$\xi = \frac{\beta_2 b}{2} \left(\frac{D_{11}}{D_{22}} \right)^{1/4} \tag{5.128}$$

For $0 \le \Upsilon \le 1$

$$S_{cr} = \frac{4\beta_1}{b} \sqrt{D_{22}(D_{12} + 2D_{66})} \tag{5.129}$$

Table 5.7. Coefficients β_1 and β_2 for clamped sides.

T	β_1	β_2
.0	18.59	—
.2	18.85	1.20
.5	19.9	1.36
1.0	22.15	1.66
2.0	18.75	1.54
3.0	17.55	1.48
5.0	16.6	1.44
10.0	15.85	1.41
20.0	15.45	—
40.0	15.25	—
∞	15.07	—

$$\xi = \beta_2 \, \frac{b}{2} \, \sqrt{\frac{D_{12} + 2D_{66}}{D_{22}}} \tag{5.130}$$

The coefficients β_1 and β_2 are tabulated in Table 5.6 for the sides $y = \pm b/2$ simply-supported and Table 5.7 for the sides $y = \pm b/2$ clamped.

5.10 FREE VIBRATION OF SIMPLY-SUPPORTED RECTANGULAR PLATES

In the previous sections of this chapter, the problems of bending and buckling of specially orthotropic plates have been considered. For these problems only the static case was considered, i.e., any variation with respect to time was neglected. In this section we consider the free vibration of a specially orthotropic simply-supported rectangular plate. The solution procedure is very similar to the approach utilized for the stability analysis of the same plates.

The governing equation for free vibration without inplane or lateral loads $(q = N_x = N_y = N_{xy} = 0)$ can be obtained from (2.83) with the result

$$D_{11} \frac{\partial^4 w}{\partial y^4} + 2(D_{12} + 2D_{66}) \frac{\partial^4 w}{\partial x^2 \partial y^2} + D_{22} \frac{\partial^4 w}{\partial y^4} + \varrho \frac{\partial^2 w}{\partial t^2} = 0 \tag{5.131}$$

The solution to Equation (5.131) must satisfy the boundary conditions at the simply-supported edges as well as the initial conditions with respect to time. For resonant frequencies, time can be removed from Equation (5.131) by considering solutions of the form

$$w = w(x,y) \, e^{i\omega t} \tag{5.132}$$

where ω is a natural frequency of vibration.

Substituting Equation (5.132) into the governing Equation (5.131), we obtain the following governing equation for $w(x,y)$:

$$D_{11} \frac{\partial^4 w}{\partial x^4} + 2(D_{12} + 2D_{66}) \frac{\partial^4 w}{\partial x^2 \partial y^2} + D_{22} \frac{\partial^4 w}{\partial y^4} - \varrho \omega^2 w = 0 \tag{5.133}$$

The simply-supported boundary conditions are
    at $x = 0$ and a

$$w(x,y) = M_x = -D_{11} \frac{\partial^2 w(x,y)}{\partial x^2} - D_{12} \frac{\partial^2 w(x,y)}{\partial y^2} = 0 \tag{5.134}$$

at $y = 0$ and b

$$w(x,y) = M_y = -D_{12}\frac{\partial^2 w(x,y)}{\partial x^2} - D_{22}\frac{\partial^2 w(x,y)}{\partial y^2} = 0 \qquad (5.135)$$

The solution to Equation (5.133) which satisfies the boundary conditions (5.134) and (5.135) is of the form

$$w(x,y) = A_{mn}\sin\frac{m\pi x}{a}\sin\frac{n\pi y}{b} \qquad (5.136)$$

where m and n are integers. Substituting (5.136) into Equation (5.133) and solving for the natural frequency, we obtain

$$\omega_{mn} = \frac{\pi^2}{R^2 b^2}\frac{1}{\sqrt{\varrho}}\sqrt{D_{11}m^4 + 2(D_{12} + 2D_{66})m^2 n^2 R^2 + D_{22}n^4 R^4} \qquad (5.137)$$

where R is the plate aspect ratio a/b. Different natural frequencies are found corresponding to different combinations of m and n. The lowest fundamental frequency is given for $m = n = 1$:

$$\omega_{11} = \frac{\pi^2}{R^2 b^2}\frac{1}{\sqrt{\varrho}}\sqrt{D_{11} + 2(D_{12} + 2D_{66})R^2 + D_{22}R^4} \qquad (5.138)$$

For an isotropic plate the fundamental frequency is

$$\omega_{11} = \frac{\pi^2}{b^2}\sqrt{\frac{D}{\varrho}}\left[\left(\frac{b}{a}\right)^3 + 1\right] \qquad (5.139)$$

In each case the characteristic mode shape corresponding to this fundamental frequency is

$$w(x,y) = \sin\frac{\pi x}{a}\sin\frac{\pi y}{b} \qquad (5.140)$$

To indicate the effects of orthotropy, consider the lowest four frequencies and the corresponding characteristic shapes of a square plate. We consider the case $D_{11}/D_{22} = 10$, $(D_{12} + 2D_{66})/D_{22} = 1$, and also consider the case of a square isotropic plate, $D_{11}/D_{22} = (D_{12} + 2D_{66})/D_{22} = 1$. The lowest four frequencies

Table 5.8. Comparisons of lowest four natural frequencies.

Mode	Orthotropic $\omega = k\pi^2/b^2 \sqrt{D_{22}/\varrho}$			Isotropic $\omega = k\pi^2/b^2 \sqrt{D/\varrho}$		
	m	n	k	m	n	k
1st	1	1	3.62	1	1	2.0
2nd	1	2	5.86	1	2	5.0
3rd	1	3	10.45	2	1	5.0
4th	2	1	13.0	2	2	8.0

are compared in Table 5.8. The corresponding characteristic shapes are illustrated in Figure 5.10 where the node lines (lines of zero deflection) are drawn for these first four resonant frequencies. Note that while the first four frequencies of the isotropic plate show no preference with respect to direction, the orthotropic plate has a lower frequency corresponding to $m = 1$, $n = 3$ than for $m = 2$, $n = 1$.

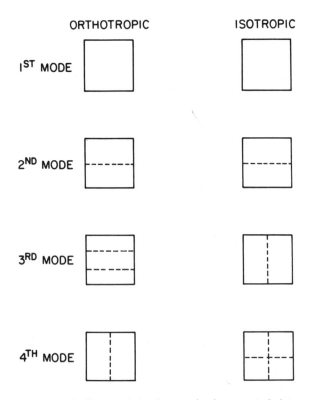

Figure 5.10. Characteristic shapes, simply-supported plates.

5.11 FREE VIBRATION OF RECTANGULAR PLATES WITH CLAMPED OR SIMPLY-SUPPORTED EDGES

The exact solution of the governing partial differential Equation (5.133) for the natural modes of free vibration of specially orthotropic plates has been obtained only for the case of two or four sides simply-supported. For any case other than four sides simply-supported, the solution to Equation (5.133) involves considerable difficulty. Thus, the use of approximate methods to determine the natural frequencies is necessary.

In Chapter 3 the appropriate energy criterion governing the free vibration of general laminated plates was formulated. Combining Equations (3.35) and (3.38) with no lateral or inplane loads ($q = N_x = N_y = N_{xy} = 0$), we obtain the energy criterion

$$U - T = \text{stationary value} \tag{5.141}$$

Substituting Equations (3.7) and (3.13) into Equation (5.141) and taking w in the form of Equation (5.132), we obtain

$$U - T = \frac{1}{2} \int_0^b \int_0^a \left[D_{11} \left(\frac{\partial^2 w}{\partial x^2} \right)^2 + 2D_{12} \frac{\partial^2 w}{\partial x^2} \frac{\partial^2 w}{\partial y^2} + D_{22} \left(\frac{\partial^2 w}{\partial y^2} \right)^2 \right.$$

$$\left. + 4D_{66} \left(\frac{\partial^2 w}{\partial x \partial y} \right)^2 - \varrho \omega^2 w \right] dx \, dy \tag{5.142}$$

The Ritz method is utilized in conjunction with Equations (5.141) and (5.142) to obtain approximate solutions for the vibration frequencies. We assume the solution in the usual form, i.e.

$$w(x,y) = \sum_{m=1}^{M} \sum_{n=1}^{N} A_{mn} X_m(x) \, Y_n(y) \tag{5.143}$$

where $X_m(x)$ and $Y_n(y)$ satisfy the boundary conditions on the edges $x = 0, a$ and $y = 0, b$, respectively. Using the assumed displacement function (5.143), the condition (5.141) reduces to the $M{\times}N$ conditions

$$\frac{\partial(U - T)}{\partial A_{mn}} = 0 \quad \begin{cases} m = 1, 2, \dots, M \\ n = 1, 2, \dots, N \end{cases} \tag{5.144}$$

Combining Equations (5.142), (5-143), and (5.144), we obtain the following MxN homogeneous algebraic equations:

$$\sum_{i=1}^{M} \sum_{j=1}^{N} \left\{ D_{11} \int_0^a \frac{d^2X_i}{dx^2} \frac{d^2X_m}{dx^2} dx \int_0^b Y_j Y_n \, dy \right.$$

$$+ D_{12} \left[\int_0^a X_m \frac{d^2X_i}{dx^2} dx \int_0^b Y_j \frac{d^2Y_n}{dy^2} dy \right.$$

$$\left. + \int_0^a X_i \frac{d^2X_m}{dx^2} dx \int_0^b Y_n \frac{d^2Y_j}{dy^2} dy \right]$$

$$+ D_{22} \int_0^a X_i X_m \, dx \int_0^b \frac{d^2Y_j}{dy^2} \frac{d^2Y_n}{dy^2} dy$$

$$+ 4D_{66} \int_0^a \frac{dX_i}{dx} \frac{dX_m}{dx} dx \int_0^b \frac{dY_j}{dy} \frac{dY_n}{dy} dy$$

$$\left. - \varrho\omega_{mn}^2 \int_0^a X_i X_m \, dx \int_0^b Y_j Y_n \, dy \right\} A_{ij} = 0$$

$$i = 1,2,\dots,M$$
$$j = 1,2,\dots,N$$

(5.145)

A comparison of Equation (5.145) to Equations (5.44) and (5.110) reveals the similarity between bending, stability, and free vibration problems as obtained by the Ritz method. The equations governing stability and free vibration are nearly identical in that each results in a set of homogeneous algebraic equations.

Since the system of Equations (5.145) is homogeneous, a nontrivial solution can be obtained only if the determinant of the coefficient matrix is zero. Just as in the stability problem, we use this condition to find our solution, which in this case is the natural frequencies of vibration, ω_{mn}.

Hearmon [8] has suggested that acceptable solutions for the natural modes of specially orthotropic plates can be obtained using a single appropriate combination of $X_m(x)$ and $Y_n(y)$ for each such mode, for plates with any combination of clamped or simply-supported edges. For opposite edges clamped, Hearmon suggests using the beam characteristic shape corresponding to a beam fixed on each end:

for edges $x = 0,a$ clamped

$$X_m(x) = \gamma_m \cos \frac{\lambda_m x}{a} - \gamma_m \cosh \frac{\lambda_m x}{a} + \sin \frac{\lambda_m x}{a} - \sinh \frac{\lambda_m x}{a} \qquad (5.146)$$

for edges $y = 0, b$ clamped

$$Y_n(y) = \gamma_n \cos \frac{\lambda_n y}{b} - \gamma_n \cosh \frac{\lambda_n y}{b} + \sin \frac{\lambda_n y}{b} - \sinh \frac{\lambda_n y}{b} \qquad (5.147)$$

where $\lambda_m, \lambda_n, \gamma_m, \gamma_n$ are found from Equations (5.50) and (5.51). For the opposite edges simply-supported, the functions can be taken as sine terms:

for the edges $x = 0,a$ simply-supported

$$X_m(x) = \sin \frac{m\pi x}{a} \qquad (5.148)$$

for the edges $y = 0,b$ simply-supported

$$Y_n(y) = \sin \frac{n\pi y}{b} \qquad (5.149)$$

When one edge is clamped and the other is simply-supported, the beam characteristic shape corresponding to a beam fixed on one end and simply-supported on the other can be used:

for the edge $x = 0$ clamped and the edge $x = a$ simply-supported

$$X_m(x) = \gamma_m \cos \frac{\lambda_m x}{a} - \gamma_m \cosh \frac{\lambda_m x}{a} + \sin \frac{\lambda_m x}{a} - \sinh \frac{\lambda_m x}{a} \qquad (5.150)$$

for the edge $y = 0$ clamped and the edge $y = b$ simply-supported

$$Y_n(y) = \gamma_n \cos \frac{\lambda_n y}{b} - \gamma_n \cosh \frac{\lambda_n y}{b} + \sin \frac{\lambda_n y}{b} - \sinh \frac{\lambda_n y}{b} \qquad (5.151)$$

where λ_m, λ_n are determined from the frequency equation [4]:

$$\tan \lambda_i - \tanh \lambda_i = 0 \qquad (5.152)$$

Table 5.9. First five roots to Equation (5.50).

λ_1	λ_2	λ_3	λ_4	λ_5
3.927	7.069	10.210	13.352	16.493

and

$$\gamma_i = \frac{\sin \lambda_i - \sinh \lambda_i}{\cosh \lambda_i - \cos \lambda_i} \tag{5.153}$$

where $i = m,n$ in both Equations (5.152) and (5.153). The first five values of λ_i which satisfy Equation (5.152) have been tabulated in Reference [4] and are given in Table 5.9.

An approximate relationship similar to the one given by Equation (5.54) can be derived for determining values of λ_i. For large values of λ_i

$$\tanh \lambda_i \approx 1 \tag{5.154}$$

and Equation (5.152) becomes

$$\tan \lambda_i = 1 \tag{5.155}$$

Roots to this relationship can be written in the form

$$\lambda_i = (i + 0.25)\pi \tag{5.156}$$

Calculating the first five values of λ_i from Equation (5.156), we find that they are identical to the exact values listed in Table 5.9.

If we restrict ourselves to the first approximation for each of the natural frequencies, then Equation (5.145) reduces to the following simple form for all boundary conditions currently under consideration:

$$\omega_{mn} = \frac{1}{a^2\sqrt{\varrho}} \sqrt{D_{11}\alpha_1^4 + 2(D_{12} + 2D_{66})R^2\alpha_2 + D_{22}R^4\alpha_3^4} \tag{5.157}$$

where α_1, α_2, and α_3 are tabulated in Table 5.10 for each combination of simple and clamped supports.

For the case of an isotropic plate, the solutions (5.157) reduce to the form

$$\omega_{mn} = \frac{1}{a^2} \sqrt{\frac{D}{\varrho} (\alpha_1^4 + 2R^2\alpha_2 + R^4\alpha_3^4)} \tag{5.158}$$

Table 5.10. Coefficients for natural frequency calculations.

Boundary Conditions	α_1	α_3	α_2	m	n
(edges: top c, left c, right c, bottom c)	4.730	4.730	151.3	1	1
	4.730	$(n+.5)\pi$	$12.30\,\alpha_3\,(\alpha_3-2)$	1	2,3,4 . . .
	$(m+.5)\pi$	4.730	$12.30\,\alpha_1\,(\alpha_1-2)$	2,3,4 . . .	1
	$(m+.5)\pi$	$(n+.5)\pi$	$(\alpha_1\,\alpha_3(\alpha_1-2)^*$ $(\alpha_3-2))$	2,3,4 . . .	2,3,4 . . .
(edges: top c, left c, right c, bottom s)	4.730	$(n+.25)\pi$	$12.30\,\alpha_3\,(\alpha_3-1)$	1	1,2,3 . . .
	$(m+.5)\pi$	$(n+.25)\pi$	$(\alpha_1\,\alpha_3\,(\alpha_1-2)^*$ $(\alpha_3-1.))$	2,3,4 . . .	2,3,4 . . .
(edges: top s, left c, right c, bottom s)	4.730	$n\pi$	$12.30\,n^2\,\pi^2$	1	1,2,3 . . .
	$(m+.5)\pi$	$n\pi$	$n^2\pi^2\alpha_1\,(\alpha_1-2)$	2,3,4 . . .	1,2,3 . . .
(edges: top s, left c, right s, bottom c)	$(m+.25)\pi$	$(n+.25)\pi$	$(\alpha_1\alpha_3(\alpha_1-1)^*$ $(\alpha_3-1))$	1,2,3 . . .	1,2,3 . . .
(edges: top s, left c, right s, bottom s)	$(m+.25)\pi$	$n\pi$	$n^2\pi^2\alpha_1(\alpha_1-1.)$	1,2,3 . . .	1,2,3 . . .
(edges: top s, left s, right s, bottom s)	$m\pi$	$n\pi$	$m^2n^2\pi^4$	1,2,3 . . .	1,2,3 . . .

BENDING OF SIMPLY-SUPPORTED RECTANGULAR PLATES

Table 5.11. Lowest six natural frequencies,
first approximation vs 25 term series.

$$\omega_{ij} = (k_{ij}/a^2)\, \sqrt{D_{22}/\varrho}$$

	k_{ij} Equation (5.157)	25 Term Series		k_{ij} Equation (5.157)	25 Term Series
ω_{11}	77.5	77.4	ω_{11}	55.3	55.2
ω_{12}	103.4	103.3	ω_{12}	80.7	80.6
ω_{13}	153.9	153.7	ω_{13}	129.8	129.6
ω_{21}	201.0	200.8	ω_{21}	163.8	163.6
ω_{22}	221.1	220.8	ω_{11}	183.3	183.1
ω_{14}	228.1	227.6	ω_{14}	201.2	200.8

From Table 5.10 we find that the fundamental frequency for the case of a square isotropic plate with clamped edges is

$$\omega_{11} = \frac{36.1}{a^2} \sqrt{\frac{D}{\varrho}} \qquad (5.159)$$

A solution using 25 terms ($M = N = 5$) of the form given in Equations (5.146) and (5.147) has been obtained by Young [3] for this fundamental frequency. Young's result is

$$\omega_{11} = \frac{35.99}{a^2} \sqrt{\frac{D}{\varrho}} \qquad (5.160)$$

and clearly in this case the approximate solution (5-158) is quite accurate.

For the orthotropic case, we consider a square plate with $D_{11}/D_{22} = 10$, $(D_{12} + 2D_{66})/D_{22} = 1.67$. The first six natural frequencies are tabulated in Table 5.11 as obtained with Equation (5.157) for the case of all edges clamped and for the case of two adjacent edges clamped and two simply-supported. Also shown in this table are the corresponding solutions obtained using 25 terms ($M = N = 5$) of the appropriate $X_m(x)$ and $Y_n(y)$ terms in Equation (5.145). Clearly the first approximation results are sufficiently accurate for most applications.

REFERENCES

1. Timoshenko, S. and S. Woinowsky-Krieger. *Theory of Plates and Shells*, McGraw-Hill (1959).
2. Lekhnitskii, S. G. *Anisotropic Plates*, Translated by S. W. Tsai and T. Cheron, Gordon and Breach (1968).
3. Young, D. "Vibrations of Rectangular Plates by the Ritz Method," *Journal of Applied Mechanics*, 17:448–453 (1950).
4. Timoshenko, S. P., D. H. Young and W. Weaver, Jr. *Vibration Problems in Engineering*, Fourth Edition, John Wiley & Sons (1974).
5. Faddeva, V. N. *Computational Methods of Linear Algebra*, Translated from Russian by C. D. Benster, N.Y.: Dover Publications, Inc. (1959).
6. Timoshenko, S. and J. M. Gere. *Theory of Elastic Stability*, McGraw-Hill (1961).
7. Seydel, E., "On the Buckling of Rectangular Isotropic or Orthogonal-Isotropic Plates by Tangential Stresses," *Ing. Archiv* (1933).
8. Hearmon, R. F. S., "The Frequency of Flexural Vibration of Rectangular Orthotropic Plates with Clamped or Supported Edges," *Transactions of ASME, Journal of Applied Mechanics*, pp. 537–540 (1959).

CHAPTER 6

Midplane Symmetric Laminates

6.1 INTRODUCTION

IN THE PREVIOUS chapter, the analysis of specially orthotropic plates in which the bending-twisting coupling stiffness terms D_{16} and D_{26} vanish was considered. As indicated, such plates are special cases of general laminated plates and occur only for laminates constructed of orthotropic or isotropic plies with the principal material axes parallel to the plate axes. In this chapter we consider symmetric laminates containing orthotropic layers in which the principal material axes are not parallel to the plate axes. Such laminates are characterized by nonvanishing bending-twisting coupling stiffness terms D_{16} and D_{26}. Since the laminates under consideration are symmetric, the bending-extensional coupling terms, B_{ij}, vanish.

Laminated plates for which bending-extensional coupling does not exist, but which do exhibit nonzero D_{16} or D_{26} terms, are an important class of plates. They are mathematically equivalent to homogeneous anisotropic plates.

The inclusion of the bending-twisting coupling terms, D_{16} and D_{26}, in the governing equations significantly increases the complexity of the analysis. As indicated earlier, one approach which is sometimes taken in the analysis of such plates is simply to neglect these coupling terms. In this chapter a variety of solutions will be presented to indicate the nature of the error that is introduced when this is done. Bending, buckling, and free vibration will be considered, and solutions will be compared to the corresponding specially orthotropic solutions developed in Chapter 5.

6.2 BENDING OF SIMPLY-SUPPORTED RECTANGULAR PLATES

Consider a rectangular plate simply-supported on the edges and subjected to the transverse load $q = q(x,y)$, Figure 6.1. For this loading in conjunction with

133

symmetric laminates, equation (2.37) in the absence of inertia terms (static loading) becomes

$$D_{11} \frac{\partial^4 w}{\partial x^4} + 4D_{16} \frac{\partial^4 w}{\partial x^3 \partial y} + 2(D_{12} + 2D_{66}) \frac{\partial^4 w}{\partial x^2 \partial y^2}$$

$$+ 4D_{26} \frac{\partial^4 w}{\partial x \partial y^3} + D_{22} \frac{\partial^4 w}{\partial y^4} = q \tag{6.1}$$

For simply-supported edges the following boundary conditions are applicable: at $x = 0$ and a

$$w = M_x = - D_{11} \frac{\partial^2 w}{\partial x^2} - 2D_{16} \frac{\partial^2 w}{\partial x \partial y} - D_{12} \frac{\partial^2 w}{\partial y^2} = 0 \tag{6.2}$$

at $y = 0$ and b

$$w = M_y = - D_{12} \frac{\partial^2 w}{\partial x^2} - 2D_{26} \frac{\partial^2 w}{\partial x \partial y} - D_{22} \frac{\partial^2 w}{\partial y^2} = 0 \tag{6.3}$$

To obtain a solution to this problem for the specially orthotropic case, we assumed the solution in terms of a series of double sine terms and expanded the loading into a double sine series. The assumed form for the solution satisfied the simply-supported boundary conditions, and when the series was substituted into the governing equations, the resulting equation contained only double sine terms. This was due to the fact that the governing equations only contained even order derivatives in x and y. Thus, by comparing like double sine terms, the coefficients in the assumed series were determined, and the problem was solved.

For the anisotropic case presently under consideration, this approach will not work. In particular, the double sine series will not satisfy the boundary conditions, Equations (6.2) and (6.3), due to the presence of the bending-twisting

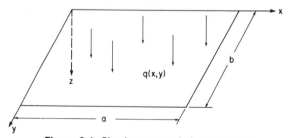

Figure 6.1. *Simply-supported plate geometry.*

coupling terms. Furthermore the governing partial differential Equation (6.1) contains odd order derivatives in x and y, in addition to the usual even order derivatives, which result in double cosine terms. Thus, like sine terms cannot be compared for the purpose of determining the coefficients in the assumed series.

Both of these difficulties can be overcome by using either the Ritz or Galerkin method to obtain an approximate solution to the problem. Using the Ritz method we substitute Equation (3.6) for midplane symmetric laminates along with Equation (3.15) into Equation (3.36), and taking into account the energy criterion (3.35), we obtain:

$$
\frac{1}{2} \int_0^b \int_0^a \left[D_{11} \left(\frac{\partial^2 w}{\partial x^2} \right)^2 + 2D_{12} \frac{\partial^2 w}{\partial x^2} \frac{\partial^2 w}{\partial y^2} + D_{22} \left(\frac{\partial^2 w}{\partial y^2} \right)^2 \right.
$$

$$
+ 4 \left(D_{16} \frac{\partial^2 w}{\partial x^2} + D_{26} \frac{\partial^2 w}{\partial y^2} \right) \frac{\partial^2 w}{\partial x \partial y} + 4D_{66} \left(\frac{\partial^2 w}{\partial x \partial y} \right)^2 \qquad (6.4)
$$

$$
\left. - 2qw \vphantom{\frac{\partial^2 w}{\partial x^2}} \right] dx\, dy = \text{stationary value}
$$

The solution is assumed to be of the variables separable form

$$
w = \sum_{m=1}^{M} \sum_{n=1}^{N} A_{mn}\, X_m(x)\, Y_n(y) \qquad (6.5)
$$

Substituting this series into the energy conditions (6.4), and applying Equation (3.42) for the determination of A_{mn}, we obtain the following $M \times N$ conditions:

$$
\sum_{i=1}^{M} \sum_{j=1}^{N} \left\{ D_{11} \int_0^a \frac{d^2 X_i}{dx^2} \frac{d^2 X_m}{dx^2}\, dx \int_0^b Y_j\, Y_n\, dy \right.
$$

$$
+ D_{12} \left[\int_0^a X_m \frac{d^2 X_i}{dx^2}\, dx \int_0^b Y_j \frac{d^2 Y_n}{dy^2}\, dy \right. \qquad (6.6)
$$

$$
+ \left. \int_0^a X_i \frac{d^2 X_m}{dx^2}\, dx \int_0^b Y_n \frac{d^2 Y_j}{dy^2}\, dy \right]
$$

$$+ D_{22} \int_0^a X_i X_m \, dx \int_0^b \frac{d^2 Y_j}{dy^2} \frac{d^2 Y_n}{dy^2} \, dy$$

$$+ 4D_{66} \int_0^a \frac{dX_i}{dx} \frac{dX_m}{dx} \, dx \int_0^b \frac{dY_j}{dy} \frac{dY_n}{dy} \, dy$$

$$+ 2D_{16} \left[\int_0^a \frac{d^2 X_i}{dx^2} \frac{dX_m}{dx} \, dx \int_0^b Y_j \frac{dY_n}{dy} \, dy \right.$$

$$+ \int_0^a \frac{dX_i}{dx} \frac{d^2 X_m}{dx^2} \, dx \int_0^b Y_n \frac{dY_j}{dy} \, dy \left. \right]$$

$$+ 2D_{26} \left[\int_0^a X_m \frac{dX_i}{dx} \, dx \int_0^b \frac{dY_j}{dy} \frac{d^2 Y_n}{dy^2} \, dy \right.$$

$$+ \int_0^a X_i \frac{dX_m}{dx} \, dx \int_0^b \frac{d^2 Y_j}{dy^2} \frac{dY_n}{dy} \, dy \left. \right] A_{mn}$$

$$= q_0 \int_0^a X_m \, dx \int_0^b Y_n \, dy \quad \begin{matrix} m = 1,2,\ldots,M \\ n = 1,2,\ldots,N \end{matrix}$$

where the transverse load is assumed to be uniform, i.e. $q(x,y) = q_0 =$ constant. Equations (6.6) form a system of MxN linear simultaneous equations for the determination of the MxN unknowns A_{mn}.

In the applications of the Ritz method in Chapter V, the functions $X_m(x)$ and $Y_n(y)$ satisfied the boundary conditions at the edges $x = 0,a$ and $y = 0,b$. As pointed out in Chapter 3, however, for the case of simple-supports in conjunction with symmetric laminates in which the D_{16} and D_{26} stiffness terms are non-zero, a solution in a variables separable form which satisfies the moment boundary conditions does not exist [1]. This difficulty can be overcome in conjunction with either the Ritz or Galerkin methods, as only the displacement boundary conditions are required to be satisfied. Thus, in the case of simple-supports, it is only necessary to choose functions in conjunction with Equation (6.5) in which $w = 0$ on the plate boundary and the slope is free to take on any value. For cases in which the moment boundary conditions are not satisfied, convergence of Equation (6.5) may be very slow.

We now consider Equation (6.5) in the form of a double sine series, i.e.

$$X_m(x) = \sin \frac{m\pi x}{a}$$

(6.7)

$$Y_n(y) = \sin \frac{n\pi y}{b}$$

If we attempt to find a first approximation using the double sine series in conjunction with Equation (6.6), we find that a one term approximation yields a solution which does not contain D_{16} and D_{26} terms. This is due to the fact that

$$\int_0^a \cos \frac{m\pi x}{a} \sin \frac{m\pi x}{a} \, dx = 0$$

$$\int_0^b \cos \frac{n\pi y}{b} \sin \frac{n\pi y}{b} \, dy = 0$$

However, this result does not indicate that the terms have an insignificant effect. For example let us consider a square plate with $D_{22}/D_{11} = 1$, $(D_{12} + 2D_{66})/D_{11} = 1.5$, and $D_{16}/D_{11} = D_{26}/D_{11} = -0.5$. The maximum deflection obtained using 49 terms ($M = N = 7$) in the series (6.5) is found to be

$$w_{max} = 0.00425 \frac{q_0 a^4}{D_{11}}$$

(6.8)

If D_{16} and D_{26} are neglected, the maximum deflection is found to be

$$w_{max} = 0.00324 \frac{q_0 a^4}{D_{11}}$$

(6.9)

and we see that the maximum deflection as computed (ignoring the coupling terms) in 24% less than the deflection computed including the coupling terms.

The above solution method can be applied to any combination of plate bending stiffness properties. It is useful, however, to consider certain representative properties such that an exact solution is available, in order that we might assess the accuracy of the Ritz method. This is of particular importance since we know that Equation (6.6) does not represent a complete set of functions because the exact solution cannot be represented by a variables separable solution. Without the

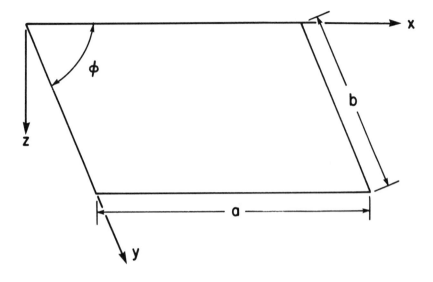

Figure 6.2. *Skew plate geometry.*

property of completeness, we are not assured that the double sine terms used in conjunction with Equation (6.6) will converge to the exact deflection. For comparative purposes an analogy between isotropic skew (parallelogram) plates and rectangular anisotropic plates is now considered. This analogy provides a few solutions which completely satisfy the boundary conditions of zero moment along the boundaries and, although the available solutions are limited in scope, they provide a guideline to assess the approximate solution under consideration.

The governing differential equation for skew (parallelogram shaped) isotropic plates under a uniform load $q(x,y) = \bar{q} =$ constant can be written for oblique coordinates x and y (Figure 6.2), in the following form [2]:

$$
\frac{\partial^4 w}{\partial x^4} - 4 \cos\phi \, \frac{\partial^4 w}{\partial x^3 \partial y} + 2(1 + 2 \cos^2\phi) \, \frac{\partial^4 w}{\partial x^2 \partial y^2}
$$
$$
- 4 \cos\phi \, \frac{\partial^4 w}{\partial x \partial y^3} + R^4 \, \frac{\partial^4 w}{\partial y^4} = \frac{\bar{q}}{D} \sin^4\phi
$$

(6.10)

If the skew plate is simply-supported along the edges $x = 0,a$ and $y = 0,b$, then the appropriate boundary conditions are at $x = 0$ and a

$$w = M_x = \frac{\partial^2 w}{\partial x^2} - 2 \cos\phi \frac{\partial^2 w}{\partial x \partial y} = 0 \tag{6.11}$$

at $y = 0$ and b

$$w = M_y = \frac{\partial^2 w}{\partial y^2} - 2 \cos\phi \frac{\partial^2 w}{\partial x \partial y} = 0 \tag{6.12}$$

As indicated in Reference [3], the governing Equation (6.1) and boundary conditions (6.2) and (6.3) for the rectangular anisotropic plate are identical to the governing Equation (6.10) and boundary conditions (6.11) and (6.12) for the skew isotropic plate if

$$D_{22} = D_{11} = D, \quad \frac{D_{12} + 2D_{66}}{D_{11}} = (1 + 2\cos^2\phi) \tag{6.13}$$

$$\frac{D_{16}}{D_{11}} = \frac{D_{26}}{D_{11}} = -\cos\phi, \quad q_0 = \bar{q} \sin^4\phi$$

It should be noted that

at $x = 0, a$

$$\frac{\partial^2 w}{\partial y^2} = 0 \tag{6.14}$$

at $y = 0, b$

$$\frac{\partial^2 w}{\partial x^2} = 0 \tag{6.15}$$

This is due to the fact that $w = 0$ along the entire boundary.

The stiffness terms described with the relationship (6.13) are equivalent to the stiffness terms of an orthotropic material with the principal axes of orthotropy at 45° to the plate edges. The orthotropic bending stiffnesses of this equivalent material as a function of ϕ can be deduced from the transformation equations for bending stiffness, which as indicated in Chapter 2, are the same as the transformation of the C_{ij}, Equations (1.35–1.48):

$$D'_{11} = D(1 + 2\cos\phi + \cos^2\phi)$$
$$D'_{22} = D(1 - 2\cos\phi + \cos^2\phi) \qquad (6.16)$$
$$D'_{12} + 2D'_{66} = D\sin^2\phi$$

where the primed terms are the equivalent orthotropic material bending stiffnesses with respect to the plate axes and with respect to the material principal axes (primed quantities), for selected values of ϕ, as listed in Table 6.1. Clearly the ratio D'_{11}/D'_{22} becomes larger for plates highly skewed, and thus the equivalent rectangular plate becomes more strongly anisotropic, i.e. the ratios $D_{16}/D_{11} = D_{26}/D_{11}$ become larger.

The purpose for citing the above analogy is to use the solutions which are available for skew isotropic plates to obtain solutions for rectangular anisotropic plates. Reference [4] presents solutions for simply-supported skew isotropic plates. These solutions were obtained by utilizing conformal mapping to map the skew plate boundaries onto a unit circle; then the problem was solved using a series which exactly satisfied the boundary conditions. These solutions have been utilized to obtain exact solutions for square anisotropic plates with the material properties given in Table 6.1 and plotted in Figure 6.3. Also shown in Table 6.2 and Figure 6.3 are solutions for the same plates obtained with 49 double sine terms ($M = N = 7$) in conjunction with the Ritz method previously described, and solutions obtained using the specially orthotropic analysis of Section 5.2, neglecting the D_{16} and D_{26} terms. For the strongly orthotropic plates, neglect of the D_{16} and D_{26} terms leads to large errors, and the predicted deflections are too small. The Ritz method also underestimates the deflection, but the error is much smaller. For example when $\phi = 60°$ ($D_{11}/D_{22} = 1$, $(D_{12} + 2D_{66})/D_{11} = 1.5$, $D_{16}/D_{11} = D_{26}/D_{11} = -0.5$) the exact solution is

$$w_{max} = 0.00452 \frac{q_0 a^4}{D_{11}} \qquad (6.17)$$

Table 6.1. Equivalent bending stiffness ratios.

Equivalent Skew Angle ϕ	D'_{22}/D'_{11}	$(D'_{12} + 2D'_{66})/D'_{11}$	D'_{16}/D'_{11}	D_{22}/D_{11}	$(D_{12} + 2D_{66})/D_{11}$	D_{16}/D_{11}
90°	1.	1.	0.	1.	1.	0.
80°	.495	.702	0.	1.	1.061	−.174
63°	.141	.376	0.	1.	1.412	−.454
60°	.111	.333	0.	1.	1.500	−.500
54°	.0675	.260	0.	1.	1.690	−.587

Table 6.2. Comparison of solutions.

$$w_{max} = (k\ a^4/D_{11})\ q_0$$

Equivalent Skew Angle ϕ	Exact Solution k	Orthotropic Solution k	Energy Solution k
90°	.00406	.00406	.00406
80°	.00411	.00394	.00408
63°	.00444	.00336	.00422
60°	.00452	.00324	.00425
54°	.00476	.00301	.00430

Comparing the results to the Ritz solution (6.8), we find that the approximate solution is 6% in error, while the specially orthotropic solution in (6.9) is in error by 28%.

The convergence of the solution given by Equation (6.17) is shown in Figure 6.4. As a basis of comparison, the orthotropic solution ($D_{16} = D_{26} = 0$) is also shown. In Figure 6.5 the convergence of the bending moment M_x at the center of

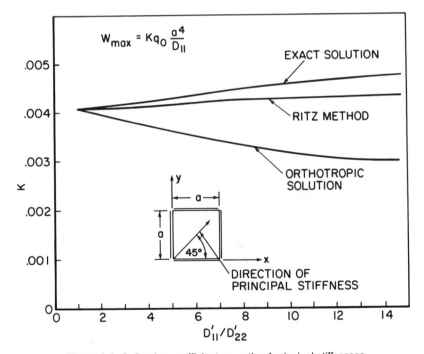

Figure 6.3. Deflection coefficient vs. ratio of principal stiffnesses.

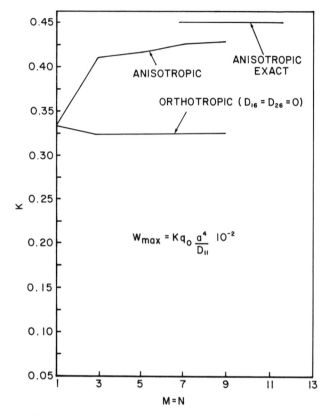

Figure 6.4. *Maximum deflection for increasing number of terms.*

the plate is illustrated. Again the corresponding orthotropic solution is also shown. The anisotropic plate solution in Figure 6.4 as represented by the Ritz method appears to be converging slowly toward the exact solution for deflection. For the bending moment illustrated in Figure 6.5, the anisotropic solution appears to be oscillating about a relative constant value, and this value seems consistent with the orthotropic solution. It is difficult, however, to generate much confidence in the results, due to the erratic nature of the solution. For both the center deflection and center moment, rapid convergence for the specially orthotropic case ($D_{16} = D_{26} = 0$) is observed.

The results in Figures 6.4 and 6.5 are consistent with a more extensive study on convergence of anisotropic plate solutions performed by Ashton [4]. In particular it was found that the Ritz method provides convergent solutions for the deflections as long as the geometrical boundary conditions are satisfied. The convergence appears to be considerably more rapid when the natural boundary conditions are also satisfied. In addition, if all of the boundary conditions are satisfied, the moments are generally also rapidly convergent, while the moments may or may not converge to a correct result if the natural boundary conditions are not satisfied.

The results presented here and discussed in more detail in Reference [4] suggest that variational methods such as Ritz and Galerkin should be acceptable for bending deflections, buckling, and free vibration of anisotropic plates (gross response) but may not be acceptable when the derivatives of the deflection are required.

6.3 BENDING OF CLAMPED RECTANGULAR PLATES

In this section the bending of a rectangular anisotropic plate with clamped edges under a uniform load is considered. The Ritz method is utilized for this analysis, and the solution procedure is identical to the approach used in Section

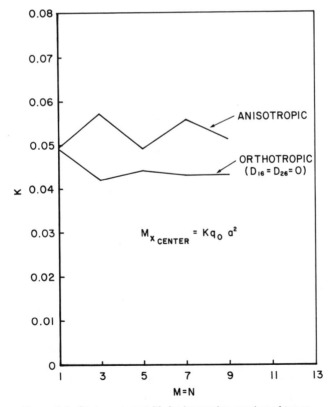

Figure 6.5. *Center moment M_x for increasing number of terms.*

5.4 for the specially orthotropic case except the D_{16} and D_{26} terms are included in the present analysis.

The following boundary conditions are applicable for fully clamped edges: at $x = 0$ and a

$$w = \frac{\partial w}{\partial x} = 0 \tag{6.18}$$

at $y = 0$ and b

$$w = \frac{\partial w}{\partial y} = 0 \tag{6.19}$$

To apply the Ritz method, we assume the solution in the variables separable series form

$$w = \sum_{m=1}^{M} \sum_{n=1}^{N} A_{mn} X_m(x) \, Y_n(y) \tag{6.20}$$

Substituting the assumed series (6.20) into the energy criterion, Equation (6.4), and differentiating with respect to each undetermined A_{mn}, we obtain MxN linear simultaneous equations in MxN unknowns. These equations are given by (6.6).

In the present applications, we select the functions $X_m(x)$ and $Y_n(y)$ as the characteristic shapes of freely vibrating beams with fixed supports:

$$X_m(x) = \gamma_m \cos \frac{\lambda_m x}{a} - \gamma_m \cosh \frac{\lambda_m x}{a} + \sin \frac{\lambda_m x}{a} - \sinh \frac{\lambda_m x}{a}$$

$$Y_n(y) = \gamma_n \cos \frac{\lambda_n y}{b} - \gamma_n \cosh \frac{\lambda_n y}{b} + \sin \frac{\lambda_n y}{b} - \sinh \frac{\lambda_n y}{b} \tag{6.21}$$

where γ_m, γ_n, λ_m, and λ_n are obtained from Equations (5.49) and Equations (5.51). These functions satisfy the boundary conditions (6.18) and (6.19) for the clamped plate.

To indicate the nature of the solution, we consider a square plate composed of an orthotropic material with $E_L/E_T = 10$, $G_{LT}/E_T = 0.25$, $\nu_{LT} = 0.3$. The maximum deflections for this plate are tabulated in Table 6.3 and plotted in Figure 6.6 for various orientations of the principal orthotropic axis of the material with respect to the plate edges (θ = angle between the x-axis and the principal material axis as defined by the laminate stacking code in Chapter 1). Also pre-

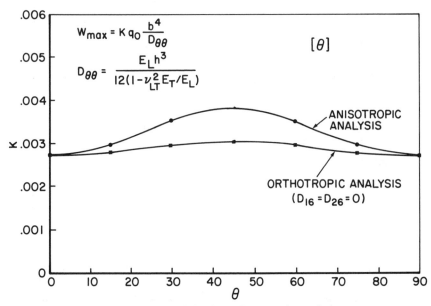

Figure 6.6. Maximum deflections of square clamped plates.

sented in Table 6.3 and Figure 6.6 are solutions for the same case neglecting D_{16} and D_{26} terms (specially orthotropic analysis). All solutions were obtained using 49 terms ($M = N = 7$) in the assumed series [5]. Clearly the D_{16} and D_{26} terms have a significant effect on some of the these solutions.

In Figure 6.7 the moments M_x and M_y along the centerline $y = b/2$ are presented for several orientations of the material axes. These moments were obtained using the calculated deflection surface from Equation (6.6) together with Equation (2.72).

In Figure 6.8 the maximum deflection is presented for a rectangular plate

Table 6.3. Maximum deflection under uniform load for anisotropic plate.

	w_{max} = kq_0 $(b^4/D_{\theta\theta})$ 10^{-3}			$D_{\theta\theta}$ = E_L $h^3/\{12(1 - v_{LT}^2 E_T/E_L)\}$				
Orientation	0°	15°	26.6°	30°	45°	60°	75°	90°
Anisotropic, a = b	2.72	2.96	—	3.52	3.83	3.52	2.96	2.72
Orthotropic, a = b	2.72	2.79	—	2.93	3.00	2.93	2.79	2.72
Anisotropic, a = 2b	18.9	17.4	14.3	—	7.89	4.49	3.10	2.75

k

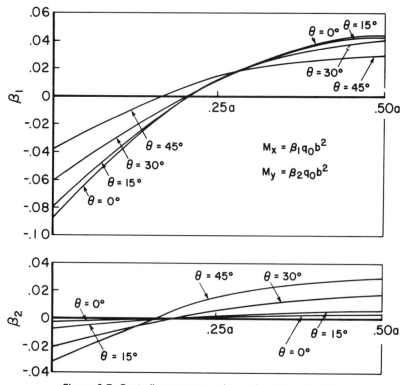

Figure 6.7. Centerline moments, clamped anisotropic plates.

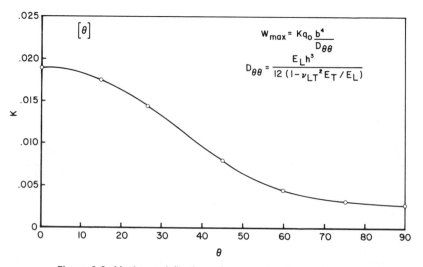

Figure 6.8. Maximum deflections of rectangular clamped plates.

Table 6.4. Convergence of Solutions.

	$w = \alpha q_0 (b^4/D_{\theta\theta}) \, 10^{-3}$		$M_x = \beta_1 \, q_0 b^2$	
	Anisotropic Solution		Orthotropic Solution	
M = N	α	β	α	β
1	3.153	.0313	3.153	.0313
3	3.800	.0427	2.937	.0366
5	3.803	.0475	3.006	.0411
7	3.832	.0494	2.997	.0429

($a = 2b$) of the same orthotropic material considered above. The influence of the orientation of the principal axis of the material is more pronounced for this case than found for the square plate.

To indicate the convergence of the solution for increasing M and N, consider a square plate composed of the same orthotropic material, with the principal axes oriented at 45 degrees to the plate edges. Table 6.4 contains the maximum deflection and maximum moment results for $M = N = 1, 3, 5$, and 7 in the assumed series. Also presented in this table are the sequential solutions for the specially orthotropic analysis. Examination of these results reveals that the specially orthotropic solution converges more rapidly than the anisotropic analysis, but results in a large error for the case considered.

6.4 STABILITY OF AN INFINITE STRIP UNDER COMPRESSION OR SHEAR

The stability of an infinite strip composed of a specially orthotropic material was considered in Chapter 5. In this section the stability of such a strip is again considered with additional generalization to include both anisotropic materials and also orthotropic materials with principal axes which are not parallel to the plate edges. The solution method was proposed by Thielemann [6], and is essentially the same as applied to the case of specially orthotropic plates in Section 5.9.

The geometry has been presented in Figure 5.9. We will consider the edges $y = \pm b/2$ to be either simply-supported or clamped. The governing differential equation for applied uniform compression in the x direction ($N_x = -N_0$) and applied uniform shear ($N_{xy} = S$) is a specialization of Equation (2.82):

$$D_{11} \frac{\partial^4 w}{\partial x^4} + 4D_{16} \frac{\partial^4 w}{\partial x^3 \partial y} + 2(D_{12} + 2D_{66}) \frac{\partial^4 w}{\partial x^2 \partial y^2} + 4D_{26} \frac{\partial^4 w}{\partial x \partial y^3} \quad (6.22)$$

$$+ D_{22} \frac{\partial^4 w}{\partial y^4} = - N_0 \frac{\partial^2 w}{\partial x^2} + 2S \frac{\partial^2 w}{\partial x \partial y}$$

The boundary conditions are
(1) Simply-supported edges at $y = \pm b/2$

$$w = M_y = - D_{12} \frac{\partial^2 w}{\partial y^2} - 2D_{26} \frac{\partial^2 w}{\partial x \partial y} - D_{22} \frac{\partial^2 w}{\partial y^2} = 0 \qquad (6.23)$$

(2) Clamped edges

$$w = \frac{\partial w}{\partial y} = 0 \qquad (6.24)$$

The solution of the governing Equation (6.22) takes the same form as found for the specially orthotropic plate, that is:

$$w = e^{2\xi x i / b} (Ae^{2\lambda_1 i y / b} + Be^{2\lambda_2 i y / b} + Ce^{2\lambda_3 i y / b} + De^{2\lambda_4 i y / b}) \qquad (6.25)$$

except in this case λ_1, λ_2, λ_3, and λ_4 are the real or complex roots of the following characteristic equation:

$$D_{22}\lambda^4 + 4D_{26}\xi\lambda^3 + 2(D_{12} + 2D_{66})\xi^2\lambda^2 + 4D_{16}\xi^3\lambda + D_{11}\xi^4$$
$$+ 2S \left(\frac{b}{2}\right)^2 \xi\lambda - N_0 \left(\frac{b}{2}\right)^2 \xi^2 = 0 \qquad (6.26)$$

If the solution (6.25) is used in conjunction with the boundary conditions (6.23) and (6.24), four homogeneous equations in the four unknowns A, B, C, and D are obtained. Since we desire non-zero solutions, the determinant of the coefficients must be zero. This condition is sufficient to determine the buckling load. For fixed edges the four conditions take the following matrix form:

$$\begin{bmatrix} \cos \lambda_1 & \cos \lambda_2 & \cos \lambda_3 & \cos \lambda_4 \\ \sin \lambda_1 & \sin \lambda_2 & \sin \lambda_3 & \sin \lambda_4 \\ \lambda_1 \cos \lambda_1 & \lambda_2 \cos \lambda_2 & \lambda_3 \cos \lambda_3 & \lambda_4 \sin \lambda_4 \\ \lambda_1 \sin \lambda_1 & \lambda_2 \sin \lambda_2 & \lambda_3 \sin \lambda_3 & \lambda_4 \sin \lambda_4 \end{bmatrix} \begin{bmatrix} A \\ B \\ C \\ D \end{bmatrix} = \begin{bmatrix} 0 \\ 0 \\ 0 \\ 0 \end{bmatrix} \qquad (6.27)$$

where the identity

$$e^{\pm i\lambda_n} = \cos \lambda_n \pm i \sin \lambda_n, \; n = 1,2,3,4$$

has been used. Equating the determinant of the coefficient matrix of Equation (6.27) to zero, we obtain the following transcendental equation governing the buckling load of clamped plates:

$$(\lambda_1 - \lambda_3)(\lambda_2 - \lambda_4)\sin(\lambda_1 - \lambda_2)\sin(\lambda_3 - \lambda_4)$$

$$- (\lambda_1 - \lambda_2)(\lambda_3 - \lambda_4)\sin(\lambda_1 - \lambda_3)\sin(\lambda_2 - \lambda_4) = 0 \qquad (6.28)$$

In the same manner four homogeneous equations may be written using the boundary conditions (6.23) for the simply-supported case, and after certain algebraic manipulations we arrive at the governing transcendental equation for simply-supported plates:

$$(\alpha_1 - \alpha_3)(\alpha_2 - \alpha_4)\sin(\lambda_1 - \lambda_2)\sin(\lambda_3 - \lambda_4)$$

$$- (\alpha_1 - \alpha_2)(\alpha_3 - \alpha_4)\sin(\lambda_1 - \lambda_3)\sin(\lambda_2 - \lambda_4) = 0 \qquad (6.29)$$

where

$$\alpha_n = D_{22}\lambda_n^2 + 2D_{66}\lambda_n\xi, \quad i = 1,2,3,4 \qquad (6.30)$$

To determine the actual buckling load for applied compressive loads or shear loads, the roots of the characteristic equation (6.26) must be determined in conjunction with either Equation (6.28) or (6.29) to determine the critical buckling load for a given value of ξ. By successively solving these equations for different values of ξ, the minimum buckling load may be determined. This procedure involves extensive algebraic transformations, and only a few cases have been solved numerically. Thielemann [6] presents a few exact solutions for the cases of pure compression and of pure shear with simply-supported edges. Reference [6] also presents approximate solutions for these same problems, obtained with the Ritz method. The approximate solutions result in relatively simple solutions, and with reference to the exact solutions the magnitude of the error can be easily assessed.

In Figure 6.9 solutions are presented for the critical compressive load for simply-supported plates composed of orthotropic material oriented at 45 degrees to the plate edges. For this case $D_{11} = D_{22}$, $D_{16} = D_{26}$, and the solutions can be presented as a function of two stiffness variables D_{16}/D_{22} and $(D_{12} + 2D_{66})/D_{22}$. The solutions shown in Figure 6.9 were obtained by both the approximate method and the exact solution of Equations (6.29) and (6.26). Comparison of the results at appropriate points indicates that all of the values presented are within graphical accuracy ($\sim 1\%$).

Examination of Figure 6.9 indicates that plates with $D_{16}/D_{22} = 0$ (specially orthotropic plates) are more buckling-resistant than plates with the ratio $D_{16}/D_{22} \neq 0$. This indicates that a specially orthotropic analysis of an anisotropic plate will, for this case, result in a predicted buckling load higher than the actual solution. For example an infinitely long plate composed of an orthotropic material with $E_L/E_T = 10$, $G_{LT}/E_T = 1$, $\nu_{LT} = 0.3$, and oriented with the principal material axes at 45 degrees to the plate axes, will have a buckling load of

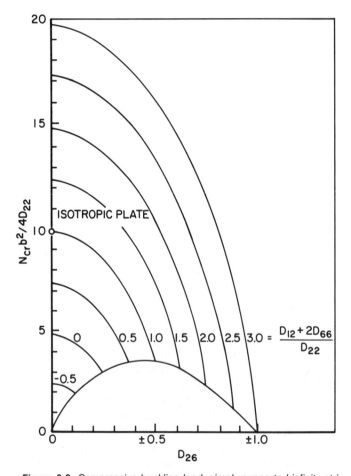

Figure 6.9. Compressive buckling load, simply-supported infinite strip.

Table 6.5. Shear buckling coefficients.

$(D_{12} + 2D_{66})/D_{22}$	$D_{16}/D_{22} = +.5$		$D_{16}/D_{22} = 0$		$D_{16}/D_{22} = -.5$		$D_{16}/D_{22} = -1.0$	
	k	L/b	k	L/b	k	L/b	k	L/b
0			8.125	1.025				
1	2.32	—	13.165	1.245	20.96	1.45		
2	8.07	1.126	17.25	1.462	24.87	1.69		
3	12.06	1.45	20.50	1.715	28.19	1.911	34.56	1.98

the following magnitude (from Figure 6.9 using $D_{16}/D_{22} = 0.71$, $(D_{12} + 2D_{66})/D_{22} = 2.5$):

$$N_{cr} = 34.8 \ D_{22}/b^2 \qquad (6.31)$$

If the specially orthotropic solutions were used for this case, the predicted buckling load would be

$$N_{cr} = 69.2 \ D_{22}/b^2 \qquad (6.32)$$

which, for this case, is twice the actual buckling resistance.

For simply-supported plates under pure shear, Thielemann has presented a range of exact solutions for 45-degree plates. These solutions are tabulated in Table 6.5. Also presented in Table 6.5 are values of the length between successive buckles (L/b). Examination of these results indicates that the actual value of the critical shear load depends upon the sign of the ratio D_{16}/D_{22}. When this ratio is positive the buckling loads are smaller than the predictions for the specially orthotropic case, while when the ratio is negative the actual buckling loads are greater than predicted with the specially orthotropic analysis. Physically, the case of positive D_{16}/D_{22} indicates that the minimum plate stiffness occurs parallel to the direction of the resolved compressive load (pure shear resolved 45 degrees), while a negative value of this ratio occurs when the maximum plate stiffness occurs parallel to the direction of the resolved compressive laod. Reversing the sign of the ratio D_{16}/D_{22} is equivalent to reversing the direction of applied shear, consequently these results show the buckling resistance to be highly dependent upon the direction of the applied shear load.

6.5 STABILITY OF SIMPLY-SUPPORTED RECTANGULAR PLATES

In Sections 5.5 and 5.7 the stability of specially orthotropic plates was considered. For uniform compression the exact solution was determined to take the form of a double sine term, while for applied shear the solutions were obtained

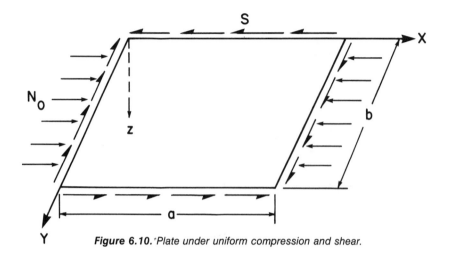

Figure 6.10. *Plate under uniform compression and shear.*

utilizing the Galerkin method in conjunction with a series composed of double sine terms. In each case the double sine terms exactly satisfied the boundary conditions. In this section the stability of a simply-supported plate is again considered. However, in the present analysis, the plate will be assumed to be composed of a midplane symmetric laminated material, and the analysis will consider the anisotropic nature of such a material.

The geometry is indicated in Figure 6.10. We will again utilize the Galerkin method as in Section 5.7. The governing differential equation is the same as utilized in the previous section, Equation (6.22). The boundary conditions are

at $x = 0,a$

$$w = M_x = -D_{11} \frac{\partial^2 w}{\partial x^2} - 2D_{16} \frac{\partial^2 w}{\partial x \partial y} - D_{12} \frac{\partial^2 w}{\partial y^2} = 0 \qquad (6.33)$$

at $y = 0,b$

$$w = M_y = -D_{12} \frac{\partial^2 w}{\partial x^2} - 2D_{26} \frac{\partial^2 w}{\partial x \partial y} - D_{22} \frac{\partial^2 w}{\partial y^2} = 0 \qquad (6.34)$$

In the previous application of the Galerkin method, we indicated that solutions could be obtained by selecting a series of functions each of which satisfies the boundary conditions, and then determine the coefficients in the series by requiring the solution to be orthogonal to each term in the series. To apply this approach directly to Equation (6.22), a series must be found which exactly satisfies the boundary conditions (6.33) and (6.34), but no such series exists.

As a result the boundary integrals in Equation (3.47) must be included in the formulation of the Galerkin equation. The double sine series

$$
w = \sum_{m=1}^{M} \sum_{n=1}^{N} A_{mn} \sin \frac{m\pi x}{a} \sin \frac{n\pi y}{b} \tag{6.35}
$$

satisfies the geometrical condition $w = 0$ on the boundary. Recognizing that

$$
\frac{\partial^2 w}{\partial x^2} = \frac{\partial^2 w}{\partial y^2} = 0 \quad \text{(on boundary)} \tag{6.36}
$$

for the assumed series (6.35), the Galerkin Equation (3.47) becomes

$$
\int_0^b \int_0^a \left[D_{11} \frac{\partial^4 w}{\partial x^4} + 4D_{16} \frac{\partial^4 w}{\partial x^3 \partial y} + 2(D_{12} + 2D_{66}) \frac{\partial^4 w}{\partial x^2 \partial y^2} + 4D_{26} \frac{\partial^4 w}{\partial x \partial y^3} \right.
$$

$$
\left. + D_{22} \frac{\partial^4 w}{\partial y^4} + N_0 \frac{\partial^2 w}{\partial x^2} - 2S \frac{\partial^2 w}{\partial x \partial y} \right] \sin \frac{m\pi x}{a} \sin \frac{n\pi y}{b} \, dx \, dy
$$

$$
\tag{6.37}
$$

$$
- 2D_{26} \int_0^a \left[(-1)^n \left(\frac{\partial^2 w}{\partial x \partial y} \right)_{y=b} - \left(\frac{\partial^2 w}{\partial x \partial y} \right)_{y=0} \right] \frac{\pi}{b} \sin \frac{m\pi x}{a} \, dx
$$

$$
- 2D_{16} \int_0^b \left[(-1)^m \left(\frac{\partial^2 w}{\partial x \partial y} \right)_{x=a} - \left(\frac{\partial^2 w}{\partial x \partial y} \right)_{x=0} \right] \frac{\pi}{a} \sin \frac{n\pi y}{b} \, dy
$$

$$
= 0 \quad \left\{ \begin{array}{l} m = 1,2,\ldots,M \\ n = 1,2,\ldots,N \end{array} \right.
$$

Substituting Equation (6.35) into Equation (6.37), and performing the resulting integration, we arrive at the following set of algebraic equations:

$$
\pi^4 [D_{11} m^4 + 2(D_{12} + 2D_{66}) m^2 n^2 R^2 + D_{22} n^4 R^4 - N_0 \frac{m^2 a^2}{\pi^2}] A_{mn}
$$

$$
- 32 \, mn \, R\pi^2 \sum_{i=1}^{M} \sum_{j=1}^{N} M_{ij} \left[(m^2 + i^2) D_{16} + (n^2 + j^2) D_{26} \right] \tag{6.38}
$$

$$
- SR \frac{a^2}{\pi^2}] \, A_{ij} = 0
$$

where

$$M_{ij} = \frac{ij}{(m^2 - i^2)(n^2 - j^2)} \quad \begin{cases} m \pm i & \text{odd} \\ n \pm j & \text{odd} \end{cases}$$

$$= 0 \quad \begin{cases} i = m, m \pm i & \text{even} \\ j = n, n \pm j & \text{even} \end{cases}$$

and $R = a/b$. Equation (6.38) yields a set of MxN homogeneous equations. As in the case of specially orthotropic plates, the resulting equations can be separated into two sets, one for $m + n$ even and a second set corresponding to $m + n$ odd. To obtain other than a trivial solution ($A_{mn} = 0$; $m = 1,2,...,M$; $n = 1,2,...,N$), the determinant of the coefficient matrix must vanish. The smallest value of N_0 and S (or a combination of N_0 and S at a fixed ratio) for which this determinant is equal to zero establishes the critical buckling load. It is significant to note that the equations represented in (6.38) are equivalent to the equations obtained by using the Ritz method with the same assumed series, and thus the discussion in Section 6.2 concerning the boundary conditions and convergence is applicable here also.

Two cases will be considered to indicate the effect of D_{16} and D_{26} terms on the buckling behavior. The first case is uniform axial compression ($S = 0$). Results are presented in Figure 6.11 for plates with properties typical of a boron/epoxy composite material, i.e. $E_T/E_L = 0.1$, $G_{LT}/E_L = 0.03$, $\nu_{LT} = 0.3$. The aspect ratio considered is $R = 1.13$. The results are presented in terms of a buckling coefficient, k, defined such that

$$k = \frac{N_{cr} a^2}{E_L h^3} \tag{6.39}$$

The bottom curve in Figure 6.11 represents the results calculated with $M = N = 7$ in Equations (6.38) for the case of unidirectional material with the principal material axis oriented at θ degrees to the plate edges. The top curve represents the orthotropic solution obtained by neglecting the D_{16} and D_{26} terms, and the curve just below the orthotropic case is the solution obtained for 20-ply laminates of the class $[\pm\theta]_{5S}$. The D_{16} and D_{26} terms are smaller for the case of alternating plies [7], and as evidenced in the figure, such "interspersed" 20-layer plates behave essentially in a specially orthotropic manner. Also presented in Figure 6.11 are experimentally determined buckling loads obtained on 11″ by 9.75″ boron/epoxy plates [8]. These experimental results clearly follow the analytical predictions quite acceptably.

As a second example we consider the case of pure shear applied to square plates with the principal material axes oriented at 45 degrees to the plate edges.

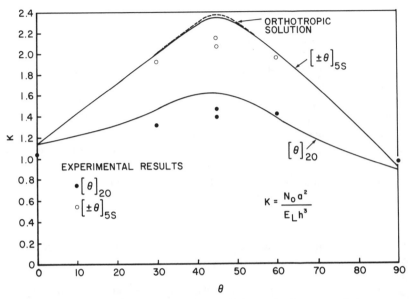

Figure 6.11. Compressive buckling coefficients, simply-supported plates.

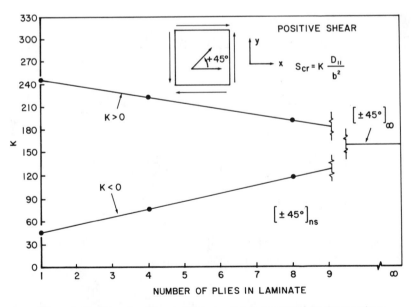

Figure 6.12. Shear buckling coefficients, simply-supported 45-degree plates.

The orthotropic material properties considered are $E_T/E_L = 0.1$, $G_{LT}/E_L = 0.025$, $\nu_{LT} = 0.3$. The shear buckling coefficient is presented in Figure 6.12 for unidirectional (all +45 degree) plates and for multidirectional laminates of the class $[\pm 45°]_{ns}$ ($n = 1, 2, \infty$). Forty-nine terms in the series (6.35) were used to obtain these results. Just as for infinitely long plates, the finite length anisotropic plates have a preferred direction of shear, and for a unidirectional all 45-degree plate the least resistant buckling direction has a buckling load only 18% of the buckling load for reversed shear. The difference between these buckling loads decreases as the number of alternating layers increases, and in the limit of an infinity of alternations, the solutions coincide and equal the specially orthotropic solution. Note also that the specially orthotropic solution can be highly conservative or nonconservative when used for an anisotropic plate.

6.6 STABILITY OF UNIFORM RECTANGULAR PLATES BY THE RITZ METHOD

In this section the stability of rectangular midplane symmetric laminated plates by the Ritz method is considered. The appropriate energy criterion is deduced from Equations (3.6) and (3.18) in conjunction with Equations (3.35) and (3.37) with $q = 0$ (no lateral load) which leads to the result:

$$\frac{1}{2} \int_0^a \int_0^b \left\{ D_{11} \left(\frac{\partial^2 w}{\partial x^2} \right)^2 + 2D_{12} \frac{\partial^2 w}{\partial x^2} \frac{\partial^2 w}{\partial y^2} + D_{22} \left(\frac{\partial^2 w}{\partial y^2} \right)^2 \right.$$

$$+ 4D_{16} \frac{\partial^2 w}{\partial x^2} \frac{\partial^2 w}{\partial x \partial y} + 4D_{26} \frac{\partial^2 w}{\partial y^2} \frac{\partial^2 w}{\partial x \partial y} + 4D_{66} \left(\frac{\partial^2 w}{\partial x \partial y} \right)^2$$

$$+ N_x \left(\frac{\partial^2 w}{\partial x^2} \right)^2 + N_y \left(\frac{\partial^2 w}{\partial y^2} \right)^2 + 2N_{xy} \frac{\partial w}{\partial x} \frac{\partial w}{\partial y} \right\} dy dx \qquad (6.40)$$

$$= \text{stationary value}$$

Following the approach of Reference [5], we assume the solution in the form of a series of products of beam characteristic shapes:

$$w = \sum_{m=1}^{M} \sum_{n=1}^{N} A_{mn} X_m(x) Y_n(y) \qquad (6.41)$$

where $X_m(x)$ and $Y_n(y)$ are the characteristic shapes of free vibration of a beam satisfying a set of the following boundary conditions on each end:

$$w = \frac{\partial w}{\partial \eta} = 0 \quad \text{(for clamped boundary)}$$

$$w = \frac{\partial^2 w}{\partial \eta^2} = 0 \quad \text{(for simply-supported boundary)} \qquad (6.42)$$

$$\frac{\partial^2 w}{\partial \eta^2} = \frac{\partial^3 w}{\partial \eta^3} = 0 \quad \text{(for a free boundary)}$$

where η is the outward normal along the boundary.

By combining the appropriate beam characteristic shapes, the geometrical boundary conditions corresponding to clamped, simply-supported, or free edges can be enforced on any edge, and the natural boundary conditions (moment and edge reaction) will be approximated by the Ritz process. For the numerical results to be presented below, the characteristic shapes corresponding to a beam clamped on both ends, Equation (6.21), and a beam simply-supported on both ends, Equation (6.7), will be used. The procedure is equally applicable to the use of the characteristic shapes of beams with free ends or combinations of supported and free ends. For the six classical beam shapes (combinations of clamped, simply-supported, and free ends), the resulting integrals are given in Reference [9].

With the assumed series (6.41) and the energy criterion (6.40), the process of minimization, Equation (3.42), leads directly to a set of linear homogeneous equations for the parameters A_{mn}:

$$
\sum_{i=1}^{M} \sum_{j=1}^{N} \left\{ D_{11} \int_0^a \frac{d^2 X_i}{dx^2} \frac{d^2 X_m}{dx^2} \, dx \int_0^b Y_j Y_n \, dy \right.
$$

$$
+ D_{12} \left[\int_0^a X_m \frac{d^2 X_i}{dx^2} \, dx \int_0^b Y_j \frac{d^2 Y_n}{dy^2} \, dy \right. \qquad (6.43)
$$

$$
\left. + \int_0^a X_i \frac{d^2 X_m}{dx^2} \, dx \int_0^b Y_n \frac{d^2 Y_j}{dy^2} \, dy \right]
$$

$$+ D_{22} \int_0^a X_i \, X_m \, dx \int_0^b \frac{d^2 Y_j}{dy^2} \, \frac{d^2 Y_n}{dy^2} \, dy$$

$$+ 4D_{66} \int_0^a \frac{dX_i}{dx} \, \frac{dX_m}{dx} \, dx \int_0^b \frac{dY_j}{dy} \, \frac{dY_n}{dy} \, dy$$

$$+ 2D_{16} \left[\int_0^a \frac{d^2 X_i}{dx^2} \, \frac{dX_m}{dx} \, dx \int_0^b Y_j \, \frac{dY_n}{dy} \, dy \right.$$

$$+ \int_0^a \frac{dX_i}{dx} \, \frac{d^2 X_m}{dx^2} \, dx \int_0^b Y_n \, \frac{dY_j}{dy} \, dy \left. \right]$$

$$+ 2D_{26} \left[\int_0^a X_m \frac{dX_i}{dx} \, dx \int_0^b \frac{dY_j}{dy} \, \frac{d^2 Y_n}{dy^2} \, dy \right.$$

$$+ \int_0^a X_i \frac{dX_m}{dx} \, dx \int_0^b \frac{d^2 Y_j}{dy^2} \, \frac{dY_n}{dy} \, dy \left. \right]$$

$$+ N_x \int_0^a \frac{dX_i}{dx} \, \frac{dX_m}{dx} \, dx \int_0^b Y_j \, Y_n \, dy$$

$$+ N_y \int_0^a X_i \, X_m \, dx \int_0^b \frac{dY_j}{dy} \, \frac{dY_n}{dy} \, dy$$

$$+ N_{xy} \left. \int_0^a X_m \frac{dX_i}{dx} \, dx \int_0^b Y_j \, \frac{dY_n}{dy} \, dy \right.$$

$$+ \int_0^a X_i \frac{dX_m}{dx} \, dx \int_0^b Y_n \, \frac{dY_j}{dy} \left. \right] \right\} A_{mn}$$

$$= 0 \qquad \begin{array}{l} i = 1,2,\ldots,M \\ j = 1,2,\ldots,N \end{array}$$

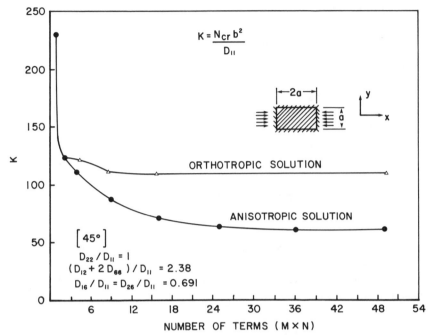

Figure 6.13. *Convergence of buckling solution.*

Setting the determinant of this system of equations to zero, an equation is obtained from which critical combinations of N_x, N_y, and N_{xy} can be determined.

To illustrate the effects of D_{16} and D_{26} on the stability of laminated plates, several examples are considered below. In each case the function $X_m(x)$ is taken as the characteristic shape of a clamped beam, and therefore the boundary conditions at $x = 0$ and $x = a$ correspond to a clamped edge. Two conditions are considered for the boundaries $y = 0$ and $y = b$. For clamped boundaries, $Y_n(y)$ is also taken as the characteristic shape of a beam clamped on both ends. For simply-supported sides $Y_n(y)$ is taken as a sine term which is the characteristic shape of a beam simply-supported on both ends.

As a first example a plate clamped on all edges with an aspect ratio $a/b = 2$ under a uniform compressive load $N_x = -N_0 = $ constant is considered. The plate is composed of a single orthotropic layer at 45 degrees to the x-axis and has properties typical of boron/epoxy ($D_{22}/D_{11} = 1$, $(D_{12} + 2D_{66})/D_{11} = 2.38$, $D_{16}/D_{11} = D_{26}/D_{11} = 0.69$). Solutions are shown in Figure 6.13 for assumed series of 1, 2, 4, 9, 16, 18, 25, 36, and 49 terms. Solutions corresponding to the specially orthotropic analysis ($D_{16} = D_{26} = 0$) are also shown. Numerical

Table 6.6. Convergence of buckling solution (Figure 6.13).

	$N_{cr} = k \, (D_{11}/b^2)$	
Terms	Anisotropic k	Orthotropic k
1	231.5	231.5
2	122.7	122.7
4	112.3	122.7
9	87.77	111.6
16	71.92	111.6
25	63.81	—
36	62.90	—
49	62.72	109.46

results are summarized in Table 6.6. Examination of these solutions reveals that the orthotropic analysis converges more rapidly, but that the solution obtained is decidedly nonconservative. The node lines of the buckled shape as obtained with the 49-term anisotropic analysis are shown in Figure 6.14. The strong effect of anisotropy is evident.

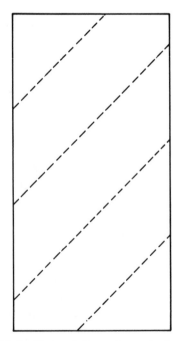

Figure 6.14. Buckled node lines, clamped anisotropic plate.

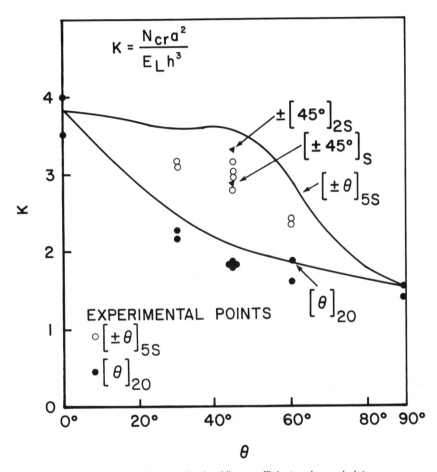

$$K = \frac{N_{cr}a^2}{E_L h^3}$$

Figure 6.15. *Compressive buckling coefficients, clamped plates.*

Results are presented in Figure 6.15 for square clamped plates under uniform compression with properties typical of boron/epoxy ($E_T/E_L = 0.09$, $G_{LT}/E_L = 0.03$, $\nu_{LT} = 0.3$). These results are given in terms of the buckling coefficient k, Equation (6.39). This figure is similar to Figure 6.11 except that in this case, the boundary conditions are full clamping on all the edges. The bottom curve corresponds to one layer plates with the principal axis of orthotropy at θ degrees to the loading axis, and the top curve corresponds to 20-layer plates of the class $[\pm\theta]_{5s}$. The effect of the D_{16} and D_{26} terms on the buckling load is again evident and important. Also plotted in Figure 6.15 are analytical results for four-layer $[\pm45°]_s$ and eight-ply $[\pm45°]_{2s}$ plates, and experimental results from Reference [10].

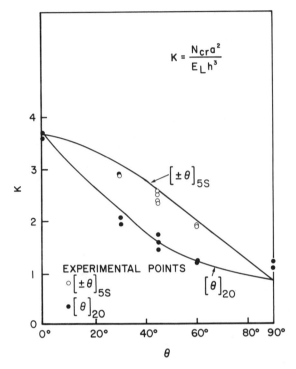

Figure 6.16. *Compressive buckling coefficients for plates simply-supported on unloaded sides.*

In Figure 6.16 we present results for uniformly compressed square plates with the loaded edges clamped and the unloaded edges simply-supported. This figure is of the same form as the previous figure. The same material properties were used for these curves, and the trends are similar. Further experimental verification from [10] is also presented. The above results indicate that the Ritz method together with the lamination model used herein allow accurate prediction of the buckling loads of laminated composite plates for a variety of boundary conditions. Furthermore the use of a specially orthotropic analysis often leads to non-conservative solutions and appreciable inaccuracies.

6.7 STABILITY OF NONUNIFORM RECTANGULAR PLATES

In the previous section the use of the Ritz method in conjunction with beam characteristic shapes has been presented for the stability analysis of uniform rectangular anisotropic plates under uniform inplane loads. Following the approach presented in Reference [9], this same analysis can easily be extended to plates with nonuniform stiffness properties and nonuniform loadings. Although the method can be applied to quite general problems with relative ease, and to all the classical boundary conditions considered in the previous section, we will illustrate the approach on a specific example for the sake of clarity. Consider a square rectangular plate with linear thickness variation in the y direction under the linearly varying compressive loading N_y, Figure 6.17. If this plate is fabricated of an orthotropic material with the principal material axes at 45 degrees to the plate edges, then the bending stiffness terms for this plate can all be written in the following form

$$D_{ij} = (D_{ij})_{y=0} (1 + 3\alpha \frac{y}{b} + 3\alpha^2 \frac{y^2}{b^2} + \alpha^3 \frac{y^3}{b^3})$$ (6.44)

where $(D_{ij})_{y=0}$ is the stiffness property at $y = 0$ as computed from the material elastic constants and the transformation Equations (1.35) to (1.47), and α is the ratio of the increase in thickness at $y = 0$. Similarly the compressive load N_y can be expressed in the form:

$$N_y = - N_0 (1 + \beta \frac{x}{a})$$ (6.45)

where N_0 is the magnitude of the compressive load at $x = 0$, and β is the ratio of the increase in compressive stress resultant at $x = a$ to the resultant at $x = 0$.

The energy criterion for anisotropic plates, Equation (6.40), can now be used together with the assumed series of beam characteristic shapes, Equation (6.41). The procedure is the same as for the case of uniform properties and loadings except the stiffness and loading terms are now functions of x and y, and thus cannot be removed from integrations. After substituting the series (6.41) into the energy criterion (6.40) and following the usual process of minimization, Equation (3.42), while using expressions (6.44) and (6.45), we obtain a system of $M \times N$ homogeneous equations for the coefficients A_{mn}:

$$\sum_{i=1}^{M} \sum_{j=1}^{N} \left\{ (D_{11})_{y=0} \int_0^a \frac{d^2 X_i}{dx^2} \frac{d^2 X_m}{dx^2} dx \int_0^b D(y) Y_j Y_n dy \right.$$

$$+ (D_{12})_{y=0} \left[\int_0^a X_m \frac{d^2 X_i}{dx^2} dx \int_0^b D(y) Y_j \frac{d^2 Y_n}{dy^2} dy \right.$$ (6.46)

$$+ \int_0^a X_i \frac{d^2 X_m}{dx^2} dx \int_0^b D(y) Y_n \frac{d^2 Y_j}{dy^2} dy \Bigg]$$

$$+ (D_{22})_{y=0} \int_0^a X_i X_m dx \int_0^b D(y) \frac{d^2 Y_j}{dy^2} \frac{d^2 Y_n}{dy^2} dy$$

$$+ 4(D_{66})_{y=0} \int_0^a \frac{dX_i}{dx} \frac{dX_m}{dx} dx \int_0^b D(y) \frac{dY_j}{dy} \frac{dY_n}{dy} dy$$

$$+ 2(D_{16})_{y=0} \Bigg[\int_0^a \frac{d^2 X_i}{dx^2} \frac{dX_m}{dx} dx \int_0^b D(y) Y_j \frac{dY_n}{dy} dy$$

$$+ \int_0^a \frac{dX_i}{dx} \frac{d^2 X_m}{dx^2} dx \int_0^b D(y) Y_n \frac{dY_j}{dy} dy \Bigg]$$

$$+ 2(D_{26})_{y=0} \Bigg[\int_0^a X_m \frac{dX_i}{dx} dx \int_0^b D(y) \frac{dY_j}{dy} \frac{d^2 Y_n}{dy^2} dy$$

$$+ \int_0^a X_i \frac{dX_m}{dx} dx \int_0^b D(y) \frac{d^2 Y_j}{dy^2} \frac{dY_n}{dy} dy \Bigg]$$

$$- N_0 \int_0^a (1 + \beta \frac{x}{a}) X_i X_m dx \int_0^b \frac{dY_j}{dy} \frac{dY_n}{dy} dy \Bigg\} A_{mn}$$

$$= 0 \qquad \begin{matrix} m = 1,2 \ldots ,M \\ n = 1,2 \ldots ,N \end{matrix}$$

where

$$D(y) = (1 + 3\alpha \frac{y}{b} + 3\alpha^2 \frac{y^2}{b^2} + \alpha^3 \frac{y^3}{b^3})$$

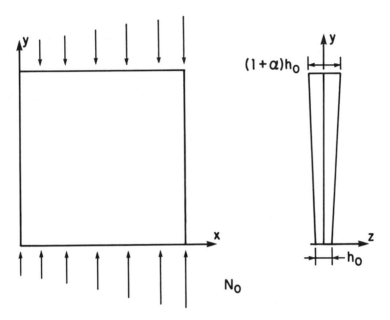

Figure 6.17. *Tapered plate geometry.*

Note that Equations (6.46) are similar to Equations (6.43) for uniform plates under uniform loads, except that the necessary integrations are somewhat more difficult. However, Reference [9] presents a simple exact means for calculating such integrals for the classical boundary conditions, and after these more general integrals are evaluated, the nonuniform problem is no more difficult than the uniform plate problems.

As a numerical example consider the plate of Figure 6.17 with $E_T/E_L = 0.1$, $G_{LT}/E_L = 0.03$, $\nu_{LT} = 0.3$, $\alpha = 1$ (the plate is twice as thick at $y = b$ as at $y = 0$) and $\beta = -2$ (inplane bending). Solutions calculated with the assumed

Table 6.7. **Buckling loads for tapered 45-degree plate under inplane bending.**

	$K = N_{cr}b^2/(D_{11})y = 0$		
M = N =	3	5	7
k, Anisotropic Analysis	1229.	923.	847.
k, Orthotropic Analysis	1679.	1484.	1470.

series (6.41), using the beam mode shape corresponding to fixed edges, are given in Table 6.7 for $m = n = 3$, 5, and 7. The equivalent specially orthotropic solutions for these same series are also given. Note again that the specially orthotropic solution is appreciably inaccurate and nonconservative.

6.8 FREE VIBRATION OF RECTANGULAR ANISOTROPIC PLATES

In Section 5.11 the Ritz method was used in conjunction with the characteristic shapes of freely vibrating beams to obtain the normal modes of specially orthotropic plates. Examples of the convergence of the solution indicate that as long as each boundary is simply-supported or clamped, a one-term approximation accurately predicts the natural frequencies of such plates, and therefore a simple equation can be utilized for such predictions. In this section we extend the analysis to include the effects of anisotropy, and use the extended formulation to investigate the effects of lamination sequence, orientations, and properties.

The energy criterion governing free vibration of midplane symmetric laminates is obtained from Equations (3.7) and (3.40) in conjunction with Equations (3.35) and (3.37) with no lateral or inplane loads ($q = N_x = N_y = N_{xy} = 0$) which leads to the result:

$$
\begin{aligned}
\frac{1}{2} \int_0^a \int_0^b \Bigg\{ & D_{11} \left(\frac{\partial^2 w}{\partial x^2} \right)^2 + 2D_{12} \frac{\partial^2 w}{\partial x^2} \frac{\partial^2 w}{\partial y^2} + D_{22} \left(\frac{\partial^2 w}{\partial y^2} \right)^2 \\
& + 4D_{66} \left(\frac{\partial^2 w}{\partial x \partial y} \right)^2 + 4D_{16} \frac{\partial^2 w}{\partial x \partial y} \frac{\partial^2 w}{\partial x^2} + 4D_{26} \frac{\partial^2 w}{\partial x \partial y} \frac{\partial^2 w}{\partial y^2} \\
& - \varrho \omega^2 w^2 \Bigg\} \, dx \, dy = \text{stationary value}
\end{aligned}
\tag{6.47}
$$

Proceeding to solve for the natural modes using the Ritz procedures, we assume the solution in the form of a series with undetermined coefficients:

$$
w = \sum_{m=1}^{M} \sum_{n=1}^{N} A_{mn} X_m(x) Y_n(y)
\tag{6.48}
$$

where $X_m(x)$ and $Y_n(y)$ satisfy (or approximate) the boundary conditions on the edges $x = 0,a$ and $y = 0,b$, respectively. Substituting the series (6.48) into the energy condition (6.47) and minimizing with respect to the A_{mn}, we obtain a set of MxN homogeneous simultaneous equations:

$$\sum_{i=1}^{M} \sum_{j=1}^{N} \left\{ D_{11} \int_0^a \frac{dX_i}{dx^2} \frac{d^2X_m}{dx^2} dx \int_0^b Y_j Y_n \, dy \right.$$

$$+ D_{12} \left[\int_0^a X_m \frac{d^2X_i}{dx^2} dx \int_0^b Y_j \frac{d^2Y_n}{dy^2} dy \right.$$

$$+ \int_0^a X_i \frac{d^2X_m}{dx^2} dx \int_0^b Y_n \frac{d^2Y_j}{dy^2} dy \left. \right]$$

$$+ D_{22} \int_0^a X_i X_m \, dx \int_0^b \frac{d^2Y_j}{dy^2} \frac{d^2Y_n}{dy^2} dy$$

$$\hspace{10cm} (6.49)$$

$$+ 4D_{66} \int_0^a \frac{dX_i}{dx} \frac{dX_m}{dx} dx \int_0^b \frac{dY_j}{dy} \frac{dY_n}{dy} dy$$

$$+ 2D_{16} \left[\int_0^a \frac{d^2X_i}{dx^2} \frac{dX_m}{dx} dx \int_0^b Y_j \frac{dY_n}{dy} dy \right.$$

$$+ \int_0^a \frac{dX_i}{dx} \frac{d^2X_m}{dx^2} dx \int_0^b Y_n \frac{dY_j}{dy} dy \left. \right]$$

$$+ 2D_{26} \left[\int_0^a X_m \frac{dX_i}{dx} dx \int_0^b \frac{dY_j}{dy} \frac{d^2Y_n}{dy^2} dy \right.$$

$$+ \int_0^a X_i \frac{dX_m}{dx} dx \int_0^b \frac{d^2Y_j}{dy^2} \frac{dY_n}{dy} dy \left. \right]$$

$$- \varrho\omega^2 \int_0^a X_i X_m\, dx \int_0^b Y_j Y_n\, dy \Bigg\} A_{mn}$$

$$= 0 \quad \begin{matrix} m = 1,2,\ldots,M \\ n = 1,2,\ldots,N \end{matrix}$$

(6.49)

Since Equations (6.49) are a set of homogeneous equations, a nontrivial solution can be obtained only if the determinant of the coefficient matrix is zero. This condition is sufficient to determine the natural frequencies of free vibration.

Following the analysis for specially orthotropic plates, we use the characteristic shapes of freely vibrating beams for the approximating functions $X_m(x)$ and $Y_n(y)$. The approximate nature of these functions for the case of simply-supported and free edges on anisotropic plates has been discussed in Sections 6.2 and 6.6, and such discussion applies here also. The integrals in Equations (6.49) are restricted to the analysis of uniform plates, the generalization to include nonuniform plate properties can be done in essentially the same manner as

Table 6.8. Natural frequencies of anisotropic plate.

a = b

Orientation θ	Mode 1 Aniso-tropic	Ortho-tropic	Mode 2 Aniso-tropic	Ortho-tropic	k Mode 3 Aniso-tropic	Ortho-tropic	Mode 4 Aniso-tropic	Ortho-tropic
0°	23.97	23.97	31.15	31.15	46.41	46.41	62.77	62.77
15°	23.10	23.80	31.52	33.31	47.65	50.30	59.46	60.17
30°	21.35	23.42	33.18	38.96	50.72	54.63	51.87	62.15
45°	20.51	23.23	35.01	46.69	47.07	46.69	52.21	74.44

a = 2b

Orientation θ	Mode 1	Mode 2	k Mode 3	Mode 4
0°	9.34	17.61	20.83	26.49
15°	9.68	17.19	22.02	26.44
26.6°	10.57	16.93	25.11	25.46
45°	13.88	17.73	23.85	31.73
60°	17.87	19.86	23.75	29.75
75°	21.27	22.32	24.69	28.94
90°	22.57	23.38	25.30	28.87

demonstrated in the previous section for the case of stability, using the general integral results of Reference [9].

A number of numerical and experimental results are presented here to illustrate the convergence of the solution, the effect of D_{16} and D_{26} terms on the solutions, and the effect of several lamination variations. The first of these examples concerns clamped plates, and for these solutions the functions $X_m(x)$ and $Yn(y)$ are given by Equations (5.150) and (5.151). We consider square and rectangular ($a = 2b$) plates composed of an orthotropic material oriented with the principal axis of orthotropy at θ degrees from the x axis. The material properties of the orthotropic layer are $E_L/E_T = 10$, $G_{LT}/E_T = 0.25$, $\nu_{LT} = 0.3$. Results obtained with $M = N = 7$ in the series (6.48) are given in Table 6.8. The first four frequencies are given in terms of k where

$$\omega = \frac{k}{b^2} \sqrt{\frac{D_{\theta\theta}}{\varrho}} \tag{6.50}$$

$$D_{\theta\theta} = \frac{E_L h^3}{(1 - \nu_{LT}^2 E_T/E_L)} = \begin{array}{l} \text{bending stiffness in} \\ \text{direction of principal} \\ \text{material axis} \end{array} \tag{6.51}$$

The corresponding specially orthotropic solutions for the square plate are also given, and the results for the square plate are presented graphically in Figure 6.18. The strong effect of the D_{16} and D_{26} terms on the natural frequencies is evident, and neglect of these effects will generally lead to predicted frequencies that are too high.

To illustrate the convergence the square plate with the orthotropic material axes oriented at 45 degrees to the plate edges has been analyzed to obtain the fundamental frequency using $M = N = 1, 3, 5$, and 7 in the assumed series. These results and the corresponding specially orthotropic results are presented in Table 6.9. Again we find that the specially orthotropic solution converges much more rapidly than the anisotropic solution, but results in appreciable inaccuracy. We were able to present a simple equation for the natural frequencies of specially orthotropic plates in Section 5.11 because of the rapid convergence of the solutions, but we cannot do this for the anisotropic plates because of the considerably slower convergence.

Experimental results have been reported for clamped plates of boron/epoxy composite material [11]. Some of the experimental results are reproduced in Figures 6.19 and 6.20. In these figures experimentally obtained Chladni figures and natural frequencies are compared to analytical predictions. The analytical

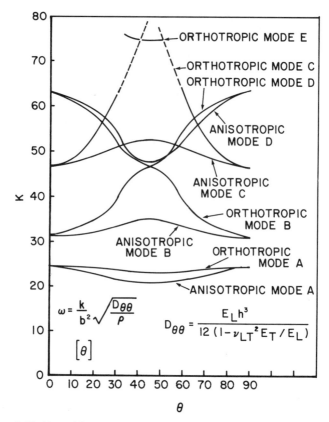

Figure 6.18. *Natural frequency variation with orientation, square anisotropic plate.*

predictions were obtained using $M = N = 7$ in the series (6.48) and the following plate properties: $E_L = 31 \times 10^6$ psi, $E_T = 2.7 \times 10^6$ psi, $G_{LT} = 0.75 \times 10^6$ psi, $\nu_{LT} = 0.28$, $h = 0.0424$ in, $a = b = 12$ in, density $= 1.92 \times 10^{-4}$ lb-sec^2/in^4. The frequencies are given in terms of cycles per second (hz). The Chladni figures illustrate the node lines of the characteristic shapes. They are

Table 6.9. Fundamental frequency of 45-degree plate.

	$\omega = k/b^2$ $(\sqrt{D_{\theta\theta}/\varrho})$			
M = N = number of terms	1	3	5	7
k, Anisotropic Analysis	23.51	20.82	20.57	20.51
k, Orthotropic Analysis	23.51	23.27	23.24	23.23

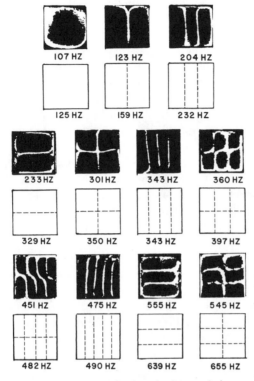

Figure 6.19. *Frequencies and mode shapes, 0-degree plate.*

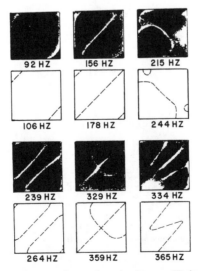

Figure 6.20. *Frequencies and mode shapes, 45-degree plate.*

obtained by letting a loose granular material migrate to the stationary positions while vibrating the plate at a natural frequency. Figure 6.19 presents the first eleven identified modes of a plate with the principal material axes parallel to the plate edges. In this configuration the plate is specially orthotropic, and the node lines are all parallel to the plate edges. Figure 6.20 presents the first six modes of a plate with the principal material axes at 45 degrees to the plate edges. In this configuration the plate is strongly anisotropic, and the node lines are skewed and curved. Clearly a specially orthotropic analysis of this plate would not be able to reasonably approximate the characteristic shapes, and the previous example has indicated the predicted frequencies would also be in considerable error. On the other hand the anisotropic analysis does an excellent job of predicting the shapes. The agreement of the natural frequencies is reasonable because the boundaries are assumed to be perfectly clamped for the analytical predictions, while this is not possible experimentally.

Further insight into the dynamic behavior of laminated plates can be obtained by considering the vibration of unrestrained rectangular plates, i.e., plates free on all edges [12]. For these results we use $M = N = 6$ in the series (6.48), and we take $X_m(x)$ and $Y_n(y)$ as the natural modes of free-free beams. In order to have a series which includes rigid body motion, the modes $m, n = 1, 2$ are taken as rigid body translation and rotation, respectively. That is,

$$X_1(x) = Y_1(y) = 1$$

$$X_2(x) = \sqrt{3}\, (1 - \frac{2x}{a})$$

$$Y_2(y) = \sqrt{3}\, (1 - \frac{2y}{b})$$

$$(6.52)$$

The remaining terms in the series are taken as the elastic vibration modes of free-free beams:

$$X_m(x) = \cosh \frac{\lambda_m x}{a} + \cos \frac{\lambda_m x}{a} - \gamma_m \sinh \frac{\lambda_m x}{a} - \gamma_m \sin \frac{\lambda_m x}{a}$$

$$Y_n(y) = \cosh \frac{\lambda_n y}{b} + \cos \frac{\lambda_n y}{b} - \gamma_n \sinh \frac{\lambda_n y}{b} - \gamma_n \sin \frac{\lambda_n y}{b}$$

$$(6.53)$$

where λ_m and λ_n are determined from the frequency equations [13]

$$\cos \lambda_i \cosh \lambda_i = 1 \qquad (6.54)$$

and

$$\gamma_i = \frac{\cosh \lambda_i - \cos \lambda_i}{\sin \lambda_i + \sinh \lambda_i} \qquad (6.55)$$

where $i = m,n$ in both Equations (6.54) and (6.55). The first five values of λ_i which satisfy Equation (6.54) have been tabulated in Table 5.2. For $i > 2$, Equation (5.54) can be utilized, i.e.

$$\lambda_i = (2i + 1)\frac{\pi}{2}, i > 2 \qquad (6.56)$$

The integrals occuring in Equation (6.49) are given in Reference [9]. A variety of results have been calculated to illustrate the effects of anisotropy on the natural frequencies of such plates. The natural frequencies are presented in terms of a parameter α defined such that

$$\omega^2 = \alpha \frac{E_T h^3}{\varrho a^4} \qquad (6.57)$$

The major Poisson's ratio, ν_{LT}, for the orthotropic laminas considered in these examples has been taken as 0.3, and all of the plates considered are square.

Table 6.10 presents the first four natural frequencies of plates composed of a single layer of orthotropic material with the principal axis of orthotropy at angles of 0, 22.5, and 45 degrees to the plate edge. The material ratios $E_L/E_T = 10$ and

Table 6.10. Effect of orientation for a single orthotropic layer.

Orientation	0°	22.5°	45°
α_1 Anisotropic	22.1	28.9	30.4
α_1 Orthotropic	22.1	46.0	65.1
α_2 Anisotropic	41.8	45.5	54.9
α_2 Orthotropic	41.8	65.6	108.8
α_3 Anisotropic	134.7	148.3	158.5
α_3 Orthotropic	134.7	318.6	203.4
α_4 Anisotropic	317.0	320.4	245.7
α_4 Orthotropic	317.0	324.2	548.4

Table 6.11. Effect of stacking sequence for a ±45 plate.

Laminate	α_1	α_2	α_3	α_4
$[\pm 45°]$	30.4	54.9	158.5	245.7
$[\pm 45°]_s$	57.9	59.8	239.6	295.4
$[\pm 45°]_{2s}$	63.9	92.7	254.4	463.2
$[\pm 45°]_\infty$ (Orthotropic)	65.1	108.8	263.4	548.4

Table 6.12. Effect of variations in E_L/E_T.

E_L/E_T	α_1	α_2	α_3	α_4
10.	30.4	54.9	158.5	245.7
7.5	29.3	50.6	151.2	194.6
5.0	27.7	46.0	140.6	142.4
2.5	24.4	40.7	84.4	129.0
1.0	16.7	35.9	54.7	112.4

$G_{LT}/E_T = 0.5$ have been used for these calculations. Also presented in this table are the corresponding orthotropic solutions. Again we find that the orthotropic solution predicts appreciably higher frequencies than those predicted with the anisotropic analysis, except for plates with the material axes coincident with the plate axes.

To illustrate the effects of laminating multiple plies at alternating $\pm \theta$ angles (with respect to the plate edges), several laminates of the class $[\pm 45°]_{ns}$ have been considered using the same material property ratios as above. The first four frequencies are given in Table 6.11 for a single layer 45-degree plate, for $[\pm 45°]_s$, $[\pm 45°]_{2s}$, and $[\pm 45°]_\infty$ (orthotropic analysis). Clearly the plate becomes effectively stiffer for alternating plies and converges on the orthotropic predictions for a large number of alternations.

The examples above consider strongly orthotropic laminates, i.e. $E_L/E_T = 10$. Table 6.12 presents results for all 45-degree plates for decreasing values of the ratio E_L/E_T; in particular for values of $E_L/E_T = 10$, 7.5, 5, 2.5, and 1, the ratio G_{LT}/ET has been varied linearly with E_L from 0.5 when $E_L/E_T = 10$ to 0.385 when $E_L/E_T = 1$. The frequencies presented in Table 6.12 make a continuous transition from the highly anisotropic case to the isotropic case.

REFERENCES

1. Wang, J. T-S. "On the Solution of Plates of Composite Materials," *Journal of Composite Materials*, 3:590–592 (1969).

2. Iyengar, K. T. S. and R. S. Srinivasan. "Clamped Skew Plate Under Uniform Normal Loading," *Journal of the Royal Aeronautical Society*, pp. 139–143 (February 1967).

3. Ashton, J. E. "An Analogy for Certain Anisotropic Plates," *Journal of Composite Materials*, 3:355–358 (1969).

4. Ashton, J. E. "Anisotropic Plate Analysis—Boundary Conditions," *Journal of Composite Materials*, 4:162–171 (1970).

5. Ashton, J. E. and M. E. Waddoups. "Analysis of Anisotropic Plates," *Journal of Composite Materials*, 3:148–165 (1969).

6. Thielemann, W. "Contributions to the Problem of the Buckling of Orthotropic Plates," *N.A.C.A. Technical Memorandum 1263* (August 1950).

7. Halpin, J. C. *Primer on Composite Materials: Analysis*, Second Edition (1984).

8. Mandell, J. F., "Experimental Investigation of the Buckling of Anisotropic Fiber Reinforced Plastic Plates," Air Force Technical Report AFML-TR-68-281 (October 1968).

9. Ashton, J. E. "Analysis of Anisotropic Plates II," *Journal of Composite Materials*, 3:470–479 (1969).

10. Ashton, J. E. and T. S. Love. "Experimental Study of the Stability of Composite Plates," *Journal of Composite Materials*, 3:230–242 (1969).

11. Ashton, J. E. and J. D. Anderson. "Natural Modes of Vibration of Boron-Epoxy Plates," *39th Shock and Vibration Bulletin*, Part 4 (April 1969).

12. Ashton, J. E., "Natural Modes of Free-Free Anisotropic Plates," *39th Shock and Vibration Bulletin*, Part 4 (April 1969).

13. Timoshenko, S. P., D. H. Young and W. Weaver, Jr. *Vibration Problems in Engineering*, Fourth Edition, John Wiley & Sons (1974).

CHAPTER 7

General Laminated Plates

7.1 INTRODUCTION

IN THE TWO previous chapters solutions have been presented for laminates exhibiting midplane symmetry with regard to the stacking sequence of the individual layers. For specially orthotropic laminates, solutions were obtained in essentially the same manner as for homogeneous isotropic plates. Laminates exhibiting D_{16} and D_{26} coupling terms, however, proved to be more formidable. Further complications are now introduced by considering the general case of laminated plates in which it is necessary to consider bending-extensional coupling.

In this chapter we consider the bending, stability, and free vibration of unsymmetrical laminated plates. The effect of the B_{ij} coupling terms is ascertained by comparing solutions to those obtained by neglecting B_{ij} in the governing equations. Thus guidelines are established for determining the circumstances under which the general laminated plate theory must be used in order to obtain reasonably accurate solutions. An approximate method for estimating the effect of bending-extensional coupling is also discussed.

7.2 BENDING OF RECTANGULAR CROSS-PLY PLATES

Consider a rectangular cross-ply composite of dimensions a,b (see Figure 7.1) having the stacking sequence $[0°/90°]_n$. For this laminate,

$$A_{16} = A_{26} = B_{12} = B_{16} = B_{26} = B_{66} = D_{16} = D_{26} = 0$$

$$A_{22} = A_{11},\ D_{22} = D_{11},\ B_{22} = -B_{11} \tag{7.1}$$

and the static case of Equations (2.32)–(2.34) becomes

$$A_{11}\frac{\partial^2 u^0}{\partial x^2} + A_{66}\frac{\partial^2 u^0}{\partial y^2} + (A_{12} + A_{66})\frac{\partial^2 v^0}{\partial x \partial y} - B_{11}\frac{\partial^3 w}{\partial x^3} = 0 \tag{7.2}$$

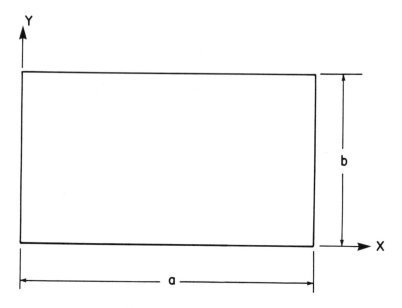

Figure 7.1. *Rectangular laminate.*

$$(A_{12} + A_{66}) \frac{\partial^2 u^0}{\partial x \partial y} + A_{66} \frac{\partial^2 v^0}{\partial x^2} + A_{11} \frac{\partial^2 v^0}{\partial y^2} + B_{11} \frac{\partial^3 w}{\partial y^3} = 0$$

$$D_{11}\left(\frac{\partial^4 w}{\partial x^4} + \frac{\partial^4 w}{\partial y^4}\right) + 2(D_{12} + 2D_{66}) \frac{\partial^4 w}{\partial x^2 \partial y^2} - B_{11}\left(\frac{\partial^3 u^0}{\partial x^3} - \frac{\partial^3 v^0}{\partial y^3}\right) = q$$

For hinged edges, free in the normal direction, the boundary conditions are:
For $x = 0,a$

$$w = 0, \ M_x = B_{11} \frac{\partial u^0}{\partial x} - D_{11} \frac{\partial^2 w}{\partial x^2} - D_{12} \frac{\partial^2 w}{\partial y^2} = 0 \qquad (7.3)$$

$$v^0 = 0, \ N_x = A_{11}\frac{\partial u^0}{\partial x} + A_{12} \frac{\partial v^0}{\partial y} - B_{11} \frac{\partial^2 w}{\partial x^2} = 0 \qquad (7.4)$$

For $y = 0,b$

$$w = 0, \ M_y = - B_{11} \frac{\partial v^0}{\partial y} - D_{12} \frac{\partial^2 w}{\partial x^2} - D_{11} \frac{\partial^2 w}{\partial y^2} = 0 \qquad (7.5)$$

$$u^0 = 0, \ N_y = A_{12} \frac{\partial u^0}{\partial x} + A_{11} \frac{\partial v^0}{\partial y} + B_{11} \frac{\partial^2 w}{\partial x^2} = 0 \qquad (7.6)$$

The transverse load q is represented by the double Fourier series:

$$q = \sum_{m=1}^{\infty} \sum_{n=1}^{\infty} q_{mn} \sin \frac{m\pi x}{a} \sin \frac{n\pi y}{b} \qquad (7.7)$$

The following displacement field [1] satisfies Equations (7.2) and the boundary conditions (7.3)–(7.7):

$$u^0 = \sum_{m=1}^{\infty} \sum_{n=1}^{\infty} A_{mn} \cos \frac{m\pi x}{a} \sin \frac{n\pi y}{b}$$

$$v^0 = \sum_{m=1}^{\infty} \sum_{n=1}^{\infty} B_{mn} \sin \frac{m\pi x}{a} \cos \frac{n\pi y}{b} \qquad (7.8)$$

$$w = \sum_{m=1}^{\infty} \sum_{n=1}^{\infty} C_{mn} \sin \frac{m\pi x}{a} \sin \frac{n\pi y}{b}$$

Substituting (7.8) into Equations (7.2), equating like Fourier terms, and solving the resulting simultaneous algebraic equations for the Fourier coefficients, we obtain the result

$$A_{mn} = q_{mn} \frac{R^3 b^3 B_{11} m}{\pi^3 D_{mn}} [A_{66} m^4 + A_{11} m^2 n^2 R^2 + (A_{12} + A_{66}) n^4 R^4]$$

$$\qquad (7.9)$$

$$B_{mn} = -q_{mn} \frac{R^4 b^3 B_{11} n}{\pi^3 D_{mn}} [(A_{12} + A_{66}) m^4 R^4 + A_{11} m^2 n^2 R^2 + A_{66} n^4 R^4]$$

$$C_{mn} = q_{mn} \frac{R^4 b^4}{\pi^4 D_{mn}} [(A_{11} m^2 + A_{66} n^2 R^2)(A_{66} m^2 + A_{11} n^2 R^2)$$

$$\qquad - (A_{12} + A_{66})^2 m^2 n^2 R^2]$$

where

$$R = a/b$$

$$D_{mn} = \{[(A_{11} m^2 + A_{66} n^2 R^2)(A_{66} m^2 + A_{11} n^2 R^2)$$

$$- (A_{12} + A_{66})m^2n^2R^2][D_{11}(m^4 + n^4R^4)$$

$$+ 2(D_{12} + 2D_{66})m^2n^2R^2] - B_{11}^2[A_{11}m^2n^2R^2(m^4 + n^4R^4)$$

$$+ 2(A_{12} + A_{66})m^4n^4R^4 + A_{66}(m^8 + n^8R^8)]\}$$

Stress and moment resultants can be determined from Equations (2.46)–(2.53). For the case $B_{11} = 0$ we find $u^0 = v^0 = 0$ and

$$w = \frac{R^4b^4}{\pi^4} \sum_{m=1}^{\infty} \sum_{n=1}^{\infty} q_{mn} \frac{\sin\dfrac{m\pi x}{a} \sin\dfrac{n\pi y}{b}}{[D_{11}(m^4 + n^4R^4) + 2(D_{12}\, 2D_{66})m^2n^2R^2]} \qquad (7.10)$$

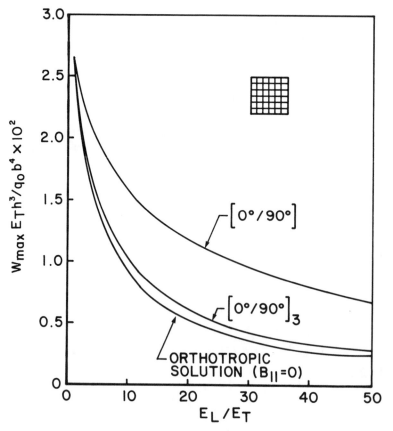

Figure 7.2. Maximum deflection of square cross-ply plate under transverse load.

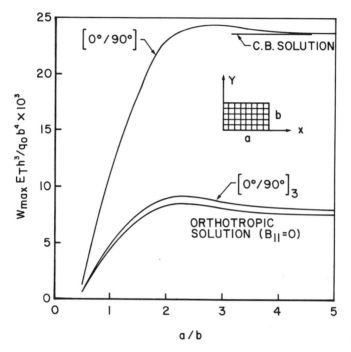

Figure 7.3. *Center deflection as a function of aspect ratio R for a cross-ply composite under uniform transverse load.*

which is the deflection equation for a homogeneous orthotropic plate (5.7) in which $D_{22} = D_{11}$. Thus, as in the case of cylindrical bending the orthotropic solution becomes the limiting case as the number of layers is increased. The effect of coupling on antisymmetric cross-ply plates is shown in Figures 7.2 and 7.3 for the following properties:

$$G_{LT}/E_T = .5, \ \nu_{LT} = .25$$

The dependence of the coupling effect on the ratio E_L/E_T is illustrated in Figure 7.2 where the maximum deflection is plotted for the load $q = q_0 \sin \pi x/a \sin \pi y/b$. Figure 7.3 shows the center deflection for a constant load q_0 where the Fourier coefficients are:

$$q_{mn} = \frac{16q_0}{\pi^2 mn} \quad (m,n \text{ odd})$$

$$q_{mn} = 0 \quad (m,n \text{ even})$$

**Table 7.1. Numerical results for simply-supported rectangular
(R = 2, b/h = 50) cross-ply plate under transverse loading.**

E_L/E_T	L^*	$w_{max}E_T/q_0b$	u^0_{max}/w_{max}	N_{xmax}/q_0b	N_{xymax}/q_0b
	2	1749	0.00720	0.06383	0.02724
40	6	676	0.00207	0.00797	0.00345
	10	630	0.00111	0.00263	0.00194
	∞	627	0	0	0
	2	1791	0.00469	0.36336	0.15507
	6	1562	0.00134	0.10562	0.05074
3.0	10	1546	0.00091	0.06273	0.02409
	∞	1540	0	0	0

*L indicates the number of layers in the plate, $\nu_{LT} = 0.25$, $G_{LT}/E_T = 0.5$.

For the results in Figure 7.3, $E_L/E_T = 40$. Equations (4.20) and (4.21) for cylindrical bending (CB) is the straight line in Figure 7.3.

The effect of coupling can be extremely severe; however, the orthotropic solution is rapidly approached as the number of plies is increased. It is interesting to note how rapidly the center deflection in Figure 7.3 approaches the maximum deflection of a cross-ply strip. Similar results are also found in homogeneous isotropic plates [2].

Table 7.1, which is taken from Reference [3], shows that the inplane forces and displacements introduced because of coupling are relatively small compared to externally applied inplane forces (such as in buckling problems) and transverse displacements, respectively. Thus, the assumption used in Chapter II concerning the validity of a linear analysis would seem justified.

7.3 BENDING OF RECTANGULAR ANGLE-PLY PLATES

Consider a rectangular angle-ply composite of the class $[\pm\theta]_n$. For this layup,

$$A_{16} = A_{26} = B_{11} = B_{12} = B_{22} = B_{66} = D_{16} = D_{26} = 0 \qquad (7.11)$$

and the static case of Equations (2.32)–(2.34) becomes

$$A_{11}\frac{\partial^2 u^0}{\partial x^2} + A_{66}\frac{\partial^2 u^0}{\partial y^2} + (A_{12} + A_{66})\frac{\partial^2 v^0}{\partial x\partial y} - 3B_{16}\frac{\partial^3 w}{\partial x^2\partial y} - B_{26}\frac{\partial^3 w}{\partial y^3} = 0$$

$$(7.12)$$

$$(A_{12} + A_{66})\frac{\partial^2 u^0}{\partial x\partial y} + A_{66}\frac{\partial^2 v^0}{\partial x^2} + A_{22}\frac{\partial^2 v^0}{\partial y^2} - B_{16}\frac{\partial^3 w}{\partial x^3} - 3B_{26}\frac{\partial^3 w}{\partial x\partial y^2} = 0$$

$$D_{11} \frac{\partial^4 w}{\partial x^4} + 2(D_{12} + 2D_{66}) \frac{\partial^4 w}{\partial x^2 \partial y^2} + D_{22} \frac{\partial^4 w}{\partial y^4} - B_{16} \left(3 \frac{\partial^3 u^0}{\partial x^2 \partial y} + \frac{\partial^3 v^0}{\partial x^3} \right)$$

$$- B_{26} \left(\frac{\partial^3 u^0}{\partial y^3} + 3 \frac{\partial^3 v^0}{\partial x \partial y^2} \right) = q$$

For hinged edges, free in the tangential direction, the boundary conditions are:
  For $x = 0, a$

$$w = 0, \ M_x = B_{16} \left(\frac{\partial u^0}{\partial y} + \frac{\partial v^0}{\partial x} \right) - D_{11} \frac{\partial^2 w}{\partial x^2} - D_{12} \frac{\partial^2 w}{\partial y^2} = 0 \qquad (7.13)$$

$$u^0 = 0, \ N_{xy} = A_{66} \left(\frac{\partial u^0}{\partial y} + \frac{\partial v^0}{\partial x} \right) - B_{16} \frac{\partial^2 w}{\partial x^2} - B_{26} \frac{\partial^2 w}{\partial y^2} = 0 \qquad (7.14)$$

  For $y = 0, b$

$$w = 0, \ M_y = B_{26} \left(\frac{\partial u^0}{\partial y} + \frac{\partial v^0}{\partial x} \right) - D_{12} \frac{\partial^2 w}{\partial x^2} - D_{22} \frac{\partial^2 w}{\partial y^2} = 0 \qquad (7.15)$$

$$v^0 = 0, \ N_{xy} = 0 \qquad (7.16)$$

The following displacement field [1] satisfies Equations (7.12) and the boundary conditions (7.13)–(7.16):

$$u^0 = \sum_{m=1}^{\infty} \sum_{n=1}^{\infty} A_{mn} \sin \frac{m \pi x}{a} \cos \frac{n \pi y}{b}$$

$$v^0 = \sum_{m=1}^{\infty} \sum_{n=1}^{\infty} B_{mn} \cos \frac{m \pi x}{a} \sin \frac{n \pi y}{b}$$

$$w = \sum_{m=1}^{\infty} \sum_{n=1}^{\infty} C_{mn} \sin \frac{m \pi x}{a} \sin \frac{n \pi y}{b} \qquad (7.17)$$

Substituting (7.17) into Equations (7.12), equating like Fourier terms, and solving the resulting simultaneous algebraic equations for the Fourier coefficients, we find

$$A_{mn} = q_{mn} \frac{R^4 b^3 n}{\pi^3 D_{mn}} \, [(A_{66}m^2 + A_{22}n^2R^2)(3B_{16}m^2 + B_{26}n^2R^2)$$

$$- m^2(A_{12} + A_{66})(B_{16}m^2 + 3B_{26}n^2R^2)]$$

$$B_{mn} = q_{mn} \frac{R^3 b^3 m}{\pi^3 D_{mn}} \, [(A_{11}m^2 + A_{66}n^2R^2)(B_{16}m^2 + 3B_{26}n^2R^2)$$

$$\hspace{6cm} (7.18)$$

$$- n^2R^2(A_{12} + A_{66})(3B_{16}m^2 + B_{26}n^2R^2)]$$

$$C_{mn} = q_{mn} \frac{R^4 b^4}{\pi^4 D_{mn}} \, [(A_{11}m^2 + A_{66}n^2R^2)(A_{66}m^2 + A_{22}n^2R^2)$$

$$- (A_{12} + A_{66})^2 \, m^2 n^2 R^2]$$

where

$$D_{mn} = \{[(A_{11}m^2 + A_{66}n^2R^2)(A_{66}m^2 + A_{22}n^2R^2)$$

$$- (A_{12} + A_{66})^2 m^2 n^2 R^2][D_{11}m^4 + 2(D_{12} + 2D_{26})m^2 n^2 R^2$$

$$+ D_{22}n^4R^4] + 2m^2 n^2 R^2(A_{12} + A_{66})(3B_{16}m^2$$

$$+ B_{26}n^2R^2)(B_{16}m^2 + 3B_{26}n^2R^2) - n^2R^2(A_{66}m^2$$

$$+ A_{22}n^2R^2)(3B_{16}m^2 + B_{26}n^2R^2)^2 - m^2(A_{11}m^2$$

$$+ A_{66}n^2R^2)(3B_{16}m^2 + B_{26}n^2R^2)^2\}$$

Stress and moment resultants can be determined from Equations (2.46)–(2.53). For the case $B_{16} = B_{26} = 0$ we find $u^0 = v^0 = 0$ and

$$w = \frac{R^4 b^4}{\pi^4} \sum_{m=1}^{\infty} \sum_{n=1}^{\infty} q_{mn} \frac{\sin \dfrac{m\pi x}{a} \sin \dfrac{n\pi y}{b}}{[D_{11}m^4 + 2(D_{12} + 2D_{66})m^2 n^2 R^2 + D_{22}n^4R^4]}$$

$$\hspace{6cm} (7.19)$$

which is the deflection equation for a homogeneous orthotropic plate (5.7). For

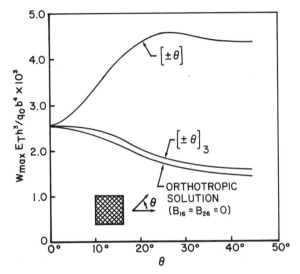

Figure 7.4. *Maximum deflection of square angle-ply plate under transverse load.*

angle-ply composites,

$$(B_{16}, B_{26}) = (Q_{16}, Q_{26}) \frac{h^2}{4n} \tag{7.20}$$

Thus, as in the case of cross-ply composites, the homogeneous orthotropic solution becomes the limiting case as the number of layers is increased. This is illustrated in Figure 7.4 for a square angle-ply plate under the transverse load

$$q = q_0 \sin \frac{\pi x}{a} \sin \frac{\pi y}{b}$$

The maximum deflection is plotted for various angle-ply orientations with

$$E_L/E_T = 40, \ G_{LT}/E_T = .5, \ \nu_{LT} = .25$$

As in the case of cross-ply composites, the effect of coupling can be very severe, but dissipates rapidly as the number of layers is increased.

7.4 BENDING OF ELLIPTIC CROSS-PLY PLATES

Cross-ply plates of the construction discussed in Section 7.2 have been investigated by Kicher [4] for elliptic plates. Several boundary conditions were

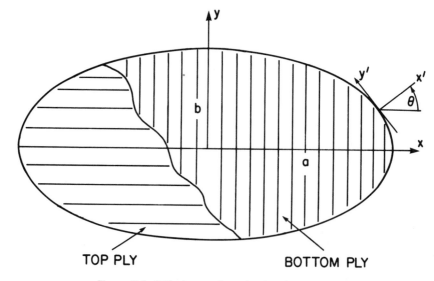

Figure 7.5. *Stiffening configuration for elliptic plate.*

considered in [4]; however, only one case is presented here. In particular we consider a special type of clamped edge such that

$$w = \frac{\partial w}{\partial x'} = N_{x'} = N_{x'y'} = 0 \qquad (7.21)$$

where x' and y' are axes normal and tangential to the boundary of the ellipse, respectively (see Figure 7.5). Integrating Equation (1.3) with respect to z and using the definition of stress resultants, Equation (2.9), we find

$$N_{x'} = N_x m^2 + N_y n^2 + 2mnN_{xy} \qquad (7.22)$$

$$N_{x'y'} = (N_y - N_x)mn + (m^2 - n^2)N_{xy} \qquad (7.23)$$

For the elliptic boundary in Figure 7.5,

$$m = \cos\theta = \frac{b^2 x}{(b^4 x^2 + a^4 y^2)^{1/2}}, \; n = \sin\theta = \frac{a^2 y}{(b^4 x^2 + a^4 y^2)^{1/2}} \qquad (7.24)$$

Since the inplane boundary conditions are of the stress type, we will consider this problem in terms of a stress function Φ and transverse deflection w. Substituting (2.59) and (7.24) into Equations (7.22) and (7.23), we obtain the following results:

$$N_{x'} = \frac{1}{b^4 x^2 + a^4 y^2} \left(b^4 x^2 \frac{\partial^2 \Phi}{\partial y^2} + a^4 y^2 \frac{\partial^2 \Phi}{\partial x^2} - 2a^2 b^2 xy \frac{\partial^2 \Phi}{\partial x \partial y} \right) \qquad (7.25)$$

$$N_{x'y'} = \frac{1}{b^4 x^2 + a^4 y^2} \left[\left(\frac{\partial^2 \Phi}{\partial x^2} - \frac{\partial^2 \Phi}{\partial y^2} \right) a^2 b^2 xy - \frac{\partial^2 \Phi}{\partial x \partial y} (b^4 x^2 - a^4 y^2) \right] \quad (7.26)$$

When

$$\frac{x^2}{a^2} + \frac{y^2}{b^2} = 1 \qquad (7.27)$$

$N_{x'}$ and $N_{x'y'}$ must vanish.

For the cross-ply composite under consideration,

$$A_{16}^* = A_{26}^* = B_{16}^* = B_{26}^* = B_{61}^* = B_{62}^* = B_{66}^* = D_{16}^* = D_{26}^* = 0 \qquad (7.28)$$

$$A_{22}^* = A_{11}^*, \ B_{21}^* = -B_{12}^*, \ B_{22}^* = -B_{11}^*, \ D_{22}^* = D_{11}^*$$

and the static form of Equations (2.65) and (2.67) becomes

$$D_{11}^* \left(\frac{\partial^4 w}{\partial x^4} + \frac{\partial^4 w}{\partial y^4} \right) + 2(D_{12}^* + 2D_{66}^*) \frac{\partial^4 w}{\partial x^2 \partial y^2} + B_{12}^* \left(\frac{\partial^4 \Phi}{\partial y^4} - \frac{\partial^4 \Phi}{\partial x^4} \right) = q \qquad (7.29)$$

$$A_{11}^* \left(\frac{\partial^4 \Phi}{\partial x^4} + \frac{\partial^4 \Phi}{\partial y^4} \right) + (2A_{12}^* + A_{66}^*) \frac{\partial^4 \Phi}{\partial x^2 \partial y^2} + B_{12}^* \left(\frac{\partial^4 w}{\partial x^4} - \frac{\partial^4 w}{\partial y^4} \right) = 0$$

For the case of a uniform load $q = q_0 = $ constant, Equations (7.29) and the boundary conditions (7.21) are satisfied by functions of the form

$$\Phi = A \left(1 - \frac{x^2}{a^2} - \frac{y^2}{b^2} \right)^2$$

$$w = B \left(1 - \frac{x^2}{a^2} - \frac{y^2}{b^2} \right)^2 \qquad (7.30)$$

Substituting (7.30) into (7.29) yields

$$3B_{12}^*(a^4 - b^4)A + [3D_{11}^*(a^4 + b^4) + 2(D_{12}^* + 2D_{66}^*)a^2b^2]B = q_0\frac{a^4b^4}{8}$$

$$- [3A_{11}^*(a^4 + b^4) + (2A_{12}^* + A_{66}^*)a^2b^2]A + 3B_{12}^*(a^4 - b^4)B = 0 \qquad (7.31)$$

Solving Equations (7.31) for the constants A and B, we obtain the following expressions:

$$A = 3B_{12}^* \frac{(a^4 - b^4)a^4b^4q_0}{8D}$$

$$\qquad (7.32)$$

$$B = [3A_{11}^*(a^4 + b^4) + (2A_{12}^* + A_{66}^*)a^2b^2] \frac{a^4b^4q_0}{8D}$$

where

$$D = [3A_{11}^*(a^4 + b^4) + (2A_{12}^* + A_{66}^*)a^2b^2][3D_{11}^*(a^4 + b^4)$$

$$+ 2(D_{12}^* + 2D_{66}^*)a^2b^2] + 9(B_{12}^*)^2(a^4 - b^4)^2$$

For circular plates, $a = b$ and Equations (7.32) become

$$A = 0$$

$$\qquad (7.33)$$

$$B = \frac{q_0a^4}{16(3D_{11}^* + D_{12}^* + 2D_{66}^*)}$$

which is the equation for a homogeneous circular plate with D_{ij} replaced by D_{ij}^*. This is an important observation and will be discussed in detail in the last section of this chapter. The moment and shear resultants are obtained from Equations (2.72) and (2.73). The inplane displacements u^0 and v^0 are determined from Equations (2.74). It should be noted that in the case of circular plates, u^0 and v^0 are nonzero even though the stress resultants vanish throughout the plate.

It can be shown that the homogeneous orthotropic solution is the limiting case of Equations (7.32) as the number of plies is increased. Thus the effect of coupling is similar to that of a rectangular cross-ply plate as discussed in Section 7.2.

7.5 STABILITY OF A RECTANGULAR ANGLE-PLY PLATE UNDER UNIFORM BIAXIAL COMPRESSION

Consider an angle-ply plate with the hinge boundary conditions discussed in

Section 7.3. Let us assume, however, that the hinge supports are such that uniform displacements along two adjacent boundaries are admissible. Thus, in addition to (7.13) and (7.15) we have the inplane boundary conditions
For $x = 0$:

$$u^0 = 0, \ N_{xy} = 0 \tag{7.34}$$

For $x = a$:

$$u^0 = c_1 = \text{constant}, \ N_{xy} = 0 \tag{7.35}$$

For $y = 0$:

$$v^0 = 0, \ N_{xy} = 0 \tag{7.36}$$

For $y = b$:

$$v^0 = c_2 = \text{constant}, \ N_{xy} = 0 \tag{7.37}$$

The initial displacement field

$$u^{0i} = \frac{c_1}{a}x, \ v^{0i} = \frac{c_2}{b}y, \ w^i = 0 \tag{7.38}$$

satisfies the boundary conditions (7.13), (7.15), (7.34)–(7.37) and Equations (7.12) for $q = 0$. The constants c_1 and c_2 are determined such that

$$N_x^i = -N_0 = \text{constant}, \ N_y^i = -kN_0, \ N_0 > 0 \tag{7.39}$$

where k is a constant. Substituting (7.38) and (7.39) into (2.46) and (2.47), we find

$$c_1 = \frac{(A_{12}k - A_{22})N_0}{(A_{11}A_{22} - A_{12}^2)}, \ c_2 = \frac{(A_{12} - kA_{11})N_0}{(A_{11}A_{22} - A_{12}^2)} \tag{7.40}$$

The postbuckling displacements are determined from Equations (2.32), (2.33), and (2.79). For the angle-ply plate currently under discussion, these equations become

$$A_{11} \frac{\partial^2 u^0}{\partial x^2} + A_{66} \frac{\partial^2 u^0}{\partial y^2} + (A_{12} + A_{66}) \frac{\partial^2 v^0}{\partial x \partial y}$$
$$- 3B_{16} \frac{\partial^3 w}{\partial x^2 \partial y} - B_{26} \frac{\partial^3 w}{\partial y^3} = 0 \tag{7.41}$$

$$(A_{12} + A_{66})\ \frac{\partial^2 u^0}{\partial x^2} + A_{66}\ \frac{\partial^2 v^0}{\partial x^2} + A_{22}\ \frac{\partial^2 v^0}{\partial y^2}$$

$$- B_{16}\ \frac{\partial^3 w}{\partial x^3} - 3B_{26}\ \frac{\partial^3 w}{\partial x \partial y^2} = 0$$

(7.42)

$$D_{11}\ \frac{\partial^4 w}{\partial x^4} + 2(D_{12} + 2D_{66})\ \frac{\partial^4 w}{\partial x^2 \partial y^2} + D_{22}\ \frac{\partial^4 w}{\partial y^4} - B_{16}\left(3\ \frac{\partial^3 u^0}{\partial x^2 \partial y} + \frac{\partial^3 v^0}{\partial x^3}\right)$$

$$- B_{26}\left(\frac{\partial^3 u^0}{\partial y^3} + 3\ \frac{\partial^3 v^0}{\partial x \partial y^2}\right) + N_0\left(\frac{\partial^2 w}{\partial x^2} + k\ \frac{\partial^2 w}{\partial y^2}\right) = 0$$

(7.43)

In order to assure that conditions (7.13), (7.15), (7.34), and (7.37) are satisfied, we require that

For $x = 0,a$:

$$w = M_x = B_{16}\left(\frac{\partial v^0}{\partial x} + \frac{\partial u^0}{\partial y}\right) - D_{11}\ \frac{\partial^2 w}{\partial x^2} - D_{12}\ \frac{\partial^2 w}{\partial y^2} = 0$$

$$u^0 = N_{xy} = A_{66}\left(\frac{\partial v^0}{\partial x} + \frac{\partial u^0}{\partial y}\right) - B_{16}\ \frac{\partial^2 w}{\partial x^2} - B_{26}\ \frac{\partial^2 w}{\partial y^2} = 0$$

(7.44)

For $y = 0,b$:

$$w = M_y = B_{26}\left(\frac{\partial v^0}{\partial x} + \frac{\partial u^0}{\partial y}\right) - D_{12}\ \frac{\partial^2 w}{\partial x^2} - D_{22}\ \frac{\partial^2 w}{\partial y^2} = 0$$

$$v^0 = N_{xy} = A_{66}\left(\frac{\partial v^0}{\partial x} + \frac{\partial u^0}{\partial y}\right) - B_{16}\ \frac{\partial^2 w}{\partial x^2} - B_{26}\ \frac{\partial^2 w}{\partial y^2} = 0$$

(7.45)

Equations (7.41)–(7.43) and the boundary conditions (7.44) and (7.45) are satisfied by the displacement field

$$u^0 = A \sin\frac{m\pi x}{a} \cos\frac{n\pi y}{b}$$

$$v^0 = B \cos\frac{m\pi x}{a} \sin\frac{n\pi y}{b}$$

(7.46)

$$w = C \sin\frac{m\pi x}{a} \sin\frac{n\pi y}{b}$$

Substituting (7.46) into (7.41)–(7.43), and collecting like coefficients leads to the following homogeneous equations (in matrix form):

$$
\begin{bmatrix}
A_{mn} & B_{mn} & C_{mn} \\
B_{mn} & D_{mn} & E_{mn} \\
C_{mn} & E_{mn} & (F_{mn} - \lambda)
\end{bmatrix}
\begin{bmatrix}
A \\
B \\
C
\end{bmatrix}
=
\begin{bmatrix}
0 \\
0 \\
0
\end{bmatrix}
\tag{7.47}
$$

where

$$A_{mn} = A_{11}m^2 + A_{66}n^2R^2$$

$$B_{mn} = (A_{12} + A_{66})mnR$$

$$C_{mn} = -\frac{n\pi}{b}(3B_{16}m^2 + B_{26}n^2R^2)$$

$$D_{mn} = A_{66}m^2 + A_{22}n^2R^2$$

$$E_{mn} = -\frac{m\pi}{Rb}(B_{16}m^2 + 3B_{26}n^2R^2)$$

$$F_{mn} = \frac{\pi^2}{R^2b^2}[D_{11}m^4 + 2(D_{12} + 2D_{66})m^2n^2R^2 + D_{22}n^4R^4]$$

$$\lambda = N_0(m^2 + kn^2R^2)$$

$$R = a/b$$

In order to obtain a non-trivial solution to (7.47), the determinant of the coefficient matrix must vanish. Thus,

$$
\begin{vmatrix}
A_{mn} & B_{mn} & C_{mn} \\
B_{mn} & D_{mn} & E_{mn} \\
C_{mn} & E_{mn} & (F_{mn} - \lambda)
\end{vmatrix}
= 0
\tag{7.48}
$$

Solving Equation (7.48) for λ we obtain the following values of N_0:

$$
N_0 = \frac{\pi^2}{R^2b^2(m^2 + kn^2R^2)}\{D_{11}m^4 + 2(D_{12} + 2D_{66})m^2n^2R^2 + D_{22}n^4R^4
$$

$$
\tag{7.49}
$$

$$
-\frac{1}{J_1}[m(B_{16}m^2 + 3B_{26}n^2R^2)J_2 + nR(3B_{16}m^2 + B_{26}n^2R^2)J_3]\}
$$

where

$$J_1 = (A_{11}m^2 + A_{66}n^2R^2)(A_{66}m^2 + A_{22}n^2R^2) - (A_{12} + A_{66})^2m^2n^2R^2$$

$$J_2 = (A_{11}m^2 + A_{66}n^2R^2)(B_{16}m^2 + 3B_{26}n^2R^2) - n^2R^2(A_{12}$$
$$+ A_{66})(3B_{16}m^2 + 3B_{26}n^2R^2)$$

$$J_3 = (A_{66}m^2 + A_{22}n^2R^2)(3B_{16}m^2 + B_{26}n^2R^2) - n^2R^2(A_{12}$$
$$+ A_{66})(B_{16}m^2 + 3B_{26}n^2R^2)$$

The buckling load is the lowest value of N_0. When $B_{16} = B_{26} = 0$, Equation (7.49) becomes

$$N_o = \frac{\pi^2[D_{11}m^4 + 2(D_{12} + 2D_{66})m^2n^2R^2 + D_{22}n^4R^4]}{R^2b^2(m^2 + kn^2R^2)} \tag{7.50}$$

which is the buckling relationship for a simply-supported orthotropic plate under uniform biaxial compression. Thus, as in the case of bending under transverse load, the orthotropic solution becomes the limiting case as the number of plies are increased.

Results are shown in Figures 7.6 and 7.7 for uniaxial compression ($k = 0$) and biaxial compression ($k = 1$), respectively, with $E_L/E_T = 40$, $G_{LT}/E_T = 0.5$, $\nu_{LT} = 0.25$, $R = 1$. Coupling is shown to severely reduce the buckling load for the two-layer plate. The effect dissipates quite rapidly, however, as the number of layers is increased.

7.6 STABILITY OF A CROSS-PLY PLATE UNDER UNIFORM SHEAR LOAD

Consider a cross-ply plate with the hinge boundary conditions discussed in Section 7.2. Let us assume, however, that the hinge supports are such that it is admissible for the middle-plane displacement v^0 to be a linear function of x (this allows the plate to undergo a uniform shear strain). Thus, in addition to (7.3), (7.5), and (7.6), we have the inplane boundary conditions

For $x = 0$:

$$v^0 = N_x = 0 \tag{7.51}$$

For $x = a$:

$$v^0 = c = \text{constant}, N_x = 0 \tag{7.52}$$

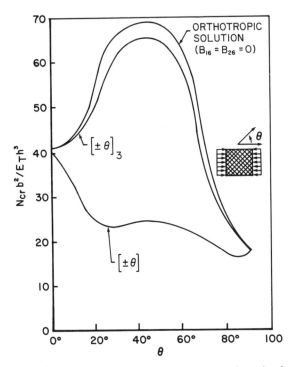

Figure 7.6. *Critical buckling load as a function of angle-ply orientation for a square plate under uniform axial compression.*

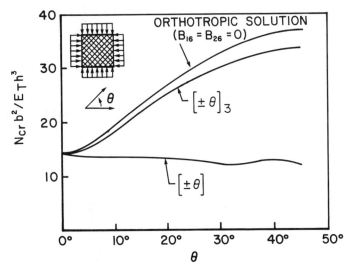

Figure 7.7. *Critical buckling load as a function of angle-ply orientation for a square plate under biaxial compression.*

The initial displacement field

$$v^{0i} = \frac{c}{a}x, \; u^{0i} = w^i = 0 \tag{7.53}$$

satisfies the boundary conditions (7.3), (7.5), (7.6), (7.52), and (7.53), and Equations (7.2) for $q = 0$. Substituting (7.52) in (2.46) and (2.47), we find

$$N_x^i = N_y^i = 0, \; N_{xy}^i = \frac{c}{a} = S = \text{constant} \tag{7.54}$$

The postbuckling displacements are determined from Equations (2.32), (2.33), and (2.79). For the cross-ply plate currently under consideration, these become

$$A_{11}\frac{\partial^2 u^0}{\partial x^2} + A_{66}\frac{\partial^2 u^0}{\partial y^2} + (A_{12} + A_{66})\frac{\partial^2 v^0}{\partial x \partial y} - B_{11}\frac{\partial^3 w}{\partial x^3} = 0 \tag{7.55}$$

$$(A_{12} + A_{66})\frac{\partial^2 u^0}{\partial x^2} + A_{66}\frac{\partial^2 v^0}{\partial x^2} + A_{11}\frac{\partial^2 v^0}{\partial y^2} + B_{11}\frac{\partial^3 w}{\partial y^3} = 0 \tag{7.56}$$

$$D_{11}\left(\frac{\partial^4 w}{\partial x^4} + \frac{\partial^4 w}{\partial y^4}\right) + 2(D_{12} + 2D_{66})\frac{\partial^4 w}{\partial x^2 \partial y^2}$$

$$- B_{11}\left(\frac{\partial^3 u^0}{\partial x^3} - \frac{\partial^3 v^0}{\partial y^3}\right) = 2S\frac{\partial^2 w}{\partial x \partial y} \tag{7.57}$$

In order to assure that conditions (7.3), (7.5), (7.6), (7.52), and (7.53) are satisfied, we require that
For $x = 0,a$:

$$w = M_x = B_{11}\frac{\partial u^0}{\partial x} - D_{11}\frac{\partial^2 w}{\partial x^2} - D_{12}\frac{\partial^2 w}{\partial y^2} = 0$$

$$\tag{7.58}$$

$$v^0 = N_x = A_{11}\frac{\partial u^0}{\partial x} + A_{12}\frac{\partial v^0}{\partial y} - B_{11}\frac{\partial^2 w}{\partial x^2} = 0$$

For $y = 0,b$:

$$w = M_y = -B_{11}\frac{\partial v^0}{\partial y} - D_{12}\frac{\partial^2 w}{\partial x^2} - D_{11}\frac{\partial^2 w}{\partial y^2} = 0$$

$$u^0 = N_y = A_{12}\frac{\partial u^0}{\partial x} + A_{11}\frac{\partial v^0}{\partial y} + B_{11}\frac{\partial^2 w}{\partial y^2} = 0$$

(7.59)

The following displacement field satisfies conditions (7.58) and (7.59) and the first two equilibrium Equations (7.55) and (7.56):

$$u^0 = \sum_{m=1}^{M} \sum_{n=1}^{N} A_{mn} \cos\frac{m\pi x}{a} \sin\frac{n\pi y}{b}$$

$$v^0 = \sum_{m=1}^{M} \sum_{n=1}^{N} B_{mn} \sin\frac{m\pi x}{a} \cos\frac{n\pi y}{b}$$

(7.60)

$$w = \sum_{m=1}^{M} \sum_{n=1}^{N} C_{mn} \sin\frac{m\pi x}{a} \sin\frac{n\pi y}{b}$$

Substituting Equations (7.60) into (7.55) and (7.56), and solving the resulting simultaneous equations, we find

$$A_{mn} = \frac{B_{11}H_{mn}m\pi}{D_{mn}Rb} C_{mn}, \quad B_{mn} = \frac{-B_{11}K_{mn}n\pi}{D_{mn}b} C_{mn}$$

(7.61)

where

$$D_{mn} = (A_{11}m^2 + A_{66}n^2R^2)(A_{66}m^2 + A_{11}n^2R^2) - (A_{12} + A_{66})^2m^2n^2R^2$$

$$H_{mn} = A_{11}m^2n^2R^2 + A_{12}n^4R^4 + A_{66}(m^4 + n^4R^4)$$

$$K_{mn} = A_{11}m^2n^2R^2 + A_{12}m^4 + A_{66}(m^4 + n^4R^4)$$

$$R = a/b$$

Equation (7.57) cannot be exactly satisfied by the displacement field (7.60). It can be approximately satisfied, however, to any desired degree of accuracy by application of the Galerkin method described in Section 3.7. For the problem currently under consideration, Equations (3.45) and (3.46) are exactly satisfied, while Equation (3.47) becomes

$$\int_0^b \int_0^a \left[D_{11}\left(\frac{\partial^4 w}{\partial x^4} + \frac{\partial^4 w}{\partial y^4}\right) + 2(D_{12} + 2D_{66})\frac{\partial^4 w}{\partial x^2 \partial y^2} - B_{11}\left(\frac{\partial^3 u^0}{\partial x^3} - \frac{\partial^3 v^0}{\partial y^3}\right) \right.$$

$$\left. - S\frac{\partial^2 w}{\partial x \partial y}\right] \sin\frac{m\pi x}{a} \sin\frac{n\pi y}{b}\, dxdy = 0 \qquad \begin{array}{l} m = 1,2\dots,M \\ n = 1,2\dots,N \end{array} \qquad (7.62)$$

Note that since the boundary conditions are exactly satisfied by Equations (7.60), the boundary integrals in Equation (3.47) vanish and are not included in Equation (7.62).

Substituting (7.60) into (7.62), taking (7.61) into account, and performing the integration leads to the following set of simultaneous equations:

$$-\lambda C_{mn} + \frac{32mnR^3}{\pi^4 M_{mn}} \sum_{i=1}^{M} \sum_{j=1}^{N} N_{ij}\, C_{ij} = 0 \qquad (7.63)$$

where

$$\lambda = \frac{1}{Sb^2}$$

$$M_{mn} = D_{11}(m^4 + n^4 R^4) + 2(D_{12} + 2D_{66})m^2 n^2 R^2 - B_{11}^2(m^2 K_{mn} + n^4 R^4 L_{mn})$$

$$N_{ij} = \frac{ij}{(m^2 - i^2)(n^2 - j^2)} \quad \begin{cases} m \pm i \quad \text{odd} \\ n \pm j \quad \text{odd} \end{cases}$$

$$= 0 \quad \begin{cases} m = i, m \pm i \quad \text{even} \\ n = j, n \pm j \quad \text{even} \end{cases}$$

Equation (7.63) results in two sets of homogeneous equations. The two sets correspond to $m + n$ odd and $m + n$ even. In each group of these equations a nontrivial solution is obtained if λ is chosen such that the determinant of the coefficient matrix vanishes. Equation (7.63) is in the form of a classical eigenvalue problem, and the critical buckling load S_{cr} corresponds to the highest eigenvalue which can be obtained by iteration [5]. For buckling of cross-ply plates under pure shear, the critical value of S does not depend on the direction of the stress. As a result the eigenvalues occur in positive and negative pairs, and

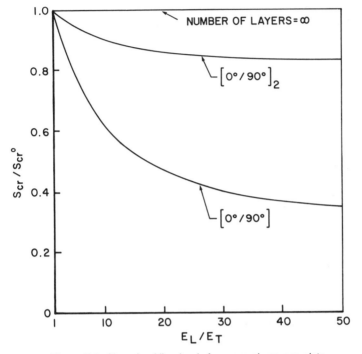

Figure 7.8. *Shear buckling loads for cross-ply square plate.*

standard iteration procedures must be modified to take this into account [6]. Convergence is established by increasing the order of the system.

When coupling is neglected ($B_{11} = 0$), Equation (7.62) reduces to (5.102), the Galerkin equation for a homogeneous orthotropic plate. The effect of coupling is illustrated in Figure 7.8 where the ratio of the coupled solution to the orthotropic solution (denoted by S_{cr}^0) is plotted as a function of E_L/E_T for a square plate with $\nu_{LT} = 0.25$ and $G_{LT}/E_T = 0.5$. Qualitatively the results are similar to those obtained in Section 7.5 for the compressive buckling of angle-ply plates. In particular, for two-layer highly anisotropic laminates, coupling reduces the buckling load by a factor of three. The effect dissipates rapidly as the number of plies is increased.

7.7 FREE-VIBRATION OF UNSYMMETRICAL LAMINATED PLATES

In this section we consider the free vibration of the cross-ply and angle-ply laminates discussed in the preceding sections of this chapter. To simplify calculations, inplane inertia terms are neglected. In order to determine the natural vibration frequencies, we consider harmonic solutions of the form

$$u^0 = e^{i\omega t} U(x,y)$$

$$v^0 = e^{i\omega t} V(x,y)$$

$$w = e^{i\omega t} W(x,y)$$

Substituting (7.64) into Equations (2.32)–(2.34) with $q = 0$ and inplane inertia neglected leads to the following equations for cross-ply plates:

$$A_{11}\frac{\partial^2 U}{\partial x^2} + A_{66}\frac{\partial^2 U}{\partial y^2} + (A_{12} + A_{66})\frac{\partial^2 V}{\partial x \partial y} - B_{11}\frac{\partial^3 W}{\partial x^3} = 0$$

$$(A_{12} + A_{66})\frac{\partial^2 U}{\partial x \partial y} + A_{66}\frac{\partial^2 V}{\partial x^2} + A_{11}\frac{\partial^2 V}{\partial y^2} + B_{11}\frac{\partial^3 W}{\partial y^3} = 0$$

$$D_{11}\left(\frac{\partial^4 W}{\partial x^4} + \frac{\partial^4 W}{\partial y^4}\right) + 2(D_{12} + 2D_{66})\frac{\partial^4 W}{\partial x^2 \partial y^2}$$

$$- B_{11}\left(\frac{\partial^3 U}{\partial x^3} - \frac{\partial^3 V}{\partial y^3}\right) - \varrho\omega^2 W = 0$$

(7.65)

For hinge supports, free in the normal direction, Equations (7.3)–(7.6) must be satisfied for all time t. Thus
For $x = 0,a$:

$$W = 0, \ B_{11}\frac{\partial U}{\partial x} - D_{11}\frac{\partial^2 W}{\partial x^2} - D_{12}\frac{\partial^2 W}{\partial y^2} = 0$$

$$V = 0, \ A_{11}\frac{\partial U}{\partial x} + A_{12}\frac{\partial V}{\partial y} - B_{11}\frac{\partial^2 W}{\partial x^2} = 0$$

(7.66)

For $y = 0,b$:

$$W = 0, \ - B_{11}\frac{\partial V}{\partial y} - D_{12}\frac{\partial^2 W}{\partial x^2} - D_{11}\frac{\partial^2 W}{\partial y^2} = 0 \qquad (7.67)$$

$$U = 0, \ A_{12} \frac{\partial U}{\partial x} + A_{11} \frac{\partial V}{\partial y} + B_{11} \frac{\partial^2 W}{\partial x^2} = 0$$

Equations (7.65–7.67) are satisfied by the displacements:

$$U = U_{mn} \cos \frac{m\pi x}{a} \sin \frac{n\pi y}{b}$$

$$V = V_{mn} \sin \frac{m\pi x}{a} \cos \frac{n\pi y}{b} \tag{7.68}$$

$$W = W_{mn} \sin \frac{m\pi x}{a} \sin \frac{n\pi y}{b}$$

Substituting Equation (7.68) into (7.65), we obtain the following homogeneous algebraic equations:

$$\begin{bmatrix} A_{mn} & B_{mn} & C_{mn} \\ B_{mn} & D_{mn} & E_{mn} \\ C_{mn} & E_{mn} & (F_{mn} - \lambda_{mn}) \end{bmatrix} \begin{bmatrix} A \\ B \\ C \end{bmatrix} = \begin{bmatrix} 0 \\ 0 \\ 0 \end{bmatrix} \tag{7.69}$$

where

$$A_{mn} = A_{11}m^2 + A_{66}n^2R^2$$

$$B_{mn} = (A_{12} + A_{66})mnR$$

$$C_{mn} = -B_{11} \frac{m^3\pi}{Rb}$$

$$D_{mn} = A_{66}m^2 + A_{11}n^2R^2$$

$$E_{mn} = \frac{B_{11}n^3R^2\pi}{R}$$

$$F_{mn} = \frac{\pi^2}{R^2b^2} [D_{11}(m^4 + n^4R^4) + 2(D_{12} + 2D_{66}m^2n^2R^2)]$$

$$\lambda_{mn} = \frac{\varrho\omega_{mn}^2R^2b^2}{\pi^2}$$

$$R = a/b$$

In order to obtain a non-trivial solution to Equation (7.69), the determinant of the coefficient matrix must vanish. Thus,

$$
\begin{vmatrix}
A_{mn} & B_{mn} & C_{mn} \\
B_{mn} & D_{mn} & E_{mn} \\
C_{mn} & E_{mn} & (F_{mn} - \lambda_{mn})
\end{vmatrix} = 0
\tag{7.70}
$$

Solving Equation (7.70) for λ yields

$$
\omega_{mn}^2 = \frac{\pi^4}{\varrho R^4 b^4} [D_{11}(m^4 + n^4 R^4) + 2(D_{12} + 2D_{66})m^2 n^2 R^2
$$

$$
- \frac{B_{11}^2}{J_3} (m^4 J_1 + n^4 R^4 J_2)]
\tag{7.71}
$$

where

$$
J_1 = A_{66}m^4 + A_{11}m^2 n^2 R^2 + (A_{12} + A_{66})n^4 R^4
$$

$$
J_2 = (A_{12} + A_{66})m^4 + A_{11}m^2 n^2 R^2 + A_{66}n^4 R^4
$$

$$
J_3 = (A_{11}m^2 + A_{66}n^2 R^2)(A_{66}m^2 + A_{11}n^2 R^2) - (A_{12} + A_{66})^2 m^2 n^2 R^2
$$

Substituting (7.64) into Equations (2.32)–(2.34) with $q = 0$ and inplane inertia neglected leads to the following equations for angle-ply plates:

$$
A_{11}\frac{\partial^2 U}{\partial x^2} + A_{66}\frac{\partial^2 U}{\partial y^2} + (A_{12} + A_{66})\frac{\partial^2 V}{\partial x \partial y} - 3B_{16}\frac{\partial^3 W}{\partial x^2 \partial y} - B_{26}\frac{\partial^3 W}{\partial y^3} = 0
$$

$$
(A_{12} + A_{66})\frac{\partial U}{\partial x \partial y} + A_{66}\frac{\partial^2 V}{\partial x^2} + A_{22}\frac{\partial^2 V}{\partial y^2} - B_{16}\frac{\partial^3 W}{\partial x^3}
$$

$$
- 3B_{26}\frac{\partial^3 W}{\partial x \partial y^2} = 0 \quad D_{11}\frac{\partial^4 W}{\partial x^4} + 2(D_{12} + 2D_{66})\frac{\partial^4 W}{\partial x^2 \partial y^2} + D_{22}\frac{\partial^4 W}{\partial y^4}
\tag{7.72}
$$

$$
- B_{16}\left(3\frac{\partial^3 U}{\partial x^2 \partial y} + \frac{\partial^3 V}{\partial x^3}\right) - B_{26}\left(\frac{\partial^3 U}{\partial y^3} + 3\frac{\partial^3 V}{\partial x \partial y^2}\right) - \varrho\omega^2 W = 0
$$

For hinge supports, free in the tangential direction, Equations (7.13)–(7.16) must be satisfied for all values of t. Thus,

For $x = 0,a$:

$$W = 0, \; B_{16}\left(\frac{\partial U}{\partial y} + \frac{\partial V}{\partial x}\right) - D_{11}\frac{\partial^2 W}{\partial x^2} - D_{12}\frac{\partial^2 W}{\partial y^2} = 0$$

(7.73)

$$U = 0, \; A_{66}\left(\frac{\partial U}{\partial y} + \frac{\partial V}{\partial x}\right) - B_{16}\frac{\partial^2 W}{\partial x^2} - B_{26}\frac{\partial^2 W}{\partial y^2} = 0$$

For $y = 0,b$:

$$W = 0, \; B_{26}\left(\frac{\partial U}{\partial y} + \frac{\partial V}{\partial x}\right) - D_{12}\frac{\partial^2 W}{\partial x^2} - D_{22}\frac{\partial^2 W}{\partial y^2} = 0$$

(7.74)

$$V = 0, \; A_{66}\left(\frac{\partial U}{\partial y} + \frac{\partial V}{\partial x}\right) - B_{16}\frac{\partial^2 W}{\partial x^2} - B_{26}\frac{\partial^2 W}{\partial y^2} = 0$$

Equations (7.72)–(7.74) are satisfied by

$$U = U_{mn}\sin\frac{m\pi x}{a}\cos\frac{n\pi y}{b}$$

$$V = V_{mn}\cos\frac{m\pi x}{a}\sin\frac{n\pi y}{b}$$

(7.75)

$$W = W_{mn}\sin\frac{m\pi x}{a}\sin\frac{n\pi y}{b}$$

Substituting Equation (7.75) into (7.72), we obtain Equation (7.69) with C_{mn} and E_{mn} replaced by C'_{mn} and E'_{mn}, respectively, where

$$C'_{mn} = -\frac{n\pi}{b}(3B_{16}m^2 + B_{26}n^2R^2)$$

(7.76)

$$E'_{mn} = -\frac{m\pi}{Rb}(B_{16}m^2 + 3B_{26}n^2R^2)$$

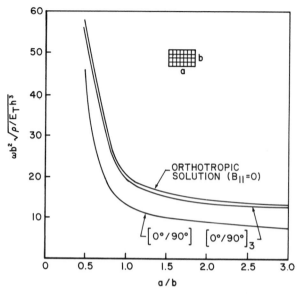

Figure 7.9. Fundamental vibration frequency for rectangular cross-ply plate.

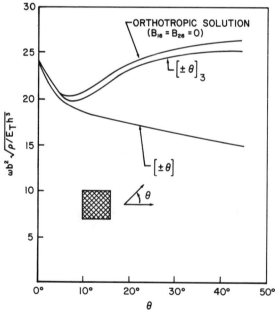

Figure 7.10. Fundamental vibration frequency as a function of angle-ply orientation for a square plate.

In a similar manner as for cross-ply laminates, the following frequency equation is obtained:

$$\omega_{mn}^2 = \frac{\pi^4}{\varrho R^4 b^4} \{D_{11}m^4 + 2(D_{12} + 2D_{66})m^2n^2R^2 + D_{22}n^4R^4$$

$$- \frac{1}{J_6}[m(B_{16}m^2 + 3B_{26}n^2R^2)J_4 + nR(3B_{16}m^2 + B_{26}n^2R^2)J_5]\} \quad (7.77)$$

where

$$J_4 = (A_{11}m^2 + A_{66}n^2R^2)(B_{16}m^2 + 3B_{26}n^2R^2) - n^2R^2(A_{12} + A_{66})(3B_{16}m^2 + B_{26}n^2R^2)$$

$$J_5 = (A_{66}m^2 + A_{22}n^2R^2)(3B_{16}m^2 + B_{26}n^2R^2) - n^2R^2(A_{12} + A_{66})(B_{16}m^2 + 3B_{26}n^2R^2)$$

$$J_6 = (A_{11}m^2 + A_{66}n^2R^2)(A_{66}m^2 + A_{22}n^2R^2) - (A_{12} + A_{66})^2m^2n^2R^2$$

When coupling is neglected ($B_{ij} = 0$) Equations (7.71) and (7.77) become

$$\omega_{mn}^2 = \frac{\pi^4}{\varrho R^4 b^4}[D_{11}m^4 + 2(D_{12} + 2D_{66})m^2n^2R^2 + D_{22}n^4R^4] \quad (7.78)$$

which is the frequency equation for the flexural vibration of a simply-supported homogeneous orthotropic plate.

Fundamental frequencies are shown in Figures 7.9 and 7.10 for cross-ply and angle-ply laminates, respectively, with

$$E_L/E_T = 40, \ G_{LT}/E_T = .5, \ \nu_{LT} = .25$$

Coupling reduces the fundamental vibration frequency. It is obvious from (7.78) that the fundamental frequency of orthotropic laminates always occurs for $m = n = 1$. However, for coupled laminates, the mode number corresponding to the lowest frequency cannot be ascertained by inspection of Equations (7.71) and (7.77). For the results presented in Figures 7.9 and 7.10, all fundamental frequencies do occur for $m = n = 1$.

7.8 THE REDUCED BENDING STIFFNESS APPROXIMATION

It is obvious from the examples presented in this chapter that bending-extensional coupling causes considerable difficulty in obtaining solutions to boundary value problems. We now consider an approximate method of solution for such problems which yields results that compare favorably with those presented in preceding sections of this chapter. Furthermore, the method reduces the problem to an equivalent anisotropic bending problem allowing the techniques developed for the analysis of orthotropic and anisotropic plates to be directly utilized. This approximate method was first discussed by Reissner and Stavsky [7].

From Equation (3.1) the strain energy U of a laminated rectangular plate can be written in the form

$$U = \frac{1}{2} \int_{-h/2}^{h/2} \int_0^b \int_0^a (\sigma_x \epsilon_x + \sigma_y \epsilon_y + \sigma_{xy} \epsilon_{xy}) dx dy dz \qquad (7.79)$$

Substituting (2.5) into (7.79), integrating with respect to z, and taking (2.9) into account, we obtain the result

$$U = \frac{1}{2} \int_0^b \int_0^a (N_x \epsilon_x^0 + N_y \epsilon_y^0 + N_{xy} \epsilon_{xy}^0 + M_x \varkappa_x + M_y \varkappa_y + M_{xy} \varkappa_{xy}) dx dy \quad (7.80)$$

Using the constitutive relation of the plate in the form of (2.64), Equation (7.80) becomes

$$U = \frac{1}{2} \int_0^b \int_0^a (A_{11}^* N_x^2 + A_{22}^* N_y^2 + A_{66}^* N_{xy}^2 + 2A_{12}^* N_x N_y + 2A_{16}^* N_x N_{xy}$$

$$+ 2A_{26}^* N_y N_{xy} + D_{11}^* \varkappa_x^2 + D_{22}^* \varkappa_y^2 + D_{66}^* \varkappa_{xy}^2 + 2D_{12}^* \varkappa_x \varkappa_y \qquad (7.81)$$

$$+ 2D_{16}^* \varkappa_x \varkappa_{xy} + 2D_{26}^* \varkappa_y \varkappa_{xy}) dx dy$$

There is no bending-extensional coupling in (7.81). This suggests that unsymmetrical laminated plate problems can be solved by using (2.64) in the uncoupled form

$$\begin{bmatrix} \epsilon^0 \\ M \end{bmatrix} = \begin{bmatrix} A^* & 0 \\ 0 & D^* \end{bmatrix} \begin{bmatrix} N \\ \varkappa \end{bmatrix} \qquad (7.82)$$

Thus, the governing equations of a symmetric plate can be utilized by substituting D_{ij}^* for the bending stiffness D_{ij}. From the previous examples presented in this chapter, it can be seen that coupling tends to reduce the effective stiffness of the plate, that is, bending deflections are increased while critical buckling loads and fundamental vibration frequencies are decreased compared to equivalent homogeneous orthotropic plates. Since $D^* = D - BA^{-1}B$, D^* represents a reduction in the bending stiffness of the plate. We will refer to D^*_{ij} as the "reduced bending stiffness." Thus, one might expect the reduced bending stiffness to give a reasonable estimate of the gross response of an unsymmetrical laminated plate. Even though Equation (7.81) is uncoupled, coupling will occur through the compatibility Equation (2.66). Hopefully the violation of compatibility will only introduce small error in the calculation of laminate response.

When $N_x = N_y = N_{xy} = 0$, a cursory examination of Equation (2.64) reveals that the reduced bending stiffness is an exact solution of the problem. This is the case for the circular plate discussed in Section 7.4. For the cylindrical bending problems in Section 4.2, the maximum bending deflection is of the form

$$\frac{A_{11}K}{(A_{11}D_{11} - B_{11}^2)} \tag{7.83}$$

where K is a constant depending on the edge conditions of the strip. If the reduced bending stiffness approximation is used, Equation (7.83) becomes

$$\frac{(A_{11} - A_{12}F)K}{A_{11}D_{11} - (B_{11}^2 + D_{11}A_{12}F - B_{12}B_{11}F)} \tag{7.84}$$

where

$$F = A_{12}/A_{22}$$

Since A_{12} involves Poisson's ratio and A_{22} involves tensile moduli, the terms in (7.84) involving F should be small compared to the other elastic stiffness terms. If these terms are neglected, Equation (7.84) reduces to (7.83). Thus the proposed approximation seems to yield satisfactory results.

If the reduced bending stiffness approximation is valid, the solution to coupled plate problems should be independent of the inplane boundary conditions. Table 7.2 shows a comparison between bending deflections obtained for hinge-free normal supports (HFN), hinge-free tangential supports (HFT), simple-supports (SS), and the reduced bending stiffness method (RBS) for cross-ply and angle-ply plates under uniform transverse load. The hinge-support results are obtained from Reference [1], while the simple-support results are obtained from a Fourier analysis presented in Reference [8]. In each one of these cases, the bending edge conditions were the same (i.e., $w = M_n = 0$ on all edges). Thus, the only difference is the imposed inplane boundary conditions. In all cases the reduced bending stiffness method agrees very well with the simple-support results. The

Table 7.2. Comparison of center deflections for 2-layer laminates under uniform transverse load.

Cross-Ply Laminates ($E_L/E_T = 40$, $\nu_{LT} = 0.25$, $G_{LT}/E_T = 0.5$) w_{center} $E_T h^3/q_0 b^4 \times 10^2$		
R	**HFN**	**SS**	**RBS**
1	1.1249	1.0874	1.0874
3	2.4364	2.4354	2.4368
5	2.3555	2.3728	2.3799

Cross-Ply Laminates ($E_L/E_T = 40$, $\nu_{LT} = 0.25$, $G_{LT}/E_T = 0.5$) w_{center} $E_T h^3/q_0 b^4 \times 10^3$		
	HFT	**SS**	**RBS**
$15°$	7.1416	8.7934	8.8268
$30°$	7.7576	8.0678	8.0723
$45°$	7.3217	7.3241	7.3245

approximate method also compares favorably with those obtained for hinge-supports except for the $\pm 15°$ angle-ply plate where the difference is greater than 20%. Similar comparison for flexural vibrations and buckling shows good agreement between the three different solutions.

Stresses in an unsymmetric laminate can also be approximated by the reduced bending stiffness method. Assuming $\Phi = 0$, Equation (2.74) yields the following middle plane displacements:

$$u^0 = - B_{11}^* \frac{\partial w}{\partial x} - 2B_{12}^* \int \frac{\partial^2 w}{\partial y^2} dx - 2B_{16}^* \frac{\partial w}{\partial y} + f(y) \qquad (7.85)$$

$$v^0 = - 2B_{26}^* \frac{\partial w}{\partial x} - B_{21}^* \int \frac{\partial^2 w}{\partial x^2} dy - B_{22}^* \frac{\partial w}{\partial y} + g(x) \qquad (7.86)$$

With the displacements determined, the stresses in each layer can be calculated from Equations (2.54)–(2.58). This approach should result in a reasonable estimate of the stresses.

It appears that the approximate uncoupling through the reduced bending stiffness method, which greatly reduces the labor involved in solving unsymmetrically laminated plate problems, should yield acceptable solutions for simply-supported plates, which represent the most important case in practical application. The applicability of this simplified method for the general case of unsymmetrical laminates remains to be demonstrated. Further discussion of this approximation can be found in Reference [9].

REFERENCES

1. Whitney, J. M. "Bending-Extensional Coupling in Laminated Plates Under Transverse Loading," *J. Composite Materials*, 3:20–28 (1969).

2. Timoshenko, S. P. and S. Woinowsky-Krieger. *Theory of Plates and Shells*, Second Edition. McGraw-Hill: New York (1961).

3. Whitney, J. M. and A. W. Leissa. "Analysis of Heterogeneous Anisotropic Plates," *Journal of Applied Mechanics*, 36:261–266 (1969).

4. Kicher, T. P. "The Analysis of Unbalanced Cross-Plied Elliptic Plates Under Uniform Pressure," *J. Composite Materials*, 3:424–432 (1969).

5. Whitney, J. M. "Shear Buckling of Unsymmetrical Cross-Ply Plates," *J. Composite Materials*, 3:359–363 (1969).

6. Faddeva, V. N. *Computational Methods of Linear Algebra*, translated from Russian by C. D. Benster, Dover Publication, Inc., N.Y. (1959).

7. Reissner, E. and Y. Stavsky. "Bending and Stretching of Certain Types of Heterogeneous Aeolotropic Elastic Plates," *Journal of Applied Mechanics*, 28:402–408 (1961).

8. Whitney, J. M. and A. W. Leissa. "Analysis of a Simply-Supported Laminated Anisotropic Rectangular Plate," *AIAA Journal*, 7:28–33 (1970).

9. Ashton, J. E. "Approximate Solutions for Unsymmetrically Laminated Plates," *J. Composite Materials*, 3:189–191 (1969).

$$\mathrm{C}_{\text{HAPTER 8}}$$

Expansional Strain Effects in Laminated Plates

8.1 INTRODUCTION

VERY OFTEN THE properties of composite materials are compromised by the environment to which they are exposed. Among the environmental factors, those which induce expansional strains (volume change in the absence of surface tractions) are of particular concern. In the case of advanced composite structures, such phenomena are primarily caused by an increase in temperature, absorption by a polymeric matrix material of a swelling agent such as water vapor, and by sudden expansion of absorbed gases in the matrix. In addition to inducing residual stresses, expansional strains can affect the gross response characteristics of a composite structure. In particular, bending deflections, buckling loads, and vibration frequencies can be modified considerably by the presence of environmentally induced strains.

In this chapter, previously derived laminated plate equations will be modified to include the effect of expansional strains. Solutions to specific boundary value problems will be presented in order to show the effect of expansional strains on the bending, buckling, and free-vibration of laminated plates.

8.2 CONSTITUTIVE EQUATIONS

Consider a thin rectangular plate of constant thickness h composed of layers of orthotropic materials bonded together (see Figure 8.1). As in Chapter 2 the plate is referred to a standard cartesian coordinate system located in the midplane (x-y) with the z-axis normal to this plane. All of the other assumptions are identical to those in Section 2.1.

In order to include the effects of expansional strains, we use the following Duhamel-Neumann form of Hooke's law [1,2]:

$$\epsilon_i = \sum_{j=1}^{6} S_{ij}\sigma_j - \bar{\epsilon}_i \qquad (i = 1, 2, \ldots, 6) \qquad (8.1)$$

209

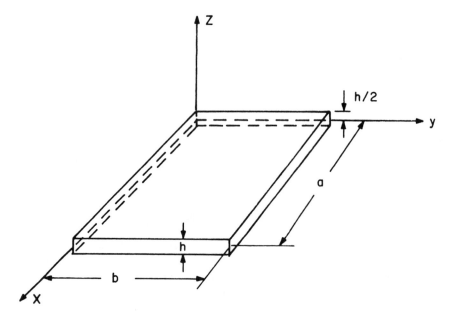

Figure 8.1. Coordinate system of plate.

where S_{ij} are the anisotropic compliances and $\bar{\epsilon}_i$ are generalized expansional strains, that is,

$$\bar{\epsilon}_i = \bar{\epsilon}_{i(\text{thermal})} + \bar{\epsilon}_{i(\text{swelling})} + \ldots \quad (8.2)$$

For thermal expansion,

$$\bar{\epsilon}_i = \alpha_i \, \Delta T(x,y,z,t) \quad (8.3)$$

where α_i are the linear coefficients of thermal expansion, and T denotes temperature. For swelling,

$$\bar{\epsilon}_i = \beta_i \, \Delta C(x,y,z,t) \quad (8.4)$$

where β_i are the linear swelling coefficients, and C is the concentration of the swelling agent. Thus, in general, $\bar{\epsilon}_i = \bar{\epsilon}_i(x,y,z,t)$ and is determined from principles of heat transfer in the case of thermal expansion and from consideration of physical chemistry concepts, such as Fick's Law [3], in the case of swelling.

The inverted form of Equation (8.1) is given by the relationship

$$\sigma_i = \sum_{j=1}^{6} C_{ij}(\epsilon_j - \bar{\epsilon}_j) \quad (i = 1,2, \ldots ,6) \tag{8.5}$$

where C_{ij} are the anisotropic stiffnesses. For plane stress, Equation (8.5) takes the form

$$\sigma_i = \sum_{j=1,2,6} Q_{ij}(\epsilon_j - \bar{\epsilon}_j) \quad (i = 1,2,6) \tag{8.6}$$

where Q_{ij} are the reduced stiffnesses for plane stress as defined in Section 1.7.

Since $\bar{\epsilon}_i$ is a strain, transformed values under a rotation in the x-y plane can be determined from Equation (1.11). Thus, using contracted notation, we obtain the following transformation relations for expansional strains under the inplane rotation shown in Figure 8.2.

$$
\begin{bmatrix} \bar{\epsilon}'_1 \\ \bar{\epsilon}'_2 \\ \bar{\epsilon}'_3 \\ \bar{\epsilon}'_4 \\ \bar{\epsilon}'_5 \\ \bar{\epsilon}'_6 \end{bmatrix}
=
\begin{bmatrix}
m^2 & n^2 & 0 & 0 & 0 & mn \\
n^2 & m^2 & 0 & 0 & 0 & -mn \\
0 & 0 & 1 & 0 & 0 & 0 \\
0 & 0 & 0 & m & -n & 0 \\
0 & 0 & 0 & n & m & 0 \\
-2mn & 2mn & 0 & 0 & 0 & (m^2 - n^2)
\end{bmatrix}
\begin{bmatrix} \bar{\epsilon}_1 \\ \bar{\epsilon}_2 \\ \bar{\epsilon}_3 \\ \bar{\epsilon}_4 \\ \bar{\epsilon}_5 \\ \bar{\epsilon}_6 \end{bmatrix}
$$

where, as in Section 1.2, $m = \cos\theta$ and $n = \sin\theta$. For the case under consideration the ply properties are known relative to the principal material axes, which are the rotated axes in Figure 8.2. Thus the inverted form of Equation (8.7) is required. In addition the properties relative to the principal material axes are ortho-

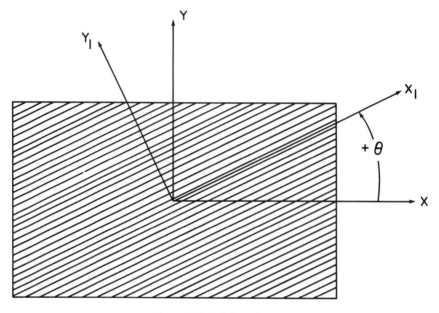

Figure 8.2. Rotation of axes.

tropic ($\bar{\epsilon}'_4 = \bar{\epsilon}'_5 = \bar{\epsilon}'_6 = 0$). Utilizing the assumption of an orthotropic material, the inverted form of Equation (8.7) is as follows:

$$
\begin{bmatrix} \bar{\epsilon}_1 \\ \bar{\epsilon}_2 \\ \bar{\epsilon}_3 \\ \bar{\epsilon}_6 \end{bmatrix}
=
\begin{bmatrix} m^2 & m^2 & 0 \\ n^2 & m^2 & 0 \\ 0 & 0 & 1 \\ 2mn & -2mn & 0 \end{bmatrix}
\begin{bmatrix} \bar{\epsilon}'_1 \\ \bar{\epsilon}'_2 \\ \bar{\epsilon}'_3 \end{bmatrix}
\tag{8.8}
$$

$$
\bar{\epsilon}_4 = \bar{\epsilon}_5 = 0
\tag{8.9}
$$

For the case of plane stress we utilize the notation $x_L - x_T$ for the principal material axes, and Equations (8.8) can be expressed in the reduced form

$$
\begin{bmatrix} \bar{\epsilon}_1 \\ \bar{\epsilon}_2 \\ \bar{\epsilon}_6 \end{bmatrix}
=
\begin{bmatrix} m^2 & n^2 \\ n^2 & m^2 \\ 2mn & -2mn \end{bmatrix}
\begin{bmatrix} \bar{\epsilon}_L \\ \bar{\epsilon}_T \end{bmatrix}
\tag{8.10}
$$

where $\bar{\epsilon}_L$ and $\bar{\epsilon}_T$ denote expansional strains parallel and transverse to the principal material axis x_L. Thus the case of an off-axis orientation leads to an anisotropic shear coupling expansional strain $\bar{\epsilon}_6$ relative to the x-y plane.

Substituting the laminate strains (2.5) into Equation (8.6), we obtain the following ply stresses:

$$
\begin{bmatrix} \sigma_x^{(k)} \\ \sigma_y^{(k)} \\ \sigma_{xy}^{(k)} \end{bmatrix}
=
\begin{bmatrix} Q_{11}^{(k)} & Q_{12}^{(k)} & Q_{16}^{(k)} \\ Q_{12}^{(k)} & Q_{22}^{(k)} & Q_{26}^{(k)} \\ Q_{16}^{(k)} & Q_{26}^{(k)} & Q_{66}^{(k)} \end{bmatrix}
\begin{bmatrix} \epsilon_x^0 + z\varkappa_x \\ \epsilon_y^0 + z\varkappa_y \\ \epsilon_{xy}^0 + z\varkappa_{xy} \end{bmatrix}
$$

$$
-
\begin{bmatrix} Q_{11}^{(k)} & Q_{12}^{(k)} & Q_{16}^{(k)} \\ Q_{12}^{(k)} & Q_{22}^{(k)} & Q_{26}^{(k)} \\ Q_{16}^{(k)} & Q_{26}^{(k)} & Q_{66}^{(k)} \end{bmatrix}
\begin{bmatrix} \bar{\epsilon}_1^{(k)} \\ \bar{\epsilon}_2^{(k)} \\ \bar{\epsilon}_6^{(k)} \end{bmatrix}
\tag{8.11}
$$

Combining Equation (8.11) with the definitions for force and moment resultants (2.9), we arrive at the laminate constitutive relations

$$
\begin{bmatrix} N_x \\ N_y \\ N_{xy} \\ \hline M_x \\ M_y \\ M_{xy} \end{bmatrix}
=
\left[\begin{array}{ccc|ccc}
A_{11} & A_{12} & A_{16} & B_{11} & B_{12} & B_{16} \\
A_{12} & A_{22} & A_{26} & B_{12} & B_{22} & B_{26} \\
A_{16} & A_{26} & A_{66} & B_{16} & B_{26} & B_{66} \\
\hline
B_{11} & B_{12} & B_{16} & D_{11} & D_{12} & D_{16} \\
B_{12} & B_{22} & B_{26} & D_{12} & D_{22} & D_{26} \\
B_{16} & B_{26} & B_{66} & D_{16} & D_{26} & D_{66}
\end{array}\right]
\begin{bmatrix} \epsilon_x^0 \\ \epsilon_y^0 \\ \epsilon_{xy}^0 \\ \hline \varkappa_x \\ \varkappa_y \\ \varkappa_{xy} \end{bmatrix}
-
\begin{bmatrix} \bar{N}_x \\ \bar{N}_y \\ \bar{N}_{xy} \\ \hline \bar{M}_x \\ \bar{M}_y \\ \bar{M}_{xy} \end{bmatrix}
$$

$$(8.12)$$

where A_{ij}, B_{ij}, and D_{ij} are as previously defined by Equation (2.27), and the expansional force and moment resultants are defined as follows:

$$
(\bar{N}_x, \bar{M}_x) = \int_{-h/2}^{h/2} (Q_{11}^{(k)} \bar{\epsilon}_1^{(k)} + Q_{12}^{(k)} \bar{\epsilon}_2^{(k)} + Q_{16}^{(k)} \bar{\epsilon}_6^{(k)})(1,z)dz
$$

$$
(\bar{N}_y, \bar{M}_y) = \int_{-h/2}^{h/2} (Q_{12}^{(k)} \bar{\epsilon}_1^{(k)} + Q_{22}^{(k)} \bar{\epsilon}_2^{(k)} + Q_{26}^{(k)} \bar{\epsilon}_6^{(k)})(1,z)dz
$$

$$
(\bar{N}_{xy}, \bar{M}_{xy}) = \int_{-h/2}^{h/2} (Q_{16}^{(k)} \bar{\epsilon}_1^{(k)} + Q_{26}^{(k)} \bar{\epsilon}_2^{(k)} + Q_{66}^{(k)} \bar{\epsilon}_6^{(k)})(1,z)dz
$$

The presence of the expansional force and moment resultants in Equation (8.12) is the only distinguishing feature between the present laminate constitutive equations and the original relationships derived in Section 2.4 and displayed in Equation (2.26).

8.3 GOVERNING EQUATIONS

The equations of motion in terms of force and moment resulants were derived in Chapter 2 with the following results:

$$\frac{\partial N_x}{\partial x} + \frac{\partial N_{xy}}{\partial y} = \varrho \frac{\partial^2 u^0}{\partial t^2} \tag{8.13}$$

$$\frac{\partial N_{xy}}{\partial x} + \frac{\partial N_y}{\partial y} = \varrho \frac{\partial^2 v^0}{\partial t^2} \tag{8.14}$$

$$\frac{\partial^2 M_x}{\partial x^2} + 2 \frac{\partial^2 M_{xy}}{\partial x \partial y} + \frac{\partial^2 M_y}{\partial y^2} + N_x^i \frac{\partial^2 w}{\partial x^2} + 2 N_{xy}^i \frac{\partial^2 w}{\partial x \partial y}$$

$$+ N_y^i \frac{\partial^2 w}{\partial y^2} + q = \varrho \frac{\partial^2 w}{\partial t^2} \tag{8.15}$$

where, as previously, u^0, v^0, and w are the midplane displacements in the x, y, and z directions, respectively. In addition, ϱ is the integral of the density as defined by Equation (2.1), q is the surface traction defined by Equation (2.15), and N_x^i, N_y^i, and N_{xy}^i are prebuckling force resultants as developed in Section 2.7.

Substituting the constitutive relations (8.12) into the equilibrium Equations (8.13)–(8.15), and taking into account the strain-displacement relations (2.6) and (2.7), we obtain the following equations of motion in terms of displacements:

$$A_{11} \frac{\partial^2 u^0}{\partial x^2} + 2A_{16} \frac{\partial^2 u^0}{\partial x \partial y} + A_{66} \frac{\partial^2 u^0}{\partial y^2} + A_{16} \frac{\partial^2 v^0}{\partial x^2} + (A_{12} + A_{66}) \frac{\partial^2 v^0}{\partial x \partial y}$$

$$+ A_{26} \frac{\partial^2 v^0}{\partial y^2} - B_{11} \frac{\partial^3 w}{\partial x^3} - 3B_{16} \frac{\partial^3 w}{\partial x^2 \partial y} - (B_{12} + 2B_{66}) \frac{\partial^3 w}{\partial x \partial y^2} \tag{8.16}$$

$$- B_{26} \frac{\partial^3 w}{\partial y^3} - \frac{\partial \bar{N}_x}{\partial x} - \frac{\partial \bar{N}_{xy}}{\partial y} = \varrho \frac{\partial^2 u^0}{\partial t^2}$$

$$A_{16} \frac{\partial^2 u^0}{\partial x^2} + (A_{12} + A_{66}) \frac{\partial^2 u^0}{\partial x \partial y} + A_{26} \frac{\partial^2 u^0}{\partial y^2} + A_{66} \frac{\partial^2 v^0}{\partial x^2} + 2A_{26} \frac{\partial^2 v^0}{\partial x \partial y}$$

$$+ A_{22} \frac{\partial^2 v^0}{\partial y^2} - B_{16} \frac{\partial^3 w}{\partial x^3} - (B_{12} + 2B_{66}) \frac{\partial^3 w}{\partial x^2 \partial y} - 3B_{26} \frac{\partial^3 w}{\partial y^2 \partial x} \qquad (8.17)$$

$$- B_{22} \frac{\partial^3 w}{\partial y^3} - \frac{\partial \overline{N}_{xy}}{\partial x} - \frac{\partial \overline{N}_y}{\partial y} = \varrho \frac{\partial^2 v^0}{\partial t^2}$$

$$D_{11} \frac{\partial^4 w}{\partial x^4} + 4D_{16} \frac{\partial^4 w}{\partial x^3 \partial y} + 2(D_{12} + 2D_{66}) \frac{\partial^4 w}{\partial x^2 \partial y^2} + 4D_{26} \frac{\partial^4 w}{\partial x \partial y^3}$$

$$+ D_{22} \frac{\partial^4 w}{\partial y^4} - B_{11} \frac{\partial^3 w}{\partial x^3} - 3B_{16} \frac{\partial^3 u^0}{\partial x^2 \partial y} - (B_{12} + 2B_{66}) \frac{\partial^3 u^0}{\partial x \partial y^2}$$

$$\qquad (8.18)$$

$$- B_{26} \frac{\partial^3 u^0}{\partial y^3} - B_{16} \frac{\partial^3 v^0}{\partial x^3} - (B_{12} + 2B_{66}) \frac{\partial^3 v^0}{\partial x^2 \partial y} - 3B_{26} \frac{\partial^3 v^0}{\partial x \partial y^2}$$

$$- B_{22} \frac{\partial^3 v^0}{\partial y^3} + \frac{\partial^2 \overline{M}_x}{\partial x^2} + 2 \frac{\partial^2 \overline{M}_{xy}}{\partial x \partial y} + \frac{\partial^2 \overline{M}_y}{\partial y^2} + \varrho \frac{\partial^2 w}{\partial t^2}$$

$$= q + N_x^i \frac{\partial^2 w}{\partial x^2} + 2N_{xy}^i \frac{\partial^2 w}{\partial x \partial y} + N_y^i \frac{\partial^2 w}{\partial y^2}$$

It should be noted that for cases in which $\overline{\epsilon}_i$ are independent of x and y, expansional force and moment resultants do not enter directly into Equations (8.16)–(8.18). Boundary constraints, however, induce inplane loads which can have an effect on bending deflections, critical buckling loads, and free vibration frequencies by entering Equation (8.18) through the prebuckled force resultants N_x^i, N_y^i, and N_{xy}^i. This is the class of problems which will be considered in this chapter.

8.4 STRAIN ENERGY

We now consider the application of energy principles to the analysis of anisotropic laminated plates which include the effect of expansional strains. If the procedure outlined in Chapter 3 is utilized, then the governing equations and boundary conditions are determined from the variational equation given by (3.19). A cursory examination of this condition, however, reveals that the expansional strains only influence the strain energy, U. All other terms in Equation (3.19) remain the same.

The strain energy in the presence of expansional strains takes the form

$$U = \frac{1}{2} \int \int \int [\sigma_x(\epsilon_x - \bar{\epsilon}_1) + \sigma_y(\epsilon_y - \bar{\epsilon}_2) + \sigma_z(\epsilon_z - \bar{\epsilon}_3) + \sigma_{yz}(\epsilon_{yz} - \bar{\epsilon}_4)$$

$$+ \sigma_{xz}(\epsilon_{xz} - \bar{\epsilon}_5) + \sigma_{xy}(\epsilon_{xy} - \bar{\epsilon}_6)]dx\,dy\,dz \qquad (8.19)$$

where the triple integral is performed over the volume of the body. We note that the strain energy is written in terms of mechanical strain, i.e., the difference between total strain and expansional strain. As readily seen from Equation (8.6), these are the strains that produce stress. Free expansion does not induce stress.

Taking into account the assumed state of plane stress along with the ply stress-strain relations, Equation (8.6), we find that Equation (8.19) takes the form

$$U = \frac{1}{2} \int \int \int [Q_{11}^{(k)}(\epsilon_x - \bar{\epsilon}_1^{(k)})^2 + 2Q_{12}^{(k)}(\epsilon_x - \bar{\epsilon}_1^{(k)})(\epsilon_y - \bar{\epsilon}_2^{(k)})$$

$$+ 2Q_{16}^{(k)}(\epsilon_x - \bar{\epsilon}_1^{(k)})(\epsilon_{xy} - \bar{\epsilon}_6^{(k)}) + 2Q_{26}^{(k)}(\epsilon_y - \bar{\epsilon}_2^{(k)})(\epsilon_{xy} \qquad (8.20)$$

$$- \bar{\epsilon}_6^{(k)}) + Q_{22}^{(k)}(\epsilon_y - \bar{\epsilon}_2^{(k)})^2 + Q_{66}^{(k)}(\epsilon_{xy} - \bar{\epsilon}_6^{(k)})^2]dx\,dy\,dz$$

Substituting the strain-displacement relations (2.5)–(2.7) into Equation (8.20) and integrating through-the-plate thickness, we obtain the double integral

$$U = \frac{1}{2} \int \int \left\{ A_{11}\left(\frac{\partial u^o}{\partial x}\right)^2 + 2A_{12}\frac{\partial u^o}{\partial x}\frac{\partial v^o}{\partial y} + A_{22}\left(\frac{\partial v^o}{\partial y}\right)^2 + 2\left(A_{16}\frac{\partial u^o}{\partial x}\right.\right.$$

$$\left. + A_{26}\frac{\partial v^o}{\partial y}\right)\left(\frac{\partial u^o}{\partial y} + \frac{\partial v^o}{\partial x}\right) + A_{66}\left(\frac{\partial u^o}{\partial y} + \frac{\partial v^o}{\partial x}\right)^2 - 2\bar{N}_x\frac{\partial u^o}{\partial x}$$

$$(8.21)$$

$$- 2\bar{N}_y\frac{\partial v^o}{\partial y} - 2\bar{N}_{xy}\left(\frac{\partial u^o}{\partial y} + \frac{\partial v^o}{\partial x}\right) - B_{11}\frac{\partial u^o}{\partial x}\frac{\partial^2 w}{\partial x^2} - 2B_{12}\left(\frac{\partial v^o}{\partial y}\frac{\partial^2 w}{\partial x^2}\right.$$

$$\left. + \frac{\partial u^o}{\partial x}\frac{\partial^2 w}{\partial y^2}\right) - B_{22}\frac{\partial v^o}{\partial y}\frac{\partial^2 w}{\partial y^2} - 2B_{16}\left[\frac{\partial^2 w}{\partial x^2}\left(\frac{\partial u^o}{\partial x} + \frac{\partial v^o}{\partial x}\right)\right.$$

$$\left. + 2\frac{\partial u^o}{\partial x}\frac{\partial^2 w}{\partial x\partial y}\right] - 2B_{26}\left[\frac{\partial^2 w}{\partial y^2}\left(\frac{\partial u^o}{\partial y} + \frac{\partial v^o}{\partial x}\right) + 2\frac{\partial v^o}{\partial y}\frac{\partial^2 w}{\partial x\partial y}\right]$$

$$- 4B_{66} \frac{\partial^2 w}{\partial x \partial y} \left(\frac{\partial u^0}{\partial y} + \frac{\partial v^0}{\partial x} \right) + D_{11} \left(\frac{\partial^2 w}{\partial x^2} \right)^2 + 2D_{12} \frac{\partial^2 w}{\partial x^2} \frac{\partial^2 w}{\partial y^2}$$

$$+ D_{22} \left(\frac{\partial^2 w}{\partial y^2} \right)^2 + 4 \left(D_{16} \frac{\partial^2 w}{\partial x^2} + D_{26} \frac{\partial^2 w}{\partial y^2} \right) \frac{\partial^2 w}{\partial x \partial y} + 4D_{66} \left(\frac{\partial^2 w}{\partial x \partial y} \right)^2$$

$$+ 2\bar{M}_x \frac{\partial^2 w}{\partial x^2} + 4\bar{M}_{xy} \frac{\partial^2 w}{\partial x \partial y} + 2\bar{M}_y \frac{\partial^2 w}{\partial y^2} \Bigg\} \, dx \, dy$$

$$+ \int \int \left[\int_{-h/2}^{h/2} f(Q_{ij}^{(k)}, \bar{\epsilon}_i^{(k)}) dz \right] dx \, dy$$

where

$$f(Q_{ij}^{(k)}, \bar{\epsilon}_i^{(k)}) = Q_{11}^{(k)} (\bar{\epsilon}_1^{(k)})^2 + 2Q_{12}^{(k)} \bar{\epsilon}_1^{(k)} \bar{\epsilon}_2^{(k)} + 2Q_{16}^{(k)} \bar{\epsilon}_1^{(k)} \bar{\epsilon}_6^{(k)}$$

$$+ 2Q_{26}^{(k)} \bar{\epsilon}_2^{(k)} \bar{\epsilon}_6^{(k)} + Q_{22}^{(k)} (\bar{\epsilon}_2^{(k)})^2 + Q_{66}^{(k)} (\bar{\epsilon}_6^{(k)})^2$$

Since $f(Q_{ij}^{(k)}, \bar{\epsilon}_i^{(k)})$ is independent of the displacements u^0, v^0, and w, the integral involving this function will vanish under the first variation, δU. Substituting Equation (8.21) into Equation (3.19) along with the expressions for T, V, and W as given by Equations (3.13), (3.15), and (3.18), respectively, we obtain the Equations of motion (8.16)–(8.18). The boundary conditions are identical to those given by Equations (3.33) and (3.34).

8.5 MIDPLANE SYMMETRIC LAMINATES

In this section we consider the effect of expansional strains on bending under transverse load, buckling, and free vibration of symmetric laminates. For the general case of midplane symmetric laminates, solutions cannot be obtained in closed form due to the presence of the bending-twisting coupling stiffness terms D_{16} and D_{26}. However, approximate solutions can be obtained by the Ritz method.

We will consider initial inplane force resultants induced by expansional strains in conjunction with boundary constraints. For the condition $\bar{\epsilon}_i^{(k)}$ = constant, the expansional force resultants are constant, and the expansional moment resultants vanish ($\bar{M}_x = \bar{M}_y = \bar{M}_{xy} = 0$). Under these conditions, Equation (8.21) is identical to Equation (3.4).

The appropriate energy criterion is deduced from Equations (3.7) and (3.40) in conjunction with Equations (3.35) and (3.37), which leads to the result:

$$\frac{1}{2} \int_0^b \int_0^a \left[D_{11} \left(\frac{\partial^2 w}{\partial x^2}\right)^2 + 2D_{12} \frac{\partial^2 w}{\partial x^2} \frac{\partial^2 w}{\partial y^2} + D_{22} \left(\frac{\partial^2 w}{\partial y}\right)^2 \right.$$

$$+ 4D_{66} \left(\frac{\partial^2 w}{\partial x \partial y}\right)^2 + 4D_{16} \frac{\partial^2 w}{\partial x \partial y} \frac{\partial^2 w}{\partial x^2} + 4D_{26} \frac{\partial^2 w}{\partial x \partial y} \frac{\partial^2 w}{\partial y^2} \qquad (8.22)$$

$$\left. + N_x \left(\frac{\partial^2 w}{\partial x^2}\right)^2 + N_y \left(\frac{\partial^2 w}{\partial y^2}\right)^2 + 2N_{xy} \frac{\partial w}{\partial x} \frac{\partial w}{\partial y} - (2q + \varrho \omega^2)w \right]$$

$$\times \; dx \, dy = \text{stationary value}$$

We assume the solution to Equation (8.22) in the form of a series with undetermined coefficients:

$$w = \sum_{m=1}^{M} \sum_{n=1}^{N} A_{mn} X_m(x) Y_n(y) \qquad (8.23)$$

where $X_m(x)$ and $Y_n(y)$ satisfy the displacement boundary conditions on the edges $x = 0, a$ and $y = 0, b$, respectively. Substituting the series (8.23) into the energy condition (8.22) and minimizing with respect to the coefficients A_{mn}, we obtain a set of $M \times N$ algebraic equations:

$$\sum_{i=1}^{M} \sum_{j=1}^{N} \left\{ D_{11} \int_0^a \frac{d^2 X_i}{dx^2} \frac{d^2 X_m}{dx^2} \, dx \int_0^b Y_j Y_n \, dy \right.$$

$$+ D_{12} \left[\int_0^a X_m \frac{d^2 X_i}{dx^2} \, dx \int_0^b Y_j \frac{d^2 Y_n}{dy^2} \, dy \right.$$

$$\qquad (8.24)$$

$$\left. + \int_0^a X_i \frac{d^2 X_m}{dx^2} \, dx \int_0^b Y_n \frac{d^2 Y_j}{dy^2} \, dy \right]$$

$$+ D_{22} \int_0^a X_i X_m \, dx \int_0^b \frac{d^2 Y_j}{dy^2} \frac{d^2 Y_n}{dy^2} \, dy$$

$$+ \, 4D_{66} \int_0^a \frac{dX_i}{dx} \, \frac{dX_m}{dx} \, dx \int_0^b \frac{dY_j}{dy} \, \frac{dY_n}{dy} \, dy$$

$$+ \, 2D_{16} \left[\int_0^a \frac{d^2X_i}{dx^2} \, \frac{dX_m}{dx} \, dx \int_0^b Y_n \frac{dY_j}{dy} \, dy \right.$$

$$+ \int_0^a \frac{dX_i}{dx} \, \frac{d^2X_m}{dx^2} \, dx \int_0^b Y_n \frac{dY_j}{dy} \, dy \left. \right]$$

$$+ \, 2D_{26} \left[\int_0^a X_m \frac{dX_i}{dx} \, dx \int_0^b \frac{dY_j}{dy} \, \frac{d^2Y_n}{dy^2} \, dy \right.$$

$$+ \int_0^a X_i \frac{dX_m}{dx} \, dx \int_0^b \frac{d^2Y_j}{dy^2} \, \frac{dY_n}{dy} \, dy \left. \right]$$

$$+ \, N_x \int_0^a \frac{dX_i}{dx} \, \frac{dX_m}{dx} \, dx \int_0^b Y_j \, Y_n \, dy$$

$$+ \, N_{xy} \left[\int_0^a X_m \frac{dX_i}{dx} \, dx \int_0^b Y_j \frac{dY_n}{dy} \, dy \right.$$

$$+ \int_0^a X_i \frac{dX_m}{dx} \, dx \int_0^b Y_n \frac{dY_j}{dy} \, dy \left. \right]$$

$$\left. - \, \varrho\omega^2 \int_0^a X_i \, X_m \, dx \int_0^b Y_j \, Y_n \, dy \right\} A_{mn}$$

$$= q \int_0^a X_m \, dx \int_0^b Y_n \, dy \quad \begin{array}{l} m = 1,2,\dots,M \\ n = 1,2,\dots,N \end{array}$$

In the absence of lateral loads ($q = 0$), Equations (8.24) form a set of homogeneous equations for which a nontrivial solution can be obtained only if the determinant of the coefficient matrix is zero. This condition is sufficient to determine the natural frequencies in either the presence or the absence of inplane force resultants. If critical buckling loads are required, then we let $\omega = 0$ in

Equation (8.24), and the lowest combination of N_x, N_y, and N_{xy} which yields a zero determinant for the coefficient matrix corresponds to the desired solution. For the case of transverse loading, the algebraic equations resulting from (8.24) are solved directly for the coefficients A_{mn}, B_{mn}, and C_{mn}.

Following the previous analysis of anisotropic plates in Chapter 6, we use the characteristic shapes of freely vibrating beams for the approximating functions $X_m(x)$ and $Y_n(y)$.

8.6 BENDING OF UNSYMMETRIC ANGLE-PLY LAMINATES

Consider a rectangular laminate of the class $[\pm\theta]_n$. For this stacking geometry $A_{16} = A_{26} = D_{16} = D_{26} = 0$, and B_{16} and B_{26} are the only nonvanishing elements of the coupling stiffness matrix B_{ij}. Again the discussion is limited to cases where $\bar{\epsilon}_i^{(k)}$ = constant. Under the condition of constant expansional strains within each layer, $\bar{N}_x$, $\bar{N}_y$, and $\bar{M}_{xy}$ are constant and $\bar{N}_{xy} = \bar{M}_x = \bar{M}_y = 0$. The balanced nature of the laminates under consideration (same number of layers at $+\theta$ and $-\theta$) leads to the vanishing of $\bar{N}_{xy}$, while the unsymmetric angle-ply layers lead to a nonvanishing $\bar{M}_{xy}$.

For bending under transverse load in the presence of initial inplane loads, Equations (8.16)–(8.18) reduce to the following for the laminates under consideration:

$$A_{11}\frac{\partial^2 u^0}{\partial x^2} + A_{66}\frac{\partial^2 u^0}{\partial y^2} + (A_{12} + A_{66})\frac{\partial^2 v^0}{\partial x \partial y} - 3B_{16}\frac{\partial^3 w}{\partial x^2 \partial y}$$

$$- B_{26}\frac{\partial^3 w}{\partial y^3} = 0 \tag{8.25}$$

$$(A_{12} + A_{66})\frac{\partial^2 u^0}{\partial x \partial y} + A_{66}\frac{\partial^2 v^0}{\partial x^2} + A_{22}\frac{\partial^2 v^0}{\partial y^2} - B_{16}\frac{\partial^3 w}{\partial x^3}$$

$$- 3B_{26}\frac{\partial^3 w}{\partial x \partial y^2} = 0 \tag{8.26}$$

$$D_{11}\frac{\partial^4 w}{\partial x^4} + 2(D_{12} + 2D_{66})\frac{\partial^4 w}{\partial x^2 \partial y^2} + D_{22}\frac{\partial^4 w}{\partial y^4} - 3B_{16}\frac{\partial^3 u^0}{\partial x^2 \partial y}$$

$$- B_{26}\frac{\partial^3 u^0}{\partial y^3} - B_{16}\frac{\partial^3 v^0}{\partial x^3} - 3B_{26}\frac{\partial^3 v^0}{\partial x \partial y^2} \tag{8.27}$$

$$= q + N_x^i \frac{\partial^2 w}{\partial x^2} + 2N_{xy}^i \frac{\partial^2 w}{\partial x \partial y} + N_y^i \frac{\partial^2 w}{\partial y^2}$$

It should be noted that any expansional strain distribution which is independent of x and y and an even function of z will produce the preceding expansional resultants as well as the ones discussed in Section 8.5 for symmetric laminates.

For a rectangular plate with N_x^i, N_y^i, and N_{xy}^i constant, solutions to Equations (8.25)–(8.27) are sought in the form of the following double Fourier series:

$$u^0 = \sum_{m=1}^{M} \sum_{n=1}^{N} A_{mn} \sin \frac{m\pi x}{a} \cos \frac{n\pi y}{b} \tag{8.28}$$

$$v^0 = \sum_{m=1}^{M} \sum_{n=1}^{N} B_{mn} \cos \frac{m\pi x}{a} \sin \frac{n\pi y}{b} \tag{8.29}$$

$$w = \sum_{m=1}^{M} \sum_{n=1}^{N} C_{mn} \sin \frac{m\pi x}{a} \sin \frac{n\pi y}{b} \tag{8.30}$$

These displacement functions satisfy the following simply-supported boundary conditions:

at $x = 0$ and a

$$u^0 = N_{xy} = A_{66} \left(\frac{\partial u^0}{\partial y} + \frac{\partial v^0}{\partial x} \right) - B_{16} \frac{\partial^2 w}{\partial x^2} - B_{26} \frac{\partial^2 w}{\partial y^2} = 0 \tag{8.31}$$

$$w = M_x = B_{16} \left(\frac{\partial u^0}{\partial y} + \frac{\partial v^0}{\partial x} \right) - D_{11} \frac{\partial^2 w}{\partial x^2} - D_{12} \frac{\partial^2 w}{\partial y^2} = 0 \tag{8.32}$$

at $y = 0$ and b

$$v^0 = N_{xy} = 0 \tag{8.33}$$

$$w = M_y = B_{26} \left(\frac{\partial u^0}{\partial y} + \frac{\partial v^0}{\partial x} \right)$$

$$- D_{12} \frac{\partial^2 w}{\partial x^2} - D_{22} \frac{\partial^2 w}{\partial y^2} = 0 \tag{8.34}$$

Combining the constitutive relations (8.12) with the strain-displacement relations (2.6) and (2.7), we obtain the following inplane force resultants:

$$N_x = -\bar{N}_x + A_{11} \frac{\partial u^0}{\partial x} + A_{12} \frac{\partial v^0}{\partial y} - 2B_{16} \frac{\partial^2 w}{\partial x \partial y} \tag{8.35}$$

$$N_y = -\bar{N}_y + A_{12} \frac{\partial u^0}{\partial x} + A_{22} \frac{\partial v^0}{\partial y} - 2B_{26} \frac{\partial^2 w}{\partial x \partial y} \tag{8.36}$$

$$N_{xy} = A_{66} \left(\frac{\partial^2 u^0}{\partial x} + \frac{\partial v^0}{\partial y} \right) - B_{16} \frac{\partial^2 w}{\partial x^2} - B_{26} \frac{\partial^2 w}{\partial y^2} \tag{8.37}$$

Substituting Equations (8.28)–(8.30) into Equations (8.35)–(8.37), we obtain the results:

$$N_x = -\bar{N}_x + \frac{\pi}{a} \sum_{m=1}^{M} \sum_{n=1}^{N} (A_{11}mA_{mn} + A_{12}nRB_{mn}$$

$$- 2B_{16} \frac{mn\pi}{b} C_{mn}) \cos \frac{m\pi x}{a} \cos \frac{n\pi y}{b} \tag{8.38}$$

$$N_y = -\bar{N}_y + \frac{\pi}{a} \sum_{m=1}^{M} \sum_{n=1}^{N} (A_{12}mA_{mn} + A_{12}nRB_{mn}$$

$$- 2B_{26} \frac{mn\pi}{b} C_{mn}) \cos \frac{m\pi x}{a} \cos \frac{n\pi y}{b} \tag{8.39}$$

$$N_{xy} = -\frac{\pi}{a} \sum_{m=1}^{M} \sum_{n=1}^{N} [A_{66}(mA_{mn} + nB_{mn})$$

$$+ \frac{\pi}{a} (B_{16}m^2 + n^2R^2)C_{mn}] \sin \frac{m\pi x}{a} \sin \frac{n\pi y}{b} \tag{8.40}$$

where R is the plate aspect ratio a/b. The expansional force resultants in Equations (8.38)–(8.40) are induced by boundary constraints, while the double Fourier series terms are a result of the lateral load q. We are interested only in the effect of inplane forces which are directly related to expansional strains. Thus, for the present case,

$$N_x^i = -\bar{N}_x, \; N_y^i = -\bar{N}_y \; N_{xy}^i = 0 \qquad (8.41)$$

In order to utilize the displacement functions (8.28)–(8.30), we must expand the lateral load, q, into the double Fourier series

$$q(x,y) = \sum_{m=1}^{M} \sum_{n=1}^{N} q_{mn} \sin \frac{m\pi x}{a} \sin \frac{n\pi y}{b} \qquad (8.42)$$

where

$$q_{mn} = \frac{4}{ab} \int_0^b \int_0^a q(x,y) \sin \frac{m\pi x}{a} \sin \frac{n\pi y}{b} \; dx \; dy \qquad (8.43)$$

For a uniform load $q = q_0 = $ constant, integration of Equation (8.43) yields

$$q_{mn} = \frac{16q_0}{\pi^2 mn} \; (m,n \text{ odd})$$

$$q_{mn} = 0 \qquad (m,n \text{ even}) \tag{8.44}$$

Substituting Equations (8.28–8.30), (8.41), and (8.42) into Equations (8.25)–(8.27), and solving for the resulting algebraic equations for the coefficients A_{mn}, B_{mn}, and C_{mn}, we obtain the following results:

$$A_{mn} = q_{mn} \frac{R^4 b^3 n}{\pi^3 D_{mn}} \; [(A_{66}m^2 + A_{22}n^2 R^2)(3B_{16}m^2 + B_{26}n^2 R^2)$$

$$- m^2(A_{12} + A_{66})(B_{16}m^2 + 3B_{26}n^2 R^2)] \tag{8.45}$$

$$B_{mn} = q_{mn} \frac{R^3 b^3 m}{\pi^3 D_{mn}} \; [(A_{11}m^2 + A_{66}n^2 R^2)(B_{16}m^2 + 3B_{26}n^2 R^2)$$

$$- n^2 R^2(A_{12} + A_{66})(3B_{16}m^2 + B_{26}n^2 R^2)] \tag{8.46}$$

$$C_{mn} = q_{mn} \frac{R^4 b^4}{\pi^4 D_{mn}} [(A_{11}m^2 + A_{66}n^2R^2)(A_{66}m^2 + A_{22}n^2R^2)$$

$$- (A_{12} + A_{66})^2 \, m^2 n^2 R^2] \tag{8.47}$$

where

$$D_{mn} = \{[(A_{11}m^2 + A_{66}n^2R^2)(A_{66}m^2 + A_{22}n^2R^2)$$

$$- (A_{12} + A_{66})^2 m^2 n^2 R^2][D_{11}m^4 + 2(D_{12} + 2D_{66})m^2 n^2 R^2$$

$$+ D_{22}n^4 R^4 - \frac{\pi^2}{a^2} (\bar{N}_x m^2 + \bar{N}_y n^2 R^2)] + 2m^2 n^2 R^2 (A_{12}$$

$$+ A_{66})(3B_{16}m^2 + B_{26}n^2R^2)(B_{16}m^2 + 3B_{26}n^2R^2)$$

$$- n^2 R^2 (A_{66}m^2 + A_{22}n^2R^2)(3B_{16}m^2 + B_{26}n^2R^2)^2$$

$$- m^2 (A_{11}m^2 + A_{66}n^2R^2)(B_{16}m^2 + 3B_{26}n^2R^2)^2\}$$

It should be noted that values of $\bar{N}_x$ and $\bar{N}_y$ which lead to the vanishing of D_{mn} represent buckling loads, i.e., A_{mn}, B_{mn}, and C_{mn} become undefined when D_{mn} vanishes. The lowest values of $\bar{N}_x$ and $\bar{N}_y$ corresponding to $D_{mn} = 0$ are the critical buckling loads. Furthermore D_{mn} represents the determinant of the coefficient matrix generated from the algebraic equations produced by substituting Equations (8.28)–(8.30) and (8.41) into Equations (8.25)–(8.27). Thus, as in the case of homogeneous plates, the lateral load q has no effect on the critical buckling load.

Equations (8.45)–(8.47) are of exactly the same form as Equations (7.18) which were derived in conjunction with the bending of unsymmetric angle-ply plates subjected to transverse loading. The distinguishing difference is the existence of the expansional force resultants $\bar{N}_x$ and $\bar{N}_y$ in the definition of D_{mn} for the case currently under consideration.

8.7 THERMAL BUCKLING

Let us first consider symmetric angle-ply laminates of the class $[\pm \theta]_s$. It has been shown by Halpin and Pagano [1] that certain orientations of highly anisotropic materials produce a negative expansional strain relative to the x-axis. In particular, high values of E_L/E_T in conjunction with lamination theory can pro-

duce negative effective expansional coefficients for a range of angle-ply orientations. For example, Reference [1] shows an experimentally measured decrease of
10% in a $[\pm 15°]_s$ fiber reinforced rubber laminate subjected to a swelling agent.
For this material $E_L/E_T = 132$. The exact value of E_L/E_T above for which
negative expansions can be demonstrated will depend on the ratio of the axial expansional strain of the fiber to the expansional strain of the matrix ($\bar{\epsilon}_{fL}/\bar{\epsilon}_m$).

As shown in Reference [1], the effective thermal expansion coefficients for
boron/epoxy and glass/epoxy symmetric angle-ply laminates are all positive.
However, graphite fibers have a negative coefficient of thermal expansion in the
axial direction which leads to a negative effective thermal expansion coefficient in
a unidirectional graphite/epoxy composite in the fiber direction ($\alpha_L < 0$). This
fact in conjunction with a relatively high value of E_L/E_T yields a range of orientations for which graphite/epoxy symmetric angle-ply laminates have a negative
coefficient of thermal expansion. From a practical standpoint this implies that
certain orientations of graphite/epoxy and angle-ply plates having inplane boundary constraints can be buckled by lowering the temperature rather than raising it.

To illustrate this phenomenon, let us consider angle-ply plates constructed of
plies of graphite/epoxy material with the following unidirectional properties:

$$E_L/E_m = 48, \quad E_T/E_m = 2.2, \quad G_{LT}/E_m = 1.3$$

$$\nu_{LT} = 0.26, \quad \alpha_L/\alpha_m = -0.013, \quad \alpha_T/\alpha_m = 0.76$$

(8.48)

where the subscript m denotes matrix properties. If the laminate is clamped at the
ends $x = 0, a$ and free on the other two edges, then the following boundary conditions are appropriate:

at $x = 0$ and a

$$w = \frac{\partial w}{\partial x} = 0$$

(8.49)

at $y = 0$ and b

$$M_y = -D_{12}\frac{\partial^2 w}{\partial x^2} - 2D_{26}\frac{\partial^2 w}{\partial x \partial y} - D_{22}\frac{\partial^2 w}{\partial y^2} = 0$$

(8.50)

$$2\frac{\partial M_{xy}}{\partial x} + \frac{\partial M_y}{\partial y} = 2D_{16}\frac{\partial^3 w}{\partial x^3} + (D_{12} + 4D_{66})\frac{\partial^3 w}{\partial x^2 \partial y}$$

$$+ 4D_{26}\frac{\partial^3 w}{\partial x \partial y^2} + D_{22}\frac{\partial^3 w}{\partial y^3} = 0$$

(8.51)

As mentioned in Section 8.5, the characteristic shapes of freely vibrating beams are used in conjunction with Equation (8.24). Thus for the present problem we utilize the Ritz method in conjunction with the following assumed mode shapes:

$$X_m(x) = \gamma_m \cos \frac{\lambda_m x}{a} - \gamma_m \cosh \frac{\lambda_m x}{a} + \sin \frac{\lambda_m x}{a} - \sinh \frac{\lambda_m x}{a} \qquad (8.52)$$

$$Y_n(y) = \cos \frac{\lambda_n y}{b} + \cosh \frac{\lambda_n y}{b}$$

$$+ \gamma_n \sin \frac{\lambda_n y}{b} + \gamma_n \sinh \frac{\lambda_n y}{b} \qquad (8.53)$$

where λ_m and λ_n are the roots of the characteristic equation

$$\cos \lambda_i \cosh \lambda_i = 1 \qquad (8.54)$$

and

$$\gamma_i = \frac{\cos \gamma_i - \cosh \lambda_i}{\sin \lambda_i + \sinh \lambda_i} \qquad (8.55)$$

where $i = m,n$ in both Equations (8.54) and (8.55). Equation (8.52) represents the natural mode shapes of a vibrating beam with fixed ends, and Equation (8.53) represents the vibration modes of a free-free beam [4]. The first five values of λ_i which satisfy Equation (8.54) have been tabulated in Table 5.2. For $i > 2$, Equation (8.54) can be utilized, i.e.,

$$\lambda_i = (2i + 1) \frac{\pi}{2}, \, i > 2 \qquad (8.56)$$

As discussed in Section 6.8, the integrals occurring in Equation (8.24) have been tabulated in Reference [5].

The inplane loads can be determined from the constitutive relations (8.12). In particular for symmetric laminates the inplane constitutive relations become

$$
\begin{bmatrix} N_x \\ N_y \\ N_{xy} \end{bmatrix} = \begin{bmatrix} A_{11} & A_{12} & A_{16} \\ A_{12} & A_{22} & A_{26} \\ A_{16} & A_{26} & A_{66} \end{bmatrix} \begin{bmatrix} \epsilon_x^0 \\ \epsilon_y^0 \\ \epsilon_{xy}^0 \end{bmatrix} - \begin{bmatrix} \bar{N}_x \\ \bar{N}_y \\ \bar{N}_{xy} \end{bmatrix} \qquad (8.57)
$$

For the symmetric angle-ply laminate under consideration, $A_{16} = A_{26} = N_{xy} = 0$. In addition, the clamped ends at $x = 0,a$ imply $\epsilon_x^0 = 0$, while the free-edges at $y = 0, b$, imply that $N_y = N_{xy} = 0$. Combining these results with Equation (8.57) we find that

$$N_x = -\frac{(A_{22}\bar{N}_x - A_{12}\bar{N}_y)}{A_{22}} \qquad (8.58)$$

Thus for the boundary constraints under consideration, the inplane load N_x is a function of both $\bar{N}_x$ and $\bar{N}_y$.

Numerical results are illustrated in Figure 8.3. For orientations in which $0° < \theta < 17.5°$, the laminate has a negative thermal expansion coefficient relative to the x-axis. Thus, in this region temperature must be decreased in order

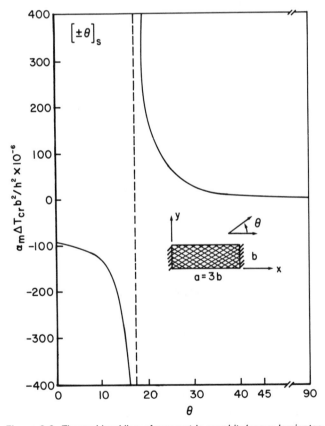

Figure 8.3. Thermal buckling of symmetric graphite/epoxy laminates.

to induce thermal buckling. At an orientation $\theta = 17.5°$ the effective thermal expansion coefficient in the x-direction vanishes, and the plate cannot buckle by either raising or lowering the temperature. For laminates with $17.5° \leq \theta \leq 90°$, the thermal expansion coefficient in the x-direction is positive, and the plate can buckle only by raising the temperature. It should be pointed out, however, that for most practical values of α_m and b/h, the critical buckling temperature is far too low to be of any significance. Thus for $0° \leq \theta \leq 17.5°$, it is virtually impossible to buckle the plate by either raising or lowering the temperature.

A cursory examination of Figure 8.3 reveals that the thermal expansion coefficient relative to the y-axis is positive when the expansion relative to the x-axis is negative and vice-versa. Thus if such laminates were constrained along all four edges, a temperature change would simultaneously induce both tension and compression in certain orientations. This combination would increase the thermal stability compared to laminates having positive thermal expansion coefficients relative to both the x and y axes.

To illustrate the mechanism discussed in the previous paragraph, we consider a simply-supported plate with the stacking geometry $[\pm\theta/0°]_s$. For simply-supported plates the following boundary conditions are appropriate:

at $x = 0$ and a

$$w = M_x = -D_{11} \frac{\partial^2 w}{\partial x^2} - 2D_{16} \frac{\partial^2 w}{\partial x \partial y} - D_{12} \frac{\partial^2 w}{\partial y^2} = 0 \qquad (8.59)$$

$$w = M_y = -D_{12} \frac{\partial^2 w}{\partial x^2} - 2D_{26} \frac{\partial^2 w}{\partial x \partial y} - D_{22} \frac{\partial^2 w}{\partial y^2} = 0 \qquad (8.60)$$

The following functions are utilized in conjunction with the Ritz method, Equation (8.24):

$$X_m(x) = \sin \frac{m\pi x}{a} \qquad (8.61)$$

$$Y_n(y) = \sin \frac{n\pi y}{b} \qquad (8.62)$$

These functions represent the natural modes of free vibration of a simply-supported beam. The simply-supported boundary conditions imply that the midplane strains vanish. Since $N_{xy} = 0$, then Equation (8.57) implies that

$$N_x = -\bar{N}_x, \quad N_y = -\bar{N}_y, \quad N_{xy} = 0 \qquad (8.63)$$

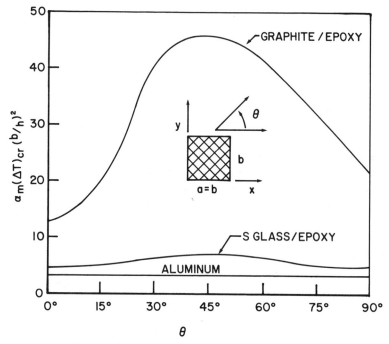

Figure 8.4. *Effect of material properties on thermal buckling.*

Numerical results are illustrated in Figure 8.4. In addition to the graphite/epoxy laminates with the ply properties as given in (8.48), we also consider glass/epoxy laminates with the following unidirectional properties:

$$E_L/E_m = 15, \quad E_T/E_m = 3.3, \quad G_{LT}/E_m = 1.2,$$

$$\nu_{LT} = 0.26, \quad \alpha_L/\alpha_m = 0.088, \quad \alpha_T/\alpha_m = 0.5 \tag{8.64}$$

As anticipated, because of the negative value of α_L/α_m, the graphite/epoxy composites have increased stability compared to glass/epoxy laminates which display positive thermal expansion coefficients relative to the x and y axes of the plate for all angle-ply orientations. Because of the low coefficient of expansion of both fibers considered, the fiber reinforced materials show increased stability compared to isotropic aluminum ($E/E_m = 20$, $\alpha/\alpha_m = 0.375$) for the plates under consideration. Analytical results show that the graphite/epoxy composites display a negative expansion coefficient in the x direction for laminates with $0° \le \theta \le 20°$. As a result, a higher buckling mode corresponding to a decrease in temperature exists in this region. As in the previous case (Figure 8.3), however, these critical temperatures are of little practical interest and are not shown in Figure 8.4.

The results in Figures 8.3 and 8.4 are normalized relative to properties of a conventional epoxy resin.

8.8 EFFECT OF SWELLING

We now consider polymeric matrix laminates which have been exposed to an increase in relative humidity for a long enough period of time for equilibrium to be obtained between the material and the environment. A polymeric matrix will often absorb moisture and swell. The amount of swelling depends on the degree of molecular crosslinking in the polymer. In general the fiber/matrix bond is subject to attack in a moisture environment, resulting in a loss in composite mechanical properties. Consideration in the present context, however, is limited to the phenomenon of swelling. As a result it is assumed that the matrix alone is affected by an increase in humidity.

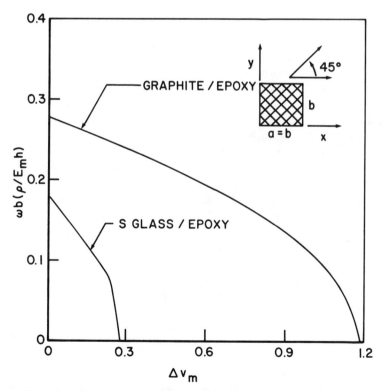

Figure 8.5. *Effect of matrix swelling on fundamental vibration frequencies.*

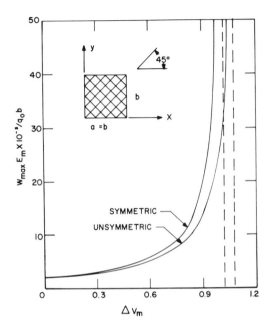

Figure 8.6. *Effect of matrix swelling on the maximum deflection of graphite/epoxy plates subjected to a uniform transverse load.*

Matrix swelling produces a gross effect which is directly analogous to thermal expansion. Thus inplane loads are induced for plates with edge constraints. As a result matrix swelling reduces fundamental vibration frequencies and increases deflection in plates subjected to lateral loads. This is illustrated in Figures 8.5 and 8.6 for both symmetric and unsymmetric simply-supported laminates.

Fundamental vibration frequencies and maximum deflection under uniform transverse load are presented as a function of the percent volume change in the matrix due to swelling in Figures 8.5 and 8.6, respectively, where

$$\Delta V_m = 100[(1 + \bar{\epsilon}_m)^3 - 1] \tag{8.65}$$

Using the micromechanics expressions developed by Shapery [5] with $\bar{\epsilon}_{fL} = \bar{\epsilon}_{fT} = 0$, we obtain the following relationships:

$$\frac{\bar{\epsilon}_L}{\bar{\epsilon}_m} = \frac{E_m}{E_L}(1 - V_f) \tag{8.66}$$

$$\frac{\bar{\epsilon}_T}{\bar{\epsilon}_m} = (1 + \nu_m - \nu_{LT}\frac{E_m}{e_L})(1 - V_f) \tag{8.67}$$

where ν_m is the Poisson ratio of the matrix and V_f denotes volume fraction of fiber. For the present examples under consideration, we assume $\nu_m = 0.35$.

The stacking geometry in Figure 8.5 is the same as considered in Figure 8.4, while results in Figure 8.6 are for symmetric and unsymmetric $\pm 45°$ laminates. In addition the plates are square $(a = b)$ and have a width-to-thickness ratio $a/h = 100$. Since we are simultaneously considering inplane loads in conjunction with either vibration frequencies or lateral loads, the ratio a/h cannot be normalized out of the solutions.

In Figure 8.5 matrix swelling reduces fundamental vibration frequencies. This is due to the fact that matrix swelling leads to positive expansional strains relative to the x and y axes for the laminates under consideration. As a result only compressive inplane loads are induced, which reduces the effective stiffness of the plate. The value of ΔV_m at which ω vanishes corresponds to the critical buckling strain. To illustrate the importance of swelling, consider that epoxy resins, although highly crosslinked, can undergo as much as a 2% volume change when immersed in water at room temperature for a long period of time. Such a volume change is of far more than sufficient magnitude to buckle the laminates shown in Figure 8.5. Thus it is obvious that long time exposure to an environment of moderate change in humidity can significantly affect laminate response. It should be pointed out that other resin systems, such as polyesters, which are not highly crosslinked are also used with glass fibers. For such cases, protection in a moisture environment becomes extremely important if phenomena such as "moisture buckling" are to be avoided.

As shown in Figure 8.6, matrix swelling in conjunction with boundary constraints induced by simple supports leads to an increase in bending deflections for both symmetric and unsymmetric $\pm 45°$ graphite/epoxy laminates. The bending deflections increase with increasing values of ΔV_m up to the critical buckling value at which point the deflection becomes unbounded. A cursory examination of Figure 8.6 reveals that matrix swelling must be less than 1.25% if buckling is to be prevented. Thus, as in the case of fundamental vibration frequencies, resin expansion in a humid environment can have serious effects on the bending of composite laminates.

REFERENCES

1. Halpin, J. C. and N. J. Pagano. "Consequences of Environmentally-Induced Dilatation in Solids," *Recent Advances in Engineering Science*, Gordon and Breach (1970).
2. Whitney, J. M. and J. E. Ashton. "Effect of Environment on the Elastic Response of Layered Composite Plates," *AIAA Journal*, 9:1708–1713 (1971).
3. Crank, J. *The Mathematics of Diffusion*, Second Edition. Oxford University Press (1975).
4. Timoshenko, S. P., D. H. Young and W. Weaver, Jr. *Vibration Problems in Engineering*. John Wiley & Sons (1974).
5. Schapery, R. A. "Thermal Expansion Coefficients of Composite Materials Based on Energy Principles," *Journal of Composite Materials*, 2:380–404 (1968).

CHAPTER 9

Laminated Cylindrical Plates

9.1 INTRODUCTION

IN THE PREVIOUS chapters we have considered bending under transverse load buckling, and free vibration of laminated plates with varying degrees of complexity. In many practical applications, such as in aircraft structures, we encounter plates having curvature in at least one direction. In the present chapter a theory is presented for the bending, buckling, and free vibration of laminated cylindrical (curved) plates. Consideration is given to plates of shallow curvature so that simplifying assumptions can be utilized in developing the governing equations.

Solutions are obtained for the bending, buckling, and free vibration of simply-supported cylindrical plates. Both symmetric and unsymmetric laminates are considered. The effect of curvature is ascertained by comparing cylindrical plate solutions to corresponding flat plate solutions.

9.2 CONSTITUTIVE EQUATIONS

Consider a cylindrical plate of constant radius R as illustrated in Figure 9.1. As in the case of flat plates, the thickness and inplane dimensions are denoted by h, a, and b, respectively. The displacements in the x, s, and z directions are denoted by u, v, and w, respectively. Assumptions are analogous to those listed in Section 2.1 for flat plates and are as follows:

1. The plate is constructed of an arbitrary number of orthotropic layers bonded together. The orthotropic axes of material symmetry, however, of an individual layer need not coincide with the x-s axes of the cylindrical plate.
2. The plate is thin, i.e., the thickness h is much smaller than the other physical dimensions.
3. The displacements u, v, and w are small compared to the plate thickness.
4. Inplane strains ϵ_x, ϵ_s, and ϵ_{xs} are small compared to unity.
5. The radius of the plate R is much larger than the thickness h.

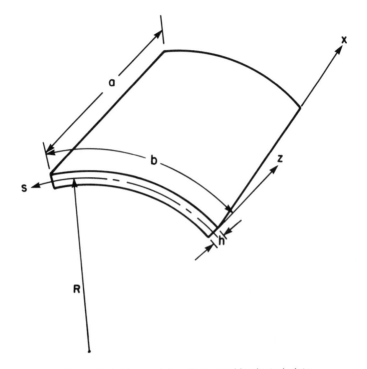

Figure 9.1. *Nomenclature for curved laminated plate.*

6. In order to include inplane force effects, nonlinear terms in the equations of motion involving products of stresses and plate slopes are retained. All other nonlinear terms are neglected.
7. Transverse shear strains ϵ_{xz} and ϵ_{sz} are negligible.
8. Tangential displacements u and v are linear functions of the z coordinate.
9. The transverse normal strain ϵ_z is negligible.
10. Each ply obeys Hooke's law.
11. The plate has constant thickness.
12. Rotatory inertia terms are negligible.
13. There are no body forces.
14. Transverse shear stresses σ_{xz} and σ_{sz} vanish on the surfaces $z = \pm\, h/2$.

The strain-displacement relations from classical theory of elasticity which are applicable to the coordinate system shown in Figure 9.1 are as follows [1]:

$$\epsilon_x = \frac{\partial u}{\partial x}, \ \epsilon_s = \frac{1}{(1 + z/R)}\frac{\partial v}{\partial s} + \frac{w}{R}, \ \epsilon_z = \frac{\partial w}{\partial z} \tag{9.1}$$

$$\epsilon_{sz} = \frac{1}{(1 + z/R)} \frac{\partial w}{\partial s} + \frac{\partial v}{\partial z} - \frac{v}{R(1 + z/R)}$$

$$\epsilon_{xz} = \frac{\partial u}{\partial z} + \frac{\partial w}{\partial x} , \quad \epsilon_{xs} = \frac{\partial v}{\partial x} + \frac{1}{(1 + z/R)} \frac{\partial u}{\partial s}$$

From assumption 7

$$u = u^o(x,s,t) + zF_1(x,s,t)$$

$$v = v^o(x,s,t) + zF_2(x,s,t)$$

(9.2)

where u^o and v^o are the axial and tangential displacements of the midplane, respectively. Substituting Equation (9.2) into the transverse shear strain-displacement relations given in (9.1), we obtain

$$\epsilon_{xz} = F_1 + \frac{\partial w}{\partial x} = 0$$

(9.3)

$$\epsilon_{sz} = \frac{1}{(1 + z/R)} \left(\frac{\partial w}{\partial s} + F_2 - \frac{v^o}{R} \right)$$

Thus

$$F_1 = - \frac{\partial w}{\partial x} , \quad F_2 = - \frac{\partial w}{\partial s} + \frac{v^o}{R}$$

(9.4)

Assumption 9 implies

$$w = w(x,s,t)$$

(9.5)

Combining Equations (9.2) and (9.4) with the strain-displacement relations (9.1), we obtain the following results:

$$\epsilon_x = \frac{\partial u^o}{\partial x} - z \frac{\partial^2 w}{\partial x^2} , \quad \epsilon_s = \frac{\partial v^o}{\partial s} + \frac{w}{R} - \frac{z}{(1 + z/R)} \frac{\partial^2 w}{\partial s^2}$$

$$\epsilon_{xs} = (1 + \frac{z}{R}) \frac{\partial v^o}{\partial s} + \frac{\partial u^o}{\partial s} - z(2 + \frac{z}{R}) \frac{\partial^2 w}{\partial x \partial s}$$

(9.6)

Since the plate is shallow ($R >> h$), z/R is small compared to unity. Thus,

$$(1 + z/R) \approx 1 \tag{9.7}$$

and Equation (9.6) can be written in the form

$$\epsilon_x = \epsilon_x^0 + z\varkappa_x$$

$$\epsilon_s = \epsilon_s^0 + z\varkappa_s \tag{9.8}$$

$$\epsilon_{xs} = \epsilon_{xs}^0 + z\varkappa_{xs}$$

where

$$\epsilon_x^0 = \frac{\partial u^0}{\partial x} , \ \epsilon_s^0 = \frac{\partial v^0}{\partial s} + \frac{w}{R} , \ \epsilon_{xs}^0 = \frac{\partial u^0}{\partial s} + \frac{\partial v^0}{\partial x}$$

$$\tag{9.9}$$

$$\varkappa_x = -\frac{\partial^2 w}{\partial x^2} , \ \varkappa_s = -\frac{\partial^2 w}{\partial s^2} , \ \varkappa_{xs} = -2\frac{\partial^2 w}{\partial x \partial s}$$

These results are of exactly the same form as Equation (2.6) with the exception of the w/R term appearing in the midplane tangential strain ϵ_s^0. The kinematic relations given by Equations (9.8) and (9.9) are equivalent to those derived by Donnell [2].

The ply constitutive relations are of the same form as Equation (2.24) and are as follows:

$$\begin{bmatrix} \sigma_x^{(k)} \\ \\ \sigma_s^{(k)} \\ \\ \sigma_{xs}^{(k)} \end{bmatrix} = \begin{bmatrix} Q_{11}^{(k)} & Q_{12}^{k} & Q_{16}^{(k)} \\ \\ Q_{12}^{(k)} & Q_{22}^{(k)} & Q_{26}^{(k)} \\ \\ Q_{16}^{(k)} & Q_{26}^{k} & Q_{66}^{(k)} \end{bmatrix} \begin{bmatrix} \epsilon_x \\ \\ \epsilon_s \\ \\ \epsilon_{xs} \end{bmatrix} \tag{9.10}$$

where Q_{ij} are the reduced stiffnesses for plane stress as defined by Equation (2.25). We define force and moment resultants in the same manner as given by Equation (2.9), i.e.

$$(N_x, N_s, N_{xs}) = \int_{-h/2}^{h/2} (\sigma_x^{(k)}, \sigma_s^{(k)}, \sigma_{xs}^{(k)}) \, dz \tag{9.11}$$

$$(M_x, M_s, M_{xs}) = \int_{-h/2}^{h/2} (\sigma_x^{(k)}, \sigma_s^{(k)}, \sigma_{xs}^{(k)}) z \ dz$$

Combining Equations (9.8) and (9.10), substituting the results into Equation (9.11), and performing the integrations, we arrive at the laminate constitutive relations which are of the same form as Equation (2.26), i.e. in abbreviated form

$$\begin{bmatrix} N \\ -- \\ M \end{bmatrix} = \begin{bmatrix} A & \vdots & B \\ --&-\vdots-&-- \\ B & \vdots & D \end{bmatrix} \begin{bmatrix} \epsilon^0 \\ -- \\ \varkappa \end{bmatrix} \tag{9.12}$$

where A_{ij}, B_{ij}, and D_{ij} are the stiffnesses defined in Equation (2.27).

9.3 GOVERNING EQUATIONS

We now consider the application of energy principles for the purpose of deriving the equations of motion. Applying the procedure discussed in Chapter 3, the governing equations and boundary conditions are determined from the variational equation given by (3.19).

The strain energy, U, for curved plates is of the same form as Equation (3.2), i.e. a state of plane stress is assumed. Thus,

$$U = \frac{1}{2} \int \int \int (Q_{11}^{(k)} \epsilon_x^2 + 2Q_{12}^{(k)} \epsilon_x \epsilon_s + 2Q_{16}^{(k)} \epsilon_x \epsilon_{xs} + 2Q_{26}^{(k)} \epsilon_s \epsilon_{xs}$$
$$+ Q_{22}^{(k)} \epsilon_s^2 + Q_{66}^{(k)} \epsilon_{xs}^2) dx \ ds \ dz \tag{9.13}$$

Substituting the kinematic relations (9.8) and (9.9) into Equation (9.13) and integrating with respect to z, we obtain the strain energy relationship

$$U = \frac{1}{2} \int \int \Bigg\{ A_{11} \left(\frac{\partial u^0}{\partial x} \right)^2 + 2A_{12} \frac{\partial u^0}{\partial x} \left(\frac{\partial v^0}{\partial s} + \frac{w}{R} \right) + A_{22} \Bigg[\frac{\partial v^0}{\partial s} \left(\frac{\partial v^0}{\partial s} \right. \Bigg.$$

$$\left. + \frac{w}{R} \right) + \left(\frac{w}{R} \right)^2 \Bigg] + 2 \Bigg[A_{16} \frac{\partial u^0}{\partial s} + A_{26} \left(\frac{\partial v^0}{\partial s} + \frac{w}{R} \right) \Bigg] \left(\frac{\partial u^0}{\partial s} + \frac{\partial v^0}{\partial x} \right)$$

$$+ A_{66} \left(\frac{\partial u^0}{\partial s} + \frac{\partial v^0}{\partial x} \right)^2 - B_{11} \frac{\partial u^0}{\partial x} \frac{\partial^2 w}{\partial x^2} - 2B_{12} \Bigg[\left(\frac{\partial v^0}{\partial s} + \frac{w}{R} \right) \frac{\partial^2 w}{\partial x^2} \tag{9.14}$$

$$+ \frac{\partial u^o}{\partial x}\frac{\partial^2 w}{\partial s^2}\Bigg] - B_{22}\left(\frac{\partial v^o}{\partial s} + \frac{w}{R}\right)\frac{\partial^2 w}{\partial s^2} - 2B_{16}\left[\frac{\partial^2 w}{\partial x^2}\left(\frac{\partial u^o}{\partial s} + \frac{\partial v^o}{\partial x}\right)\right.$$

$$+ 2\frac{\partial u^o}{\partial x}\frac{\partial^2 w}{\partial x \partial s}\Bigg] - 2B_{16}\left[\frac{\partial^2 w}{\partial x^2}\left(\frac{\partial u^o}{\partial s} + \frac{\partial v^o}{\partial x}\right) + 2\frac{\partial u^o}{\partial x}\frac{\partial^2 w}{\partial x \partial s}\right]$$

$$- 2B_{26}\left[\frac{\partial^2 w}{\partial s^2}\left(\frac{\partial u^o}{\partial s} + \frac{\partial v^o}{\partial x}\right) + 2\left(\frac{\partial v^o}{\partial s} + \frac{w}{R}\right)\frac{\partial^2 w}{\partial x \partial s}\right]$$

$$- 4D_{66}\frac{\partial^2 w}{\partial x \partial s}\left(\frac{\partial u^o}{\partial x} + \frac{\partial v^o}{\partial s}\right) + D_{11}\left(\frac{\partial^2 w}{\partial x^2}\right)^2 + 2D_{12}\frac{\partial^2 w}{\partial x^2}\frac{\partial^2 w}{\partial s^2}$$

$$+ D_{22}\left(\frac{\partial^2 w}{\partial s^2}\right)^2 + 4\left(D_{16}\frac{\partial^2 w}{\partial x^2} + D_{26}\frac{\partial^2 w}{\partial s^2}\right)\frac{\partial^2 w}{\partial x \partial s}$$

$$+ 4D_{66}\left(\frac{\partial^2 w}{\partial x \partial s}\right)^2\Bigg\} \, dx \, ds$$

Combining Equations (9.2) and (9.4), and substituting the results into Equation (3.10), we obtain the kinetic energy

$$T = \frac{1}{2}\int\int\int \varrho_0^{(k)}\left\{\left(\frac{\partial u^o}{\partial t} - z\frac{\partial^2 w}{\partial x \partial t}\right)^2 + \left[\left(1 + \frac{z}{R}\right)\frac{\partial v^o}{\partial t}\right.\right. \tag{9.15}$$

$$\left.\left. - z\frac{\partial^2 w}{\partial s \partial t}\right]^2 + \left(\frac{\partial w}{\partial t}\right)^2\right\} dx \, ds \, dz$$

where, as previously, $\varrho_0^{(k)}$ denotes the density of the kth layer. Integrating Equation (9.15) with respect to z, recognizing the approximation given by Equation (9.7), and neglecting time derivatives of plate slopes in accordance with assumption 12, we arrive at the expression

$$T = \frac{1}{2}\int\int \varrho\left[\left(\frac{\partial u^o}{\partial t}\right)^2 + \left(\frac{\partial v^o}{\partial t}\right)^2 + \left(\frac{\partial w}{\partial t}\right)^2\right] dx \, ds \tag{9.16}$$

where ϱ is the integral of the density through the plate thickness as defined by Equation (2.12).

The potential energy due to surface pressure is of the same form as Equation (3.14). Thus,

$$W = - \int \int p \, w \, dx \, ds \qquad (9.17)$$

where the surface pressure p is analogous to the transverse load q, i.e.

$$p = \sigma_z(h/2) - \sigma_z(-h/2) \qquad (9.18)$$

Thus for internal pressure $p = -\sigma_z(-h/2)$ and for external pressure $p = +\sigma_z(h/2)$.

The potential energy due to inplane loads takes the same form as Equation (3.16) and is given by

$$V = \int \int (N_x^i \epsilon_x' + N_s^i \epsilon_s' + N_{xs}^i \epsilon_{xs}') dx \, ds \qquad (9.19)$$

where N_x^i, N_s^i, and N_{xs}^i represent initial inplane force resultants applied to the plate in a prebuckled state. As in Equation (3.16), ϵ_x', ϵ_s', and ϵ_{xs}' are midplane strains associated with large deformation analysis. Considering the Green strain tensor in terms of the coordinate system shown in Figure 9.1 and retaining only the nonlinear terms involving plate slopes, we find [3]

$$\epsilon_x' = \frac{1}{2} \left(\frac{\partial w}{\partial w} \right)^2$$

$$\epsilon_y' = \frac{1}{2} \left(\frac{\partial w}{\partial s} \right)^2 \qquad (9.20)$$

$$\epsilon_{xs}' = \frac{\partial w}{\partial x} \frac{\partial w}{\partial s}$$

Combining Equations (9.19) and (9.20), we have the result

$$V = \frac{1}{2} \int \int \left[N_x^i \left(\frac{\partial w}{\partial x} \right)^2 + N_s^i \left(\frac{\partial w}{\partial s} \right)^2 + N_{xs}^i \left(\frac{\partial w}{\partial s} \right)^2 \right] dx \, ds \qquad (9.21)$$

The first variation of the strain energy, Equation (9.14), is as follows:

$$
\delta U = \int_0^b \int_0^a \left\{ \left[A_{11} \frac{\partial u^o}{\partial x} + A_{12} \left(\frac{\partial v^o}{\partial s} + \frac{w}{R} \right) + A_{16} \left(\frac{\partial u^o}{\partial s} + \frac{\partial v^o}{\partial x} \right) \right. \right.
$$

$$
\left. - B_{11} \frac{\partial^2 w}{\partial x^2} - B_{12} \frac{\partial^2 w}{\partial s^2} - 2B_{16} \frac{\partial^2 w}{\partial x \partial s} \right] \frac{\partial}{\partial x} (\delta u^o)
$$

$$
+ \left[A_{12} \frac{\partial u^o}{\partial x} + A_{22} \left(\frac{\partial v^o}{\partial s} + \frac{w}{R} \right) + A_{26} \left(\frac{\partial u^o}{\partial s} + \frac{\partial v^o}{\partial x} \right) \right.
$$

$$
\left. - B_{12} \frac{\partial^2 w}{\partial x^2} - B_{22} \frac{\partial^2 w}{\partial s^2} - 2B_{26} \frac{\partial^2 w}{\partial x \partial s} \right] \left[\frac{\partial}{\partial s} (\delta v^o) + \frac{\delta w}{R} \right]
$$

$$
+ \left[A_{16} \frac{\partial u^o}{\partial x} + A_{26} \left(\frac{\partial v^o}{\partial s} + \frac{w}{R} \right) + A_{66} \left(\frac{\partial u^o}{\partial s} + \frac{\partial v^o}{\partial x} \right) \right.
$$

$$
\left. - B_{16} \frac{\partial^2 w}{\partial x^2} - B_{26} \frac{\partial^2 w}{\partial s^2} - 2B_{66} \frac{\partial^2 w}{\partial x \partial s} \right] \left[\frac{\partial}{\partial s} (\delta u^o) \right. \qquad (9.22)
$$

$$
\left. + \frac{\partial}{\partial x} (\delta v^o) \right] - \left[B_{11} \frac{\partial u^o}{\partial x} + B_{12} \left(\frac{\partial v^o}{\partial s} + \frac{w}{R} \right) + B_{16} \left(\frac{\partial u^o}{\partial s} \right. \right.
$$

$$
\left. + \frac{\partial v^o}{\partial x} \right) - D_{11} \frac{\partial^2 w}{\partial x^2} - D_{12} \frac{\partial^2 w}{\partial s^2} - 2D_{16} \frac{\partial^2 w}{\partial x \partial s} \right] \frac{\partial^2}{\partial x^2} (\delta w)
$$

$$
- \left[B_{12} \frac{\partial u^o}{\partial x} + B_{22} \left(\frac{\partial v^o}{\partial s} + \frac{w}{R} \right) + B_{26} \left(\frac{\partial u^o}{\partial s} + \frac{\partial v^o}{\partial x} \right) \right.
$$

$$
\left. - D_{12} \frac{\partial^2 w}{\partial x^2} - D_{22} \frac{\partial^2 w}{\partial s^2} - 2D_{26} \frac{\partial^2 w}{\partial x \partial s} \right] \frac{\partial^2}{\partial s^2} (\delta w)
$$

$$
- 2 \left[B_{16} \frac{\partial u^o}{\partial x} + B_{26} \left(\frac{\partial v^o}{\partial s} + \frac{w}{R} \right) + B_{66} \left(\frac{\partial u^o}{\partial s} + \frac{\partial v^o}{\partial x} \right) \right.
$$

$$
\left. \left. - D_{16} \frac{\partial^2 w}{\partial x^2} - D_{26} \frac{\partial^2 w}{\partial x \partial s} - 2D_{66} \frac{\partial^2 w}{\partial x \partial s} \right] \frac{\partial^2}{\partial x \partial s} (\delta w) \right\} dx\, ds
$$

Comparing terms in this equation with the laminate constitutive relations, Equation (9.12), we find that

$$\delta U = \int_0^b \int_0^a \left\{ N_x \frac{\partial}{\partial x} (\delta u^o) + N_s \left[\frac{\partial}{\partial s} (\delta v^o) + \frac{\delta w}{R} \right] + N_{xs} \left[\frac{\partial}{\partial s} (\delta u^o) \right. \right.$$

$$\left. + \frac{\partial}{\partial x} (\delta v^o) \right] - M_x \frac{\partial^2}{\partial x^2} (\delta w) - M_s \frac{\partial^2}{\partial s^2} (\delta w) \qquad (9.23)$$

$$\left. - 2M_{xs} \frac{\partial^2}{\partial x \partial s} (\delta w) \right\} dx \, ds$$

Applying Green's theorem [4] in conjunction with integration by parts, we can transform Equation (9.23) into the form

$$\delta U = - \int \int \left[\left(\frac{\partial N_x}{\partial x} + \frac{\partial N_{xs}}{\partial s} \right) \delta u^o + \left(\frac{\partial N_{xs}}{\partial x} + \frac{\partial N_s}{\partial s} \right) \delta v^o \right.$$

$$\left. + \left(\frac{\partial^2 M_x}{\partial x^2} + 2 \frac{\partial^2 M_{xs}}{\partial x \partial s} + \frac{\partial^2 M_s}{\partial s^2} - \frac{N_s}{R} \right) \delta w \right] dx \, ds$$

$$\qquad (9.24)$$

$$- \int_{S_s} \left[N_{xs} \delta u^o + N_s \delta v^o + M_y \frac{\partial}{\partial y} (\delta w) \right.$$

$$\left. - \left(2 \frac{\partial M_{xs}}{\partial x} + \frac{\partial M_s}{\partial s} \right) \delta w \right] dx + \int_{S_x} \left[N_x \delta u^o + N_{xs} \delta v^o \right.$$

$$\left. + M_x \frac{\partial}{\partial x} (\delta w) - \left(2 \frac{\partial M_{xs}}{\partial s} + \frac{\partial M_x}{\partial x} \right) \delta w \right] dy$$

where S_x and S_s are defined along the edges $x = $ constant and $s = $ constant, respectively.

Performing similar operations with Equation (9.21), we obtain

$$\delta V = - \int \int \left(N_x^i \frac{\partial^2 w}{\partial x^2} + 2N_{xs}^i \frac{\partial^2 w}{\partial x \partial s} + N_s^i \frac{\partial^2 w}{\partial s^2} \right) \delta w \, dx \, ds \qquad (9.25)$$

$$- \int_{S_s} \left(N_{xs}^i \frac{\partial w}{\partial x} + N_s^i \frac{\partial w}{\partial s} \right) \delta w \, dw$$

$$+ \int_{S_x} \left(N_x^i \frac{\partial w}{\partial x} + N_{xs}^i \frac{\partial w}{\partial y} \right) \delta w \, dy$$

The first variation of Equation (9.16) leads to a relationship of the same form as Equation (3.28).

$$\delta T = - \int \int \varrho \left(\frac{\partial^2 u^o}{\partial t^2} \delta u^o + \frac{\partial^2 v^o}{\partial t^2} \delta v^o + \frac{\partial^2 w}{\partial t^2} \delta w \right) dx \, ds \quad (9.26)$$

The first variation of Equation (9.17) yields

$$\delta W = - \int \int p \delta w \, dx \, ds \quad (9.27)$$

Substituting Equations (9.24–9.27) into Equation (3.19), we obtain the following results:

$$\int_{t_1}^{t_o} \left\{ \int \int \left[\left(\frac{\partial N_x}{\partial x} + \frac{\partial N_{xs}}{\partial s} - \varrho \frac{\partial^2 u^o}{\partial t^2} \right) \delta u^o + \left(\frac{\partial N_{xs}}{\partial x} + \frac{\partial N_s}{\partial s} - \varrho \frac{\partial^2 v^o}{\partial t^2} \right) \delta v^o \right. \right.$$

$$+ \left(\frac{\partial^2 M_x}{\partial x^2} + 2 \frac{\partial M_{xs}}{\partial x \partial s} + \frac{\partial^2 M_s}{\partial s^2} - \frac{N_s}{R} + N_x^i \frac{\partial^2 w}{\partial x^2} + 2 N_{xs}^i \frac{\partial^2 w}{\partial x \partial s} + N_s \frac{\partial^2 w}{\partial s^2} \right.$$

$$\left. + p - \varrho \frac{\partial^2 w}{\partial t^2} \right) \delta w \right] dx \, ds - \int_{S_x} \left[N_x \delta u^o + N_{xs} \delta v^o + M_x \frac{\partial}{\partial x} (\delta w) \right. \quad (9.28)$$

$$- \left(\frac{\partial M_x}{\partial x} + 2 \frac{\partial M_{xs}}{\partial s} + N_x^i \frac{\partial w}{\partial x} + N_{xs}^i \frac{\partial w}{\partial s} \right) \delta w \right] ds + \int_{S_s} \left[N_{xs} \delta u^o \right.$$

$$+ N_s \delta v^o + M_y \frac{\partial}{\partial s} (\delta w) - \left(2 \frac{\partial M_{xs}}{\partial x} + \frac{\partial M_s}{\partial y} + N_{xs}^i \frac{\partial w}{\partial x} \right.$$

$$\left. \left. + N_s^i \frac{\partial w}{\partial s} \right) \delta w \right] dx \right\} dt = 0$$

The surface integral in Equation (9.28) will vanish if the following equations of motion are satisfied:

$$\frac{\partial N_x}{\partial x} + \frac{\partial N_{xs}}{\partial s} = \varrho \frac{\partial^2 u^0}{\partial t^2} \tag{9.29}$$

$$\frac{\partial N_{xs}}{\partial x} + \frac{\partial N_s}{\partial s} = \varrho \frac{\partial^2 v^0}{\partial t^2} \tag{9.30}$$

$$\frac{\partial^2 M_x}{\partial x^2} + 2\frac{\partial^2 M_{xs}}{\partial x \partial s} + \frac{\partial^2 M_s}{\partial s^2} - \frac{N_s}{R} + N_x^i \frac{\partial^2 w}{\partial x^2} + 2N_{xs}^i \frac{\partial^2 w}{\partial x \partial s}$$
$$+ N_s^i \frac{\partial^2 w}{\partial s^2} + p = \varrho \frac{\partial^2 w}{\partial t^2} \tag{9.31}$$

Comparing these equations to Equations (2.21) and (2.22) for a flat plate, we find the only difference is the existence of the N_s/R term in Equation (9.31). Thus, if we let $R \rightarrow \infty$, then $N_s/R \rightarrow 0$ and Equations (9.29–9.31) reduce to the equilibrium equations of a flat plate.

The line integrals in Equation (9.28) define the natural boundary conditions which are of the same form as Equations (3.33) and (3.34). In particular

$$N_x \delta u^0 = 0 \text{ on } S_x$$

$$N_{xs} \delta v^0 = 0 \text{ on } S_x$$

$$M_x \frac{\partial}{\partial x}(\delta w) = 0 \text{ on } S_x$$

$$\left(\frac{\partial M_x}{\partial x} + 2\frac{\partial M_{xs}}{\partial y} + N_x^i \frac{\partial w}{\partial x} + N_{xs}^i \frac{\partial w}{\partial y} \right) \delta w = 0 \tag{9.32}$$

$$N_{xs} \delta u^0 = 0 \text{ on } S_s$$

$$N_s \delta v^0 = 0 \text{ on } S_s \tag{9.33}$$

$$M_s \frac{\partial}{\partial y}(\delta w) = 0 \text{ on } S_s$$

$$\left(2\frac{\partial^2 M_{xs}}{\partial x} + \frac{\partial M_s}{\partial s} + N_{xs}^i\frac{\partial w}{\partial x} + N_s^i\frac{\partial w}{\partial s}\right)\delta w = 0 \text{ on } S_s$$

Combining the kinematic relations, Equations (9.8) and (9.9) with the constitutive relations, Equations (9.12), and substituting the results into the equations of motion (9.29–9.31), we obtain the following displacement equations:

$$A_{11}\frac{\partial^2 u^0}{\partial x^2} + 2A_{16}\frac{\partial^2 u^0}{\partial x\partial s} + A_{66}\frac{\partial^2 u^0}{\partial s^2} + A_{16}\frac{\partial^2 v^0}{\partial x^2}$$

$$+ (A_{12} + A_{66})\frac{\partial^2 v^0}{\partial x\partial s} + A_{26}\frac{\partial^2 v^0}{\partial s^2} + \frac{A_{12}}{R}\frac{\partial w}{\partial x}$$

$$+ \frac{A_{26}}{R}\frac{\partial w}{\partial s} - B_{11}\frac{\partial^3 w}{\partial x^3} - 3B_{16}\frac{\partial^3 w}{\partial x^2\partial s}$$ (9.34)

$$- (B_{12} + 2B_{66})\frac{\partial^3 w}{\partial x\partial s^2} - B_{26}\frac{\partial^3 w}{\partial s^3} = \varrho\frac{\partial^2 u^0}{\partial t^2}$$

$$A_{16}\frac{\partial^2 u^0}{\partial x^2} + (A_{12} + A_{66})\frac{\partial^2 u^0}{\partial x\partial s} + A_{26}\frac{\partial^2 u^0}{\partial s^2} + A_{66}\frac{\partial^2 v^0}{\partial x^2}$$

$$+ 2A_{26}\frac{\partial^2 v^0}{\partial x\partial s} + A_{22}\frac{\partial^2 v^0}{\partial s^2} + \frac{A_{26}}{R}\frac{\partial w}{\partial x} + \frac{A_{22}}{R}\frac{\partial w}{\partial s}$$ (9.35)

$$- B_{16}\frac{\partial^3 w}{\partial x^3} - (B_{12} + 2B_{66})\frac{\partial^3 w}{\partial x^2\partial s} - 3B_{26}\frac{\partial^3 w}{\partial x\partial s^2}$$

$$- B_{22}\frac{\partial^3 w}{\partial s^3} = \varrho\frac{\partial^2 v^0}{\partial t^2}$$

$$\frac{A_{12}}{R}\frac{\partial u^0}{\partial x} + \frac{A_{26}}{R}\frac{\partial u^0}{\partial s} - B_{11}\frac{\partial^3 u^0}{\partial x^3} - 3B_{16}\frac{\partial^3 u^0}{\partial x^2\partial s}$$ (9.36)

$$- (B_{12} + 2B_{66})\frac{\partial^3 u^0}{\partial x\partial s^2} - B_{26}\frac{\partial^3 u^0}{\partial s^3} + \frac{A_{26}}{R}\frac{\partial v^0}{\partial x}$$

$$+ \frac{A_{22}}{R} \frac{\partial v^0}{\partial s} - B_{16} \frac{\partial^3 v^0}{\partial x^3} - (B_{12} + 2B_{66}) \frac{\partial^3 v^0}{\partial x^2 \partial s}$$

$$- 3B_{26} \frac{\partial^3 v^0}{\partial x \partial s^2} - B_{22} \frac{\partial^3 v^0}{\partial s^3} + D_{11} \frac{\partial^4 w}{\partial x^4} + 4D_{16} \frac{\partial^4 w}{\partial x^3 \partial s}$$

$$+ 2(D_{12} + 2D_{66}) \frac{\partial^4 w}{\partial x^2 \partial s^2} + 4D_{26} \frac{\partial^4 w}{\partial x \partial s^3} + D_{22} \frac{\partial^4 w}{\partial s^4}$$

$$= p + N_x^i \frac{\partial^2 w}{\partial x^2} + 2N_{xs}^i \frac{\partial^2 w}{\partial x \partial s} + N_s^i \frac{\partial^2 w}{\partial s^2} - \varrho \frac{\partial^2 w}{\partial t^2}$$

Due to the presence of w/R in the midplane strain ϵ_s^0, bending-extensional coupling exists for both symmetric and unsymmetric laminates. Equations (2.35–2.37) for flat plates can be recovered by allowing $R \to \infty$ in Equations (9.34–9.36).

9.4 SIMPLY-SUPPORTED ORTHOTROPIC PLATES

In Chapter 5 we considered symmetric laminates ($B_{ij} = 0$) of a special orthotropic class ($D_{16} = D_{26} = 0$). We now define a broader class of orthotropic laminates which will be considered in this section. In particular since the inplane and bending problems for cylindrical plates are coupled due to curvature, we do not limit our discussion to symmetric laminates. Rather, we only require the vanishing of the shear coupling terms. Thus in this section we consider bending under pressure loading, stability under biaxial loading, and free vibration of simply-supported plates with stacking geometry such that $A_{16} = A_{26} = B_{16} = B_{26} = D_{16} = D_{26} = 0$. The simply-supported boundary conditions are as follows:

at $x = 0$ and a

$$v^0 = N_x = A_{11} \frac{\partial u^0}{\partial x} + A_{12} \left(\frac{\partial v^0}{\partial y} + \frac{w}{R} \right) - B_{11} \frac{\partial^2 w}{\partial x^2} - B_{12} \frac{\partial^2 w}{\partial y^2} = 0$$

$$(9.37)$$

$$w = M_x = B_{11} \frac{\partial u^0}{\partial x} + B_{12} \left(\frac{\partial v^0}{\partial s} + \frac{w}{R} \right) - D_{11} \frac{\partial^2 w}{\partial x^2} - D_{12} \frac{\partial^2 w}{\partial s^2} = 0$$

$$(9.38)$$

at $s = 0$ and b

$$u^0 = N_y = A_{12} \frac{\partial u^0}{\partial x} + A_{22} \left(\frac{\partial v^0}{\partial s} + \frac{w}{R} \right) - B_{12} \frac{\partial^2 w}{\partial x^2} - B_{22} \frac{\partial^2 w}{\partial s^2} = 0$$

$$(9.39)$$

$$w = M_y = B_{12} \frac{\partial u^0}{\partial x} + B_{22} \left(\frac{\partial v^0}{\partial s} + \frac{w}{R} \right) - D_{12} \frac{\partial^2 w}{\partial x^2} - D_{22} \frac{\partial^2 w}{\partial s^2} = 0$$

$$(9.40)$$

For the case of stability under biaxial loading, we consider the following pre-buckling problem:

$$N_x^i = - N_0 = \text{constant}$$

$$N_s^i = Rp_0, \; N_{xs}^i = 0$$

$$(9.41)$$

where $N_0 > 0$ and $p_0 = $ constant. A simplified solution can be found which leads to Equation (9.41). In particular let us assume the following prebuckling displacements:

$$u^{0i} = c_1 x, \; v^{0i} = c_2 s, \; w = 0 \qquad (9.42)$$

Substituting these relationships into Equations (9.6), we obtain the following results:

$$\epsilon_x^0 = c_1, \; \epsilon_s^0 = c_2$$

$$\epsilon_{xs}^0 = \varkappa_x = \varkappa_y = \varkappa_{xs} = 0$$

$$(9.43)$$

The constants c_1 and c_2 are determined from the constitutive relations, Equations (9.12), in conjunction with Equation (9.41), which leads to the result

$$c_1 = - \frac{(A_{22}N_0 + A_{12}Rp_0)}{(A_{11}A_{22} - A_{12}^2)}$$

$$c_2 = \frac{(A_{12}N_0 + A_{11}Rp_0)}{(A_{11}A_{22} - A_{12}^2)}$$

$$(9.44)$$

Note that Equations (9.41–9.43) are compatible with the equilibrium Equations (9.29–9.31) in the absence of inplane load and inertia effects. The following

boundary conditions are satisfied by Equations (9.42):
at $x = 0$ and a

$$w = N_{xs} = 0$$

(9.45)

$$N_x = -N_0, \quad M_x = \frac{[(B_{12}A_{22} - B_{11}A_{22})N_0 + (B_{12}A_{11} - B_{11}A_{12})Rp]}{(A_{11}A_{22} - A_{12}^2)}$$

at $y = 0$ and b

$$w = N_{xs} = 0$$

(9.46)

$$N_s = Rp_0, \quad M_s = \frac{[(B_{22}A_{12} - B_{12}A_{22})N_0 + (B_{22}A_{11} - B_{12}A_{12})Rp]}{(A_{11}A_{22} - A_{12}^2)}$$

The moment boundary conditions in Equations (9.45) and (9.46) are required in conjunction with unsymmetric laminates in order to assure that $\varkappa_x = \varkappa_s = 0$.
For the orthotropic laminates under consideration in conjunction with the in-plane loads, Equations (9.41), the displacement Equations (9.34–9.36) become

$$A_{11} \frac{\partial^2 u^0}{\partial x^2} + A_{66} \frac{\partial^2 u^0}{\partial s^2} + (A_{12} + A_{66}) \frac{\partial^2 v^0}{\partial x \partial s} + \frac{A_{12}}{R} \frac{\partial w}{\partial x}$$

(9.47)

$$- B_{11} \frac{\partial^3 w}{\partial x^3} - (B_{12} + 2B_{66}) \frac{\partial^3 w}{\partial x \partial s^2} = \varrho \frac{\partial^2 u^0}{\partial t^2}$$

$$(A_{12} + A_{66}) \frac{\partial^2 u^0}{\partial x \partial s} + A_{66} \frac{\partial^2 v^0}{\partial x^2} + A_{22} \frac{\partial^2 v^0}{\partial s^2} + \frac{A_{22}}{R} \frac{\partial w}{\partial s}$$

(9.48)

$$- (B_{12} + 2B_{66}) \frac{\partial^3 w}{\partial x^2 \partial s} - B_{22} \frac{\partial^3 w}{\partial s^3} = \varrho \frac{\partial^2 v^0}{\partial t^2}$$

$$\frac{A_{12}}{R} \frac{\partial u^0}{\partial x} - B_{11} \frac{\partial^3 u^0}{\partial x^3} - (B_{12} + 2B_{66}) \frac{\partial^3 u^0}{\partial x \partial s^2} + \frac{A_{22}}{R} \frac{\partial v^0}{\partial s}$$

(9.49)

$$- (B_{12} + 2B_{66}) \frac{\partial^3 v^0}{\partial x^2 \partial s} - B_{22} \frac{\partial^3 v^0}{\partial s^3} + D_{11} \frac{\partial^4 w}{\partial x^4}$$

$$+ 2(D_{12} + 2D_{66}) \frac{\partial^4 w}{\partial x^2 \partial s^2} + D_{22} \frac{\partial^4 w}{\partial s^4} = p - N_0 \frac{\partial^2 w}{\partial x^2}$$

$$+ Rp \frac{\partial^2 w}{\partial s^2} - \varrho \frac{\partial^2 w}{\partial t^2}$$

For static problems in the absence of inplane load effects, we consider pressure loadings which can be expanded in the double Fourier series

$$p(x,s) = \sum_{m=1}^{\infty} \sum_{n=1}^{\infty} p_{mn} \sin \frac{m\pi x}{a} \sin \frac{n\pi s}{b} \qquad (9.50)$$

where

$$p_{mn} = \frac{4}{ab} \int_0^a \int_0^b p(x,y) \sin \frac{m\pi x}{a} \sin \frac{n\pi s}{b} \qquad (9.51)$$

A solution to Equation (9.47–9.49) which satisfies the boundary conditions (9.37–9.40) is of the form

$$u^0 = \sum_{m=1}^{\infty} \sum_{n=1}^{\infty} A_{mn} e^{i\omega mnt} \cos \frac{m\pi x}{a} \sin \frac{n\pi s}{b} \qquad (9.52)$$

$$v^0 = \sum_{m=1}^{\infty} \sum_{n=1}^{\infty} B_{mn} e^{i\omega mnt} \sin \frac{m\pi x}{a} \cos \frac{n\pi s}{b} \qquad (9.53)$$

$$w = \sum_{m=1}^{\infty} \sum_{n=1}^{\infty} C_{mn} e^{i\omega mnt} \sin \frac{m\pi x}{a} \sin \frac{n\pi s}{b} \qquad (9.54)$$

Substituting Equations (9.52–9.54) into Equations (9.47–9.49), we arrive at the following algebraic equations in matrix form:

$$\begin{bmatrix} (H_{11mn} - \lambda_{1mn}) & H_{12mn} & H_{13mn} \\ H_{12mn} & (H_{22mn} - \lambda_{1mn}) & H_{23mn} \\ H_{13mn} & H_{23mn} & (H_{33mn} - \lambda_{1mn} - \lambda_{2mn}) \end{bmatrix} \begin{bmatrix} A_{mn} \\ B_{mn} \\ C_{mn} \end{bmatrix} = \begin{bmatrix} 0 \\ 0 \\ p_{mn} a^2/\pi^2 \end{bmatrix}$$

$$(9.55)$$

where

$$H_{11mn} = A_{11}m^2 + A_{66}n^2R^2$$

$$H_{12mn} = (A_{12} + A_{66})mnR$$

$$H_{13mn} = -\frac{m\pi}{a}[A_{12}\frac{a^2}{\pi^2R} + B_{11}m^2 + (B_{12} + 2B_{66})n^2R^2]$$

$$H_{22mn} = A_{66}m^2 + A_{22}n^2R^2$$

$$H_{23mn} = -\frac{nR\pi}{a}[A_{22}\frac{a^2}{\pi^2R} + (B_{12} + 2B_{66})m^2 + B_{22}n^2R^2]$$

$$H_{33mn} = \frac{\pi^2}{a^2}[D_{11}m^4 + 2(D_{12} + 2D_{66})m^2n^2R^2 + D_{22}n^4R^4]$$

$$\lambda_{1mn} = \frac{\varrho\omega_{mn}^2}{\pi^2a^2}$$

$$\lambda_{2mn} = N_0m^2 - Rp_0n^2R^2$$

We now consider three separate cases involving equations (9.55).

Case 1: Bending Under Internal Pressure Loading

This case involves static loading for which $\omega_{mn} = 0$ and Equations (9.52–9.54) reduce to the following:

$$u^0 = \sum_{m=1}^{\infty}\sum_{n=1}^{\infty} A_{mn}\cos\frac{m\pi x}{a}\sin\frac{n\pi s}{b} \qquad (9.56)$$

$$v^0 = \sum_{m=1}^{\infty}\sum_{n=1}^{\infty} B_{mn}\sin\frac{m\pi x}{a}\cos\frac{n\pi s}{b} \qquad (9.57)$$

$$w = \sum_{m=1}^{\infty}\sum_{n=1}^{\infty} C_{mn}\sin\frac{m\pi x}{a}\sin\frac{n\pi s}{b} \qquad (9.58)$$

These displacement functions are of the same form as presented by Timoshenko and Woinowsky-Krieger for homogeneous, isotropic curved plates [5]. For the case of uniform internal pressure loading, $p = p_0 = $ constant, Equation (9.52) in the absence of inplane loads ($\lambda_{2mn} = 0$) yields

$$p = \frac{16}{\pi^2} p_0 \sum_{m = 1,3,5, \ldots}^{\infty} \sum_{n = 1,3,5, \ldots}^{\infty} \frac{1}{mn} \sin \frac{m\pi x}{a} \sin \frac{n\pi s}{b} \quad (9.59)$$

Solving Equation (9.55), we obtain the result

$$A_{mn} = \frac{16 p_0 a^2}{\pi^4 mn F_{mn}} (H_{12mn} H_{23mn} - H_{22mn} H_{13mn}) \quad (9.60)$$

$$B_{mn} = \frac{16 p_0 a^2}{\pi^4 mn F_{mn}} (H_{12mn} H_{13mn} - H_{11mn} H_{23mn}) \quad (9.61)$$

$$C_{mn} = \frac{16 p_0}{\pi^4 mn F_{mn}} (H_{11mn} H_{22mn} - H_{12mn}^2) \quad (9.62)$$

where

$$F_{mn} = H_{11mn} H_{22mn} H_{33mn} + 2 H_{12mn} H_{13mn} H_{23mn} - H_{22mn} H_{13mn}^2$$

$$- H_{33mn} H_{12mn}^2 - H_{11mn} H_{23mn}^2$$

For unsymmetric laminates in which $R \rightarrow \infty$, Equations (9.60–9.62) reduce to Equations (6.39) for a flat plate. In the case of symmetric laminates in which $R \rightarrow \infty$, we find that

$$A_{mn} = B_{mn} = 0$$

$$C_{mn} = \frac{16 p_0 a^4}{\pi^6 mn [D_{11} m^4 + 2(D_{12} + 2D_{66}) m^2 n^2 R^2 + D_{22} n^4 R^4]} \quad (9.63)$$

and Equation (9.58) becomes identical to Equation (5.7) for a specially orthotropic plate subjected to a uniform transverse load.

Force and moment resultants can be obtained by substituting Equations (9.56–9.58) into the constitutive relations (9.12). Rapid convergence is obtained for the displacements, while the force and moment resultants converge rather slowly.

Case 2: Stability Under Combined Axial Load and Internal Pressure

This case again involves static loading with individual buckling modes of the same form as Equations (9.56–9.58), i.e.

$$u^0 = A_{mn} \cos \frac{m\pi x}{a} \sin \frac{n\pi s}{b} \qquad (9.64)$$

$$v^0 = B_{mn} \sin \frac{m\pi x}{a} \cos \frac{n\pi s}{b} \qquad (9.65)$$

$$w = C_{mn} \sin \frac{m\pi x}{a} \sin \frac{n\pi s}{b} \qquad (9.66)$$

Equations (9.55) in the absence of pressure loading ($p_{mn} = 0$), i.e. the pressure loading is included in the prebuckled solution, reduce to a set of homogeneous algebraic equations. Thus if we write the pressure loading in the form $Rp_0 = kN_0$, where k is a constant, then a nontrivial solution to Equation (9.55) can be obtained by choosing values of N_0 such that the determinant of the coefficient matrix vanishes. This leads to the following result:

$$N_0 = \frac{F_{mn}}{(m^2 - kn^2R^2)(H_{11mn}H_{22mn} - H_{12mn}^2)} \qquad (9.67)$$

For the case of symmetric laminates in which $R \to \infty$, Equation (9.67) reduces to an expression of the same form as Equation (5.72) for specially orthotropic plates.

$$N_0 = \frac{\pi^2[D_{11}m^4 + 2(D_{12} + 2D_{66})m^2n^2R^2 + D_{22}n^4R^4]}{a^2(m^2 - kn^2R^2)} \qquad (9.68)$$

External pressure can be considered in Equation (9.67) by considering values of $k < 0$. The critical buckling load corresponds to values of m and n which produce the lowest value of N_0.

Case 3: Free-Vibration

For this case the individual modes of free vibration are of the same form as Equations (9.52–9.54), i.e.

$$u^0 = A_{mn} e^{i\omega_{mn}t} \cos \frac{m\pi x}{a} \sin \frac{n\pi s}{b} \qquad (9.69)$$

$$v^0 = B_{mn} \, e^{i\omega mnt} \sin \frac{m\pi x}{a} \cos \frac{n\pi s}{b} \tag{9.70}$$

$$w = C_{mn} \, e^{i\omega mnt} \sin \frac{m\pi x}{a} \sin \frac{n\pi s}{b} \tag{9.71}$$

Equations (9.55) in the absence of inplane loads ($\lambda_{2mn} = 0$) and pressure loading ($p_{mn} = 0$) yield a homogeneous set of algebraic equations. In order to avoid a trivial solution, we choose ω_{mn} such that the determinant of the coefficient matrix in Equation (9.55) vanishes. Neglecting inplane inertia terms, we find

$$\omega_{mn} = \pi a \sqrt{\frac{F_{mn}}{\varrho(H_{11mn}H_{22}mn - H_{12mn}^2)}} \tag{9.72}$$

For laminates in which $R \to \infty$, Equation (9.72) reduces to Equation (6.100) in the case of unsymmetric laminates and to Equation (5.137) for specially orthotropic plates, i.e.

$$\omega_{mn} = \frac{\pi^2}{a^2\sqrt{\varrho}} \sqrt{D_{11}m^4 + 2(D_{12} + 2D_{66})m^2n^2R^2 + D_{22}n^4R^4} \tag{9.73}$$

9.5 STABILITY OF SIMPLY-SUPPORTED PLATES UNDER COMBINED LOADING

Consideration is now given to the stability of curved plates under combined uniform axial load, internal pressure, and inplane shear load [6]. In this section we also broaden the scope of laminate geometries to be considered. In particular we require that only the inplane stiffnesses and the coupling stiffnesses be orthotropic, i.e. $A_{16} = A_{26} = B_{16} = B_{26} = 0$. For these laminates, Equations (9.47) and (9.48) in the absence of inertia terms are valid. However, considering Equation (9.36) we see that equation (9.49) takes the following form:

$$\frac{A_{12}}{R} \frac{\partial u^0}{\partial x} - B_{11} \frac{\partial^3 u^0}{\partial x^3} - (B_{12} + 2B_{66})\frac{\partial^3 u^0}{\partial x \partial s^2} + \frac{A_{22}}{R} \frac{\partial v^0}{\partial s}$$

$$- (B_{12} + 2B_{66})\frac{\partial^3 v^0}{\partial x^2 \partial s} - B_{22} \frac{\partial^3 v^0}{\partial s^3} + D_{11} \frac{\partial^4 w}{\partial x^4} + 4D_{16} \frac{\partial^4 w}{\partial x^3 \partial s}$$

$$+ 2(D_{12} + 2D_{66}) \frac{\partial^4 w}{\partial x^2 \partial s^2} + 4D_{26} \frac{\partial^4 w}{\partial x \partial s^3} + D_{22} \frac{\partial^4 w}{\partial s^4} \tag{9.74}$$

$$= N_x \frac{\partial^2 w}{\partial x^2} + 2N_{xs} \frac{\partial^2 w}{\partial x \partial s} + N_s \frac{\partial^2 w}{\partial s^2}$$

Thus, we have the additional complexity of nonvanishing bending-twisting coupling stiffness terms D_{16} and D_{26}.

If we consider the same simply-supported boundary conditions as in the previous section, then Equations (9.37) and (9.39) are applicable. The bending deflections, w, vanish on the boundary as required by Equations (9.38) and (9.40). The bending moments as defined in these boundary conditions must be modified to reflect the nonvanishing D_{16} and D_{26} terms. In particular, the bending moment conditions in Equations (9.38) and (9.40) become

at $x = 0$ and a:

$$M_x = B_{11} \frac{\partial u^0}{\partial x} + B_{12} \left(\frac{\partial v^0}{\partial s} + \frac{w}{R} \right) - D_{11} \frac{\partial^2 w}{\partial x^2} - 2D_{16} \frac{\partial^2 w}{\partial x \partial s}$$

$$- D_{12} \frac{\partial^2 w}{\partial s^2} = 0$$

(9.75)

at $y = 0$ and b:

$$M_y = B_{12} \frac{\partial u^0}{\partial x} + B_{22} \left(\frac{\partial v^0}{\partial s} + \frac{w}{R} \right) - D_{12} \frac{\partial^2 w}{\partial x^2} - 2D_{26} \frac{\partial^2 w}{\partial x \partial s}$$

$$- D_{22} \frac{\partial^2 w}{\partial s^2} = 0$$

(9.76)

Solutions to Equations (9.47), (9.48), and (9.74) are assumed to be of the same form as Equations (9.56–9.58). As in Section 9.4, these displacement functions exactly satisfy Equations (9.47) and (9.48). However, Equation (9.74) and the moment boundary conditions (9.75) and (9.76) are not exactly satisfied due to the presence of the D_{16} and D_{26} stiffness terms. Equations (9.56–9.58) satisfy the in-plane boundary conditions (9.37) and (9.39) along with the requirement that w vanish on the boundary. Thus, our choice of displacement functions satisfies all of the geometric boundary conditions. This allows us to utilize the Galerkin method for obtaining an approximate solution to Equation (9.74) which is compatible with the moment boundary conditions as given by Equations (9.75) and (9.76).

We now proceed in a manner similar to the approach utilized in Section 7.6 for determining the critical load associated with unsymmetric cross-ply plates subjected to a uniform shear load. Substituting Equations (9.56–9.58) into Equations

(9.47) and (9.48), solving the resulting algebraic Equations for A_{mn} and B_{mn} in terms of C_{mn}, we find that

$$A_{mn} = \frac{(H_{23mn}H_{12mn} - H_{22mn}H_{13mn})}{(H_{11mn}H_{22mn} - H_{12mn}^2)} C_{mn}$$

(9.77)

$$B_{mn} = \frac{(H_{13mn}H_{12mn} - H_{11mn}H_{13mn})}{(H_{11mn}H_{22mn} - H_{12mn}^2)} C_{mn}$$

where the H_{ijmn} are defined in equation (9.55).

Combining the constitutive relations (9.12) with Equation (9.28), we obtain the following variational form of Equation (9.74)

$$\int_0^b \int_0^a \left[\frac{A_{12}}{R} \frac{\partial u^0}{\partial x} - B_{11} \frac{\partial^3 u^0}{\partial x^3} - (B_{12} + 2B_{66}) \frac{\partial^3 u^0}{\partial x \partial s^2} + \frac{A_{22}}{R} \frac{\partial v^0}{\partial s} \right.$$

$$- (B_{12} + 2B_{66}) \frac{\partial^3 v^0}{\partial x^2 \partial s} - B_{22} \frac{\partial^3 v^0}{\partial s^3} + D_{11} \frac{\partial^4 w}{\partial x^4} + 4D_{16} \frac{\partial^4 w}{\partial x^3 \partial s}$$

$$+ 2(D_{12} + 2D_{66}) \frac{\partial^4 w}{\partial x^2 \partial s^2} + 4D_{26} \frac{\partial^4 w}{\partial x \partial s^3} + D_{22} \frac{\partial^4 w}{\partial s^4} - N_x^i \frac{\partial^2 w}{\partial x^2}$$

(9.78)

$$\left. - 2N_{xs}^i \frac{\partial^2 w}{\partial x \partial s} - N_s^i \frac{\partial^2 w}{\partial s^2} \right] \delta w \, dx \, dy + \int_0^b \left\{ M_x(0,s) \frac{\partial}{\partial s} [\delta w(0,s)] \right.$$

$$\left. - M_x(a,s) \frac{\partial}{\partial x} [\delta w(a,s)] \right\} ds + \int_0^a \left\{ M_s(x,0) \frac{\partial}{\partial s} [\delta w(x,0)] \right.$$

$$\left. - M_s(x,b) \frac{\partial}{\partial s} [\delta w(x,b)] \right\} dx = 0$$

The line integrals are included in Equation (9.78) due to the fact that the moment boundary conditions (9.75) and (9.76) are not satisfied by Equations (9.56–9.58).

If we take the variation of Equation (9.58) with respect to the undetermined coefficients C_{mn}, we find

$$\delta w = \sum_{m=1}^{\infty} \sum_{n=1}^{\infty} \sin \frac{m \pi x}{a} \sin \frac{n \pi s}{b} \delta C_{mn}$$

(9.79)

$$\frac{\partial}{\partial x}(\delta w) = \frac{\pi}{a} \sum_{m=1}^{\infty} \sum_{n=1}^{\infty} m \cos \frac{m\pi x}{a} \sin \frac{n\pi s}{b} \delta C_{mn} \qquad (9.80)$$

$$\frac{\partial}{\partial y}(\delta w) = \frac{\pi}{b} \sum_{m=1}^{\infty} \sum_{n=1}^{\infty} n \sin \frac{m\pi x}{a} \cos \frac{n\pi s}{b} \delta C_{mn} \qquad (9.81)$$

Substituting Equations (9.79–9.81) into Equation (9.78) and recognizing that the only nonvanishing terms in the moment boundary conditions (9.75) and (9.76) are those involving the D_{16} and D_{26} stiffness coefficients, we arrive at the following Galerkin Equation:

$$\int_0^a \int_0^b \left[\frac{A_{12}}{R} \frac{\partial u^0}{\partial x} - B_{11} \frac{\partial^3 u^0}{\partial x^3} - (B_{12} + 2B_{66}) \frac{\partial^3 u^0}{\partial x \partial s^2} + \frac{A_{22}}{R} \frac{\partial v^0}{\partial s} \right.$$

$$- (B_{12} + 2B_{66}) \frac{\partial^3 v^0}{\partial x^2 \partial s} - B_{22} \frac{\partial^3 v^0}{\partial s^3} + D_{11} \frac{\partial^4 w}{\partial x^4} + 4D_{16} \frac{\partial^4 w}{\partial x^3 \partial s}$$

$$+ 2(D_{12} + 2D_{66}) \frac{\partial^4 w}{\partial x^2 \partial s^2} + 4D_{26} \frac{\partial^4 w}{\partial x \partial s^3} + D_{22} \frac{\partial^4 w}{\partial s^4} - N_x^i \frac{\partial^2 w}{\partial x^2}$$

$$\left. - 2N_{xs}^i \frac{\partial^2 w}{\partial x \partial s} - N_s^i \frac{\partial^2 w}{\partial s^2} \right] \sin \frac{m\pi x}{a} \sin \frac{n\pi s}{b} \, dx \, dy \qquad (9.82)$$

$$- 2D_{16} \int_0^b \left[(-1)^m \left(\frac{\partial^2 w}{\partial x \partial s} \right)_{x=a} - \left(\frac{\partial^2 w}{\partial x \partial s} \right)_{x=0} \right] \frac{\pi}{a} \sin \frac{n\pi s}{b} \, dy$$

$$- 2D_{26} \int_0^a \left[(-1)^n \left(\frac{\partial^2 w}{\partial x \partial s} \right)_{y=b} - \left(\frac{\partial^2 w}{\partial x \partial s} \right)_{y=0} \right] \frac{\pi}{b} \sin \frac{m\pi x}{a} \, dx$$

$$= 0 \begin{cases} m = 1,2,\ldots,\infty \\ n = 1,2,\ldots,\infty \end{cases}$$

Substituting Equations (9.56–9.58) into Equation (9.82), taking Equations (9.77) into account, and performing the resulting integration, we arrive at the following set of algebraic Equations:

$$(F_{mn} - N_x^i m^2 - N_s^i n^2 R^2) C_{mn} - 32 \frac{mnR}{a^2} \sum_{i=1}^{\infty} \sum_{j=1}^{\infty} M_{ij} \left[(m^2 + i^2) D_{16} \right.$$

$$(9.83)$$

$$+ (n^2 + j^2)D_{26} - N_{xs}^i \frac{a^2}{\pi^2} \Bigg] C_{ij} = 0$$

where F_{mn} is defined in Equations (9.60–9.62). Truncating the series (9.56–9.58) at $m = M$ and $n = N$, Equations (9.83) reduce to a set of $M \times N$ homogeneous equations. These equations can be divided into two groups, one corresponding to $m + n$ even and one corresponding to $m + n$ odd. If we consider uniform initial loads of the form

$$N_x^i = -k_1 N_0, \quad N_s^i = Rp = k_2 N_0, \quad N_{xs}^i = k_3 N_0 \qquad (9.84)$$

where $N_0 > 0$ and k_1, k_2, and k_3 are prescribed constants, then a nontrivial solution to Equations (9.83) can be obtained by choosing N_0 such that the determinant of the coefficient matrix vanishes. The lowest value of N_0 which produces a zero determinant is the critical buckling load.

The initial displacements in Equation (9.42) need to be modified slightly to accommodate the initial shear load. In particular we assume that

$$u^{0i} = c_1 x + c_3 y \qquad (9.85)$$

From the strain-displacement relations (9.8) and (9.9) in conjunction with the constitutive relations (9.12), we find that

$$c_3 = \frac{N_{xs}^i}{A_{66}} = \frac{k_3 N_0}{A_{66}} \qquad (9.86)$$

To illustrate the effect of stacking sequence and plate geometry on critical buckling loads, we consider numerical results involving the following ply properties:

$$E_L/E_T = 14, \quad G_{LT}/E_T = 0.5, \quad \nu_{LT} = 0.25 \qquad (9.87)$$

These are typical properties of current graphite/epoxy unidirectional composites. In order to compare results from the same thickness plates, we consider 12-ply laminates with various stacking sequences. In addition numerical results are for square plates ($a/b = 1$) with $a/h = 300$. Three loading conditions are considered.

Case I: Shear Loading

For this case $k_1 = k_2 = 0$, $k_3 = 1$, and $N_0 = S$. First we consider symmetric angle-ply laminates with the stacking geometries $[45_3/-45_3]_s$ and $[(\pm 45)_3]_s$.

The first stacking geometry behaves as a 4-ply laminate. Since these laminates are symmetric, $B_{ij} = 0$. The inplane properties are orthotropic ($A_{16} = A_{26} = 0$), while the bending stiffnesses are anisotropic ($D_{16}, D_{26} \neq 0$). For the class of laminates $[(\pm 45)_n]_s$, the bending-twisting coupling terms are of the form

$$D_{16} = D_{26} = \frac{h^3 Q_{16}(45°)}{16n} \qquad (9.88)$$

where n is the number of repeating $\pm 45°$ layers above the midplane. Thus as discussed in Chapter 6, the effect of the bending-twisting coupling terms dissipates with an increasing number of units of $\pm 45°$ layers.

Convergence of the solution is shown in Table 9.1, while the effect of plate curvature is illustrated in Figure 9.2. Note that bending-twisting coupling effects increase with increasing curvature. This is especially noticeable in the case of the four-layer composite. The presence of bending-twisting coupling results in different critical buckling loads for positive and negative shear.

Now consider two 12-layer unsymmetric cross-ply laminates with the stacking geometries $[0_6/90_6]$ and $[0_3/90_3]_2$ under inplane shear loading. The first stacking geometry behaves as a two-layer laminate and the second as a four-ply laminate. This again allows two stacking sequences of the same class of laminate, i.e. the [0/90] unsymmetric class of laminates, to be studied without changing laminate thickness. The only bending-extensional coupling terms that do not vanish are B_{11} and B_{22}. As discussed in Chapter 4, the bending-extensional coupling terms for this class of laminates are of the form

$$B_{11} = -B_{22} = \frac{(E_L - E_T)}{8(1 - \nu_{LT}^2 E_T / E_L)n} \qquad (9.89)$$

where n is the total number of repeating units of $0°/90°$ layers. Numerical results are presented in Figure 9.3 for the ply properties given in Equation (9.87). The

Table 9.1. Convergence of Galerkin Method.

		Shear Buckling $[45_3/-45_3]_s$ $a/b = 1$, $a/R = 90$, $\lambda = S_{cr} a^2 / E_T h^3$	
M	N	$|\lambda| < 0$	$|\lambda| > 0$
2	2	475.6	518.4
4	4	105.6	208.5
6	6	92.27	207.3
8	8	91.40	207.2
10	10	91.04	207.2

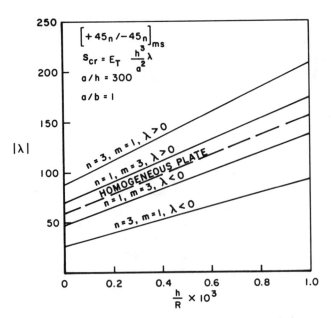

Figure 9.2. Shear buckling of symmetric angle-ply curved plates.

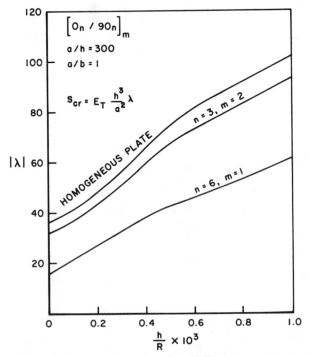

Figure 9.3. Shear buckling of unsymmetric cross-ply curved plates.

effect of bending-extensional coupling is seen to dissipate rapidly with an increasing number of layers.

Note that the homogeneous solutions in Figures 9.2 and 9.3 are obtained by allowing $n \to \infty$ in Equations (9.88) and (9.89).

Case 2: Combined Axial Compression and Inplane Shear

In this case $k_2 = 0$, $k_3 = 1$, while k_1 is a variable. Numerical results are shown in Figure 9.4 for symmetric laminates of the class $[0/\pm 45]_2$. The ply properties are those of Equation (9.87). Critical shear buckling loads are shown for increasing values of axial compression load. As in Figure 9.3 we let $N_0 = S$. Critical shear buckling loads are shown for increasing axial compression loads. The axial loads are normalized by the critical buckling load, N_{cr}, associated with the same laminate under uniaxial compression. As one might anticipate, the laminate with angle-plies on the outer surfaces produces greater shear stability. It also produces the largest difference in critical buckling load between positive and negative shear.

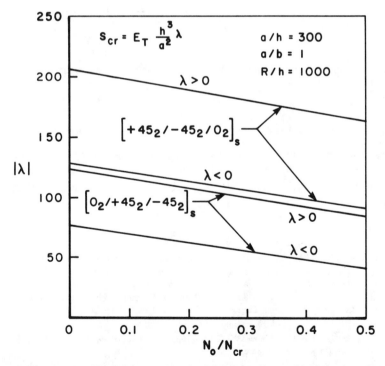

Figure 9.4. *Shear buckling under combined axial compression and inplane shear for symmetric angle-ply curved plates.*

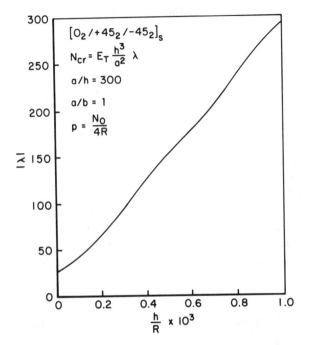

Figure 9.5. Buckling under combined axial compression and internal pressure for angle-ply curved plates.

Case 3: Combined Axial Compression and Internal Pressure

We now consider the case where $k_1 = 1$, $k_2 = 0.25$, and $k_3 = 0$. Numerical results are shown in Figure 9.5 for the $[0/\pm45]_s$ class of laminates with the ply properties as given by Equation (9.87). As can be seen the critical buckling load increases rapidly with decreasing radius of curvature.

REFERENCES

1. Timoshenko, S. and J. N. Goodier. *Theory of Elasticity*, McGraw-Hill (1951).
2. Dong, S. B., K. S. Pister and R. L. Taylor. "On the Theory of Laminated Anisotropic Shells and Plates," *Journal of the Aerospace Sciences*, 29:969–975 (1962).
3. Ambartsumyan, S. A. *Theory of Anisotropic Plates*. J. E. Ashton, ed. Translated from the first Russian Edition by T. Cheron. Lancaster, PA:Technomic Publishing Co. (1970).
4. Taylor, A. E. *Advanced Calculus*, Ginn and Company (1955).
5. Timoshenko, S. and S. Woinowsky-Krieger. *Theory of Plates and Shells*, Second Edition, McGraw-Hill Book Company, Inc. (1959).
6. Whitney, J. M. "Buckling of Anisotropic Laminated Cylindrical Plates," *AIAA Journal*, 2:1641–1645 (1984).

CHAPTER 10

Shear Deformation in Laminated Plates

10.1 LIMITATIONS OF LAMINATED PLATE THEORY

I<small>N THE PREVIOUS</small> chapters some experimental evidence has been presented which supports the classical approach to laminated plate analysis. Pagano [1,2,3] has obtained exact solutions, based on theory of elasticity, for laminated plates subjected to bending loads. Numerical results presented in conjunction with these exact solutions generally confirm the basic assumptions of laminated plate theory for reasonably high width-to-thickness ratios. For composites having a high ratio of inplane Young's moduli to interlaminar shear moduli and ratios of inplane dimensions to thickness less than ten, significant departure from classical theory was observed. In particular, maximum plate deflections were shown to be considerably larger than predicted by classical laminated plate theory. As a result it is appropriate to develop a higher order laminated plate theory which can be applied to moderately thick plates.

In this chapter we consider a higher order theory which includes the effects of transverse shear deformation and rotatory inertia. Solutions to the theory are also presented for the purpose of assessing the effect of transverse shear deformation on the behavior of laminated plates. The approach presented here is an extension of theories developed by Reissner [4] and Mindlin [5] for homogeneous, isotropic plates to laminates consisting of an arbitrary number of bonded anisotropic layers. Such an extension was originally due to Yang, Norris, and Stavsky [6] with some later modification by Whitney and Pagano [7].

No discussion on layered plate analysis is complete without considering sandwich plates. Thus the governing equations applicable to the analysis of sandwich plates are derived in this chapter. Bending under lateral loads, buckling under inplane loads, and flexural vibration frequencies are considered.

10.2 CONSTITUTIVE EQUATIONS

The governing equations are derived in terms of the same x,y,z coordinate system used in Chapter 2 for the original laminated plate theory. The basic assumptions for the shear deformation theory remain the same as listed in Section 2.1 with two exceptions being assumption 6 concerning the neglect of the interlaminar shear strains ϵ_{xz} and ϵ_{yx}, and assumption 11 in which rotatory inertia terms are neglected. In the present theory both assumptions are removed. Thus the displacements are now assumed to be of the form

$$u = u^0(x,y,t) + z\,\psi_x(x,y,t)$$

$$v = v^0(x,y,t) + z\,\psi_y(x,y,t) \tag{10.1}$$

$$w = w(x,y,t)$$

Using Equation (10.1) in conjunction with the strain-displacement relations (1.10), we obtain the following results:

$$\epsilon_x = \epsilon_x^0 + z\,\varkappa_x$$

$$\epsilon_y = \epsilon_y^0 + z\,\varkappa_y \tag{10.2}$$

$$\epsilon_{xy} = \epsilon_x^0 + z\,\varkappa_x$$

where the midplane strains are defined in the usual manner, Equation (2.6), and

$$\varkappa_x = \frac{\partial \psi_x}{\partial x}\,,\ \varkappa_y = \frac{\partial \psi_y}{\partial y}\,,\ \varkappa_{xy} = \frac{\partial \psi_x}{\partial y} + \frac{\partial \psi_y}{\partial x} \tag{10.3}$$

In addition, the interlaminar shear strains are given by the relationships

$$\epsilon_{xz} = \frac{\partial u}{\partial z} + \frac{\partial w}{\partial x} = \psi_x + \frac{\partial w}{\partial x}$$

$$\tag{10.4}$$

$$\epsilon_{yz} = \frac{\partial v}{\partial z} + \frac{\partial w}{\partial y} = \psi_y + \frac{\partial w}{\partial y}$$

Using Equation (10.2) in conjunction with the assumption of plane stress within each ply, Equation (2.24), and the definition of force and moment resultants, Equations (2.9), we obtain a constitutive relation of exactly the same form as given by Equation (2.26), i.e., in abbreviated notation,

$$\left[\begin{array}{c} N \\ \hline M \end{array}\right] = \left[\begin{array}{c|c} A & B \\ \hline B & D \end{array}\right] \left[\begin{array}{c} \epsilon \\ \hline \varkappa \end{array}\right] \tag{10.5}$$

where the stiffness terms A_{ij}, B_{ij}, and D_{ij} are defined by Equation (2.27). Applying the definition of shear force resultants, Equation (2.9), we obtain an additional constitutive relation involving transverse shear. Following Reissner [4] and Mindlin [5] we introduce a parameter k in this constitutive relation for transverse shear. Thus we obtain

$$\begin{bmatrix} Q_y \\ Q_x \end{bmatrix} = k \begin{bmatrix} A_{44} & A_{45} \\ A_{45} & A_{55} \end{bmatrix} \begin{bmatrix} \epsilon_{yz} \\ \epsilon_{xz} \end{bmatrix} \tag{10.6}$$

where ϵ_{yz} and ϵ_{xz} are defined in Equation (10.4) and

$$A_{ij} = \int_{-h/2}^{h/2} C_{ij} dz \quad (i,j = 4,5) \tag{10.7}$$

with the C_{ij} terms denoting anisotropic stiffnesses. Values associated with the factor k are discussed in Section 10.4.

10.3 GOVERNING EQUATIONS

The equations of motion in terms of stress and moment resultants, including in-plane effects, are obtained by integrating the equations of motion from classical theory of elasticity as previously done in Chapter 2 for classical laminated plate theory. Substituting the first of Equations (10.1) into Equation (2.10) and performing the integrations, we obtain the result

$$\frac{\partial N_x}{\partial x} + \frac{\partial N_{xy}}{\partial y} = \varrho \frac{\partial^2 u^0}{\partial t^2} + H \frac{\partial^2 \psi_x}{\partial t^2} \tag{10.8}$$

where

$$(\varrho, H) = \int_{-h/2}^{h/2} \varrho_0 (1, z) \, dz$$

In a similar manner, integration of Equation (2.8) with respect to z in conjunction with the second of Equations (10.1) yields

$$\frac{\partial N_{xy}}{\partial x} + \frac{\partial N_y}{\partial y} = \varrho \frac{\partial^2 v^0}{\partial t^2} + H \frac{\partial^2 \psi_y}{\partial t^2} \tag{10.9}$$

Substituting the first of Equations (1.10) into Equation (2.17) and performing the indicated integrations, we find

$$\frac{\partial M_x}{\partial x} + \frac{\partial M_{xy}}{\partial y} - Q_x = H\frac{\partial^2 u^0}{\partial t^2} + I\frac{\partial^2 \psi_x}{\partial t^2} \tag{10.10}$$

where

$$I = \int_{-h/2}^{h/2} \varrho_0\, z^2\, dz$$

A similar operation performed in conjunction with the second of Equations (2.8) yields

$$\frac{\partial M_{xy}}{\partial x} + \frac{\partial M_y}{\partial y} - Q_y = H\frac{\partial^2 v^0}{\partial t^2} + I\frac{\partial^2 \psi_y}{\partial t^2} \tag{10.11}$$

The last equilibrium equation is identical to Equation (2.16) with inplane force effects in terms of initial force resultants:

$$\frac{\partial Q_x}{\partial x} + \frac{\partial Q_y}{\partial y} + N_x^i\frac{\partial^2 w}{\partial x^2} + 2N_{xy}^i\frac{\partial^2 w}{\partial x \partial y} + N_y^i\frac{\partial^2 w}{\partial y^2} + q = \varrho\frac{\partial^2 w}{\partial t^2} \tag{10.12}$$

where, as previously,

$$q = \sigma_z(h/2) - \sigma_z(-h/2) \tag{10.13}$$

Equations (10.7)–(10.11) constitute the equations of motion in terms of force and moment resultants. Note that for symmetric laminates $H = 0$. In addition, for laminates constructed of the same unidirectional material, $H = 0$, i.e., the laminate behaves as a homogeneous material relative to density.

Substituting the constitutive relations (10.5) into Equations (10.8)–(10.12), and taking Equation (10.3) into account, we obtain the following governing equations:

$$A_{11}\frac{\partial^2 u^0}{\partial x^2} + 2A_{16}\frac{\partial^2 u^0}{\partial x \partial y} + A_{66}\frac{\partial^2 u^0}{\partial y^2} + A_{16}\frac{\partial^2 v^0}{\partial x^2} + (A_{12} + A_{66})\frac{\partial^2 v^0}{\partial x \partial y}$$

$$+ A_{26}\frac{\partial^2 v^0}{\partial y^2} + B_{11}\frac{\partial^2 \psi_x}{\partial x^2} + 2B_{16}\frac{\partial^2 \psi_x}{\partial x \partial y} + B_{66}\frac{\partial^2 \psi_x}{\partial y^2} + B_{16}\frac{\partial^2 \psi_y}{\partial x^2}$$

$$+ (B_{12} + B_{66}) \frac{\partial^2 \psi_y}{\partial x \partial y} + B_{26} \frac{\partial^2 \psi_y}{\partial y^2} = \varrho \frac{\partial^2 u^0}{\partial t^2} + H \frac{\partial^2 \psi_x}{\partial t^2} \qquad (10.14)$$

$$A_{16} \frac{\partial^2 u^0}{\partial x^2} + (A_{12} + A_{66}) \frac{\partial^2 u^0}{\partial x \partial y} + A_{26} \frac{\partial^2 u^0}{\partial y^2} + A_{66} \frac{\partial^2 v^0}{\partial x^2} + 2A_{26} \frac{\partial^2 v^0}{\partial x \partial y}$$

$$(10.15)$$

$$+ A_{22} \frac{\partial^2 v^0}{\partial y^2} + B_{16} \frac{\partial^2 \psi_x}{\partial x^2} + (B_{12} + B_{66}) \frac{\partial^2 \psi_x}{\partial x \partial y} + B_{26} \frac{\partial^2 \psi_x}{\partial y^2} + B_{66} \frac{\partial^2 \psi_y}{\partial x^2}$$

$$+ 2B_{26} \frac{\partial^2 \psi_y}{\partial x \partial y} + B_{22} \frac{\partial^2 \psi_y}{\partial y^2} = \varrho \frac{\partial^2 v^0}{\partial t^2} + H \frac{\partial^2 \psi_y}{\partial t^2}$$

$$B_{11} \frac{\partial^2 u^0}{\partial x^2} + 2B_{16} \frac{\partial^2 u^0}{\partial x \partial y} + B_{66} \frac{\partial^2 u^0}{\partial y^2} + B_{16} \frac{\partial^2 v^0}{\partial x^2} + (B_{12} + B_{66}) \frac{\partial^2 v^0}{\partial x \partial y}$$

$$+ B_{26} \frac{\partial^2 v^0}{\partial y^2} + D_{11} \frac{\partial^2 \psi_x}{\partial x^2} + 2D_{16} \frac{\partial^2 \psi_x}{\partial x \partial y} + D_{66} \frac{\partial^2 \psi_x}{\partial y^2} + D_{16} \frac{\partial^2 \psi_y}{\partial x^2}$$

$$(10.16)$$

$$+ (D_{12} + D_{66}) \frac{\partial^2 \psi_y}{\partial x \partial y} + D_{26} \frac{\partial^2 \psi_y}{\partial y^2} - k[A_{55}(\psi_x + \frac{\partial w}{\partial x}) + A_{45}(\psi_y$$

$$+ \frac{\partial w}{\partial y})] = H \frac{\partial^2 u^0}{\partial t^2} + I \frac{\partial^2 \psi_x}{\partial t^2}$$

$$B_{16} \frac{\partial^2 u^0}{\partial x^2} + (B_{12} + B_{66}) \frac{\partial^2 u^0}{\partial x \partial y} + B_{26} \frac{\partial^2 u^0}{\partial y^2} + B_{66} \frac{\partial^2 v^0}{\partial x^2} + 2B_{26} \frac{\partial^2 v^0}{\partial x \partial y}$$

$$+ B_{22} \frac{\partial^2 v^0}{\partial y^2} + D_{16} \frac{\partial^2 \psi_x}{\partial x^2} + (D_{12} + D_{66}) \frac{\partial^2 \psi_x}{\partial x \partial y} + D_{26} \frac{\partial^2 \psi_x}{\partial y^2} + D_{66} \frac{\partial^2 \psi_y}{\partial x^2}$$

$$(10.17)$$

$$+ 2D_{26} \frac{\partial^2 \psi_y}{\partial x \partial y} + D_{22} \frac{\partial^2 \psi_y}{\partial y^2} - k[A_{45}(\psi_x + \frac{\partial w}{\partial x}) + A_{44}(\psi_y + \frac{\partial w}{\partial y})]$$

$$= H \frac{\partial^2 v^0}{\partial t^2} + I \frac{\partial^2 \psi_y}{\partial t^2}$$

$$k[A_{55} \left(\frac{\partial \psi_x}{\partial x} + \frac{\partial^2 w}{\partial x^2} \right) + A_{45} \left(\frac{\partial \psi_x}{\partial y} + \frac{\partial \psi_y}{\partial x} + 2 \frac{\partial^2 w}{\partial x \partial y} \right) + A_{44} \left(\frac{\partial \psi_y}{\partial y} \right) \qquad (10.18)$$

$$+ \frac{\partial^2 w}{\partial y^2} \Bigg) \Bigg] + q + N_x^i \frac{\partial^2 w}{\partial x^2} + 2N_{xy}^i \frac{\partial^2 w}{\partial x \partial y} + N_y^i \frac{\partial^2 w}{\partial y^2} = \varrho \frac{\partial^2 w}{\partial t^2}$$

For symmetric laminates, $B_{ij} = H = 0$, and Equations (10.13) and (10.14) reduce to the plane stress Equations (2.35) and (2.36), respectively, while Equations (10.15) and (10.16) reduce to the following:

$$D_{11} \frac{\partial^2 \psi_x}{\partial x^2} + 2D_{16} \frac{\partial^2 \psi_x}{\partial x \partial y} + D_{66} \frac{\partial^2 \psi_x}{\partial y^2} + D_{16} \frac{\partial^2 \psi_y}{\partial x^2} + (D_{12} + D_{66}) \frac{\partial^2 \psi_y}{\partial x \partial y}$$

$$(10.19)$$

$$+ D_{26} \frac{\partial^2 \psi_y}{\partial y^2} - k[A_{55}(\psi_x + \frac{\partial w}{\partial x}) + A_{45}(\psi_y + \frac{\partial w}{\partial y})] = I \frac{\partial^2 \psi_x}{\partial t^2}$$

$$D_{16} \frac{\partial^2 \psi_x}{\partial x^2} + (D_{12} + D_{66}) \frac{\partial^2 \psi_x}{\partial x \partial y} + D_{26} \frac{\partial^2 \psi_x}{\partial y^2} + D_{66} \frac{\partial^2 \psi_y}{\partial x^2} + 2D_{26} \frac{\partial^2 \psi_y}{\partial x \partial y}$$

$$(10.20)$$

$$+ D_{22} \frac{\partial^2 \psi_y}{\partial y^2} - k[A_{45}(\psi_x + \frac{\partial w}{\partial x}) + A_{44}(\psi_y + \frac{\partial w}{\partial y})] = I \frac{\partial^2 \psi_y}{\partial t^2}$$

Equation (10.18) remains unchanged. For homogeneous, isotropic plates, Equations (10.18)–(10.20) reduce to Mindlin's plate theory [5].

Ply stresses can be obtained by combining Equations (2.24) with the kinematic relations (2.6), (10.2), and (10.3), with the following results in matrix notation:

$$
\begin{bmatrix} \sigma_x^{(k)} \\ \sigma_y^{(k)} \\ \sigma_{xy}^{(k)} \end{bmatrix} =
\begin{bmatrix} Q_{11}^{(k)} & Q_{12}^{(k)} & Q_{16}^{(k)} \\ Q_{12}^{(k)} & Q_{22}^{(k)} & Q_{26}^{(k)} \\ Q_{16}^{(k)} & Q_{26}^{(k)} & Q_{66}^{(k)} \end{bmatrix}
\begin{bmatrix} \dfrac{\partial u^o}{\partial x} \\ \dfrac{\partial v^o}{\partial y} \\ \dfrac{\partial u^o}{\partial y} + \dfrac{\partial v^o}{\partial x} \end{bmatrix}
$$

$$(10.21)$$

$$
+ z
\begin{bmatrix} Q_{11}^{(k)} & Q_{12}^{(k)} & Q_{16}^{(k)} \\ Q_{12}^{(k)} & Q_{22}^{(k)} & Q_{26}^{(k)} \\ Q_{16}^{(k)} & Q_{26}^{(k)} & Q_{66}^{(k)} \end{bmatrix}
\begin{bmatrix} \dfrac{\partial \psi_x}{\partial x} \\ \dfrac{\partial \psi_x}{\partial y} \\ \dfrac{\partial \psi_x}{\partial y} + \dfrac{\partial \psi_y}{\partial x} \end{bmatrix}
$$

Since ϵ_{xz} and ϵ_{yz}, as given by Equation (10.4), are independent of the z coordinate, then the interlaminar shear stresses σ_{xz} and σ_{yz} will also be independent of z if calculated from the constitutive relations (1.26). In particular,

$$
\begin{bmatrix} \sigma_{yz}^{(k)} \\ \\ \sigma_{xz}^{(k)} \end{bmatrix} = \begin{bmatrix} C_{44}^{(k)} & C_{45}^{(k)} \\ \\ C_{45}^{(k)} & C_{55}^{(k)} \end{bmatrix} \begin{bmatrix} \psi_y + \dfrac{\partial w}{\partial y} \\ \\ \psi_x + \dfrac{\partial w}{\partial x} \end{bmatrix}
\tag{10.22}
$$

Equations (10.22) yield shear stresses which are uniform through-the-thickness of each ply and discontinuous at the ply interfaces. Such a distribution is clearly unrealistic. Thus the procedure outlined in Chapter 2 for calculating interlaminar stresses should be utilized also in conjunction with the shear deformation theory presented here. In particular, the equilibrium Equations (2.8) in conjunction with Equations (10.21) can be integrated to obtain more realistic interlaminar stresses.

The proper boundary conditions to be utilized with the present shear deformation theory can be obtained from variational principles in the same manner as outlined in Chapter 3 for classical laminated plate theory. The details are not presented here, but the procedure leads to boundary conditions which require prescribing one member of each pair of the following five quantities:

$$
u_n^0, N_n; \quad u_s^0, N_{ns}; \quad \psi_n, M_n; \quad \psi_s, M_{ns}; \quad w, Q_n
\tag{10.23}
$$

where n and s denote coordinates normal and tangential to the plate edge, respectively. Thus for the general case of a laminated plate which includes transverse shear deformation, five boundary conditions are prescribed. In this case the Kirchhoff boundary conditions of classical laminated plate theory are replaced by the last two conditions listed in Equation (10.23). For stability problems the shear force boundary condition is replaced by the following:

$$
Q_n + N_n^i \frac{\partial w}{\partial n} + N_{ns}^i \frac{\partial w}{\partial s}
\tag{10.24}
$$

The following boundary conditions correspond to those listed in Section 2.8 for the Kirchhoff theory:

(1) Simply-supported

$$
N_n = N_{ns} = M_n = \psi_s = w = 0
\tag{10.25}
$$

(2) Hinged-free in the normal direction

$$N_n = u_s^0 = M_n = \psi_s = w = 0$$

(3) Hinged-free in the tangential direction

$$u_n^0 = N_{ns} = M_n = \psi_s = w = 0 \qquad (10.26)$$

(4) Clamped

$$u_n^0 = u_s^0 = \psi_n = \psi_s = w = 0 \qquad (10.27)$$

(5) Free

$$N_n = N_{ns} = M_n = M_{ns} = Q_n = 0 \qquad (10.28)$$

10.4 DETERMINATION OF THE k PARAMETER

Various values of k have been used for homogeneous, isotropic plates. For example, Reissner [4], Mindlin [5], and Uflyand [8] used values of 5/6, $\pi^2/12$, and 2/3, respectively. A similar range of values, depending on the material properties, can be calculated from numerical results presented in Reference [6] for a two-layer isotropic plate. Comparison of results derived from exact solutions to dynamic elasticity equations and solutions to the shear deformation plate theory provide the basis for the determination of k in References [5] and [6]. The transverse shear stresses as derived from equilibrium conditions in conjunction with static bending loads provide the basis for determining values of k in References [4] and [8]. The classical values of k derived from static bending of homogeneous plates are considered in conjunction with the shear deformation theory presented in this chapter.

Integrating the static form of the first relationship in Equation (2.8), we find that for a homogeneous plate

$$\sigma_{xz} = -\frac{\partial}{\partial x} \int_{-h/2}^{z} \sigma_x \, d\xi - \frac{\partial}{\partial y} \int_{-h/2}^{z} \sigma_{xy} \, d\xi \qquad (10.29)$$

For the bending of a homogeneous, orthotropic plate the constitutive relations in Equation (10.21) reduce to the form

$$\begin{bmatrix} \sigma_x \\ \sigma_y \\ \sigma_{xy} \end{bmatrix} = z \begin{bmatrix} Q_{11} & Q_{12} & 0 \\ Q_{12} & Q_{22} & 0 \\ 0 & 0 & Q_{66} \end{bmatrix} \begin{bmatrix} \varkappa_x \\ \varkappa_y \\ \varkappa_{xy} \end{bmatrix} \qquad (10.30)$$

Substituting Equation (10.30) into Equation (10.29) and performing the integration, we obtain the following result:

$$\sigma_{xz} = \frac{h^2}{8} \left[\frac{\partial}{\partial x} (Q_{11}\varkappa_x + Q_{12}\varkappa_y) + \frac{\partial}{\partial y} (Q_{66}\varkappa_{xy}) \right](1 - \frac{z^2}{h^2}) \qquad (10.31)$$

It can readily be seen from the constitutive relations (10.5) that for a homogeneous, orthotropic plate

$$(Q_{11}\varkappa_x + Q_{12}\varkappa_y) = \frac{12}{h^3} M_x$$

$$\qquad (10.32)$$

$$Q_{66}\varkappa_{xy} = \frac{12}{h^3} M_{xy}$$

Equation (10.30) can now be written in the form

$$\sigma_{xz} = \frac{3}{2h} \left(\frac{\partial M_x}{\partial x} + \frac{\partial M_{xy}}{\partial y} \right) (1 - \frac{z^2}{h^2}) \qquad (10.33)$$

Using the equilibrium Equation (2.19) in conjunction with Equation (10.33), we obtain the transverse shear stress distribution

$$\sigma_{xz} = \frac{3}{2h} (1 - 4\frac{z^2}{h^2}) Q_x \qquad (10.34)$$

A similar procedure with the static form of the second relationship in Equation (2.8) yields

$$\sigma_{yz} = \frac{3}{2h} (1 - 4\frac{z^2}{h^2}) Q_y \qquad (10.35)$$

Two approaches to determining k will now be considered. The first method is identical to the approach in Timoshenko beam theory [9].

For a homogeneous, orthotropic plate the shear stress-strain relations (10.22) take the form

$$\sigma_{xz} = C_{55}\epsilon_{xz}$$

$$\qquad (10.36)$$

$$\sigma_{yz} = C_{44}\epsilon_{yz}$$

Substituting the plate constitutive relations (10.6) into Equation (10.35), we obtain the result

$$\sigma_{xz} = \frac{Q_x}{kh}, \ \sigma_{yz} = \frac{Q_y}{kh} \tag{10.37}$$

The shear stresses at the center of the plate as given by Equations (10.34) and (10.35) are

$$\sigma_{xz} = \frac{3Q_x}{2h}, \ \sigma_{yz} = \frac{3Q_y}{2h} \tag{10.38}$$

If k is chosen such that Equations (10.37) and (10.38) are identical, then

$$k = \frac{2}{3} \tag{10.39}$$

The second approach involves a procedure which is the same as utilized in Reissner plate theory [4] and later applied to homogeneous, orthotropic plates by Medwadowski [10].

The total strain energy, U, due to transverse shear is given by the relationship

$$U = \frac{1}{2} \int_{-h/2}^{h/2} \int_0^b \int_0^a (\tau_{xz}\epsilon_{xz} + \tau_{yz}\epsilon_{yz}) dx \, dy \, dz \tag{10.40}$$

Substituting the plate constitutive relations (10.6) into Equation (10.40), for a homogeneous, orthotropic plate we obtain the result

$$U = \frac{1}{2k} \int_0^b \int_0^a \left[\int_{-h/2}^{h/2} (\frac{Q_x}{A_{55}} \tau_{xz} + \frac{Q_y}{A_{44}} \tau_{yz}) dz \right] dx \, dy \tag{10.41}$$

or

$$U = \frac{1}{2k} \int_0^b \int_0^a \left(\frac{Q_x^2}{A_{55}} + \frac{Q_y^2}{A_{44}} \right) dx \, dy \tag{10.42}$$

For a homogeneous, orthotropic material, Equation (10.40) can also be written in the form

$$U = \frac{h}{2} \int_0^b \int_0^a \int_{-h/2}^{h/2} \left(\frac{\tau_{xz}^2}{A_{55}} + \frac{\tau_{yz}^2}{A_{44}} \right) dz \, dx \, dy \tag{10.43}$$

Substituting the stress distributions given by Equations (10.34) and (10.35) into Equation (10.43) and integrating with respect to z, we obtain the result

$$U = \frac{3}{5} \int_0^b \int_0^a \left(\frac{Q_x^2}{A_{55}} + \frac{Q_y^2}{A_{44}} \right) dx \, dy \tag{10.44}$$

Now, if k is chosen such that Equations (10.42) and (10.44) are identical, then

$$k = \frac{5}{6} \tag{10.45}$$

The procedures utilized here can also be readily applied to laminated, anisotropic plates for the determination of k. However for the general case, results will not appear in closed form. Furthermore, values of k will depend on ply properties and stacking geometries of the laminate. Examples of applications to laminates can be found in References [11] and [12]. In both of these cases two k factors were applied: k_1 associated with ϵ_{xz}, and k_2 associated with ϵ_{yz}.

10.5 CYLINDRICAL BENDING OF ORTHOTROPIC LAMINATES

Consider a laminate composed of an arbitrary number of orthotropic layers such that the various axes of material symmetry are parallel to the x-y axes of the plate ($A_{16} = A_{26} = B_{16} = B_{26} = D_{16} = D_{26} = 0$). In addition, let us assume a state of plane strain in which the plate is of infinite length in the y direction, uniformly supported along the edges $x = 0, a$ and subjected to the loads $q = q(x)$, $N_y^i = N_{xy}^i = 0$, and $N_x^i = N_x^i(x)$. Under these conditions the deflected surface is cylindrical, i.e.,

$$v^0 = \psi_y = 0, \quad u^0 = u^0(x,t), \quad \psi_x = \psi_x(x,t), \quad w = w(x,t) \tag{10.46}$$

and Equations (10.14)–(10.18) reduce to the following:

$$A_{11} \frac{\partial^2 u^0}{\partial x^2} + B_{11} \frac{\partial^2 \psi_x}{\partial x^2} = \varrho \frac{\partial^2 u^0}{\partial t^2} + H \frac{\partial^2 \psi_x}{\partial t^2} \tag{10.47}$$

$$B_{11} \frac{\partial^2 u^0}{\partial x^2} + D_{11} \frac{\partial^2 \psi_x}{\partial x^2} - kA_{55}(\psi_x + \frac{\partial w}{\partial x}) = H \frac{\partial^2 u^0}{\partial t^2} + I \frac{\partial^2 \psi_x}{\partial t^2} \tag{10.48}$$

$$kA_{55} \left(\frac{\partial \psi_x}{\partial x} + \frac{\partial^2 w}{\partial x^2} \right) + q + N_x^i \frac{\partial^2 w}{\partial x^2} = \varrho \frac{\partial^2 w}{\partial t^2} \tag{10.49}$$

For simple supports the following boundary conditions are applicable:
at $x = 0$ and a

$$N_x = A_{11} \frac{\partial u^o}{\partial x} + B_{11} \frac{\partial \psi_x}{\partial x} = 0$$

$$w = M_x = B_{11} \frac{\partial u^o}{\partial x} + D_{11} \frac{\partial \psi_x}{\partial x} = 0$$

(10.50)

We now consider bending under lateral loads, stability under inplane loads, and free-vibration.

Case I: Bending Under Lateral Loads ($N_x^i = 0$)

Recognizing that almost any load function can be expressed in the form of a Fourier series, we let

$$q(x) = q_0 \sin \frac{m\pi x}{a}$$

(10.51)

Solutions to the static form of Equations (10.47)–(10.49), which satisfy the boundary conditions (10.50), are of the form

$$u^o = A_m \cos \frac{m\pi x}{a}$$

$$\psi_x = B_m \cos \frac{m\pi x}{a}$$

$$w = C_m \sin \frac{m\pi x}{a}$$

(10.52)

Substituting Equations (10.52) into Equations (10.47)–(10.49) and solving the resulting algebraic equations, we find

$$A_m = \frac{B_{11}\, a^3\, q_0}{m^3 \pi^3 (A_{11}D_{11} - B_{11}^2)}$$

$$B_m = \frac{-A_{11}\, a^3\, q_0}{m^3 \pi^3 (A_{11}D_{11} - B_{11}^2)}$$

(10.53)

$$C_m = \frac{q_0\, a^2}{m^2 \pi^2} \left[\frac{A_{11}\, a^2}{m^2 \pi^2 (A_{11}D_{11} - B_{11}^2)} + \frac{1}{kA_{55}} \right]$$

The maximum deflection occurs at the center of the plate ($x = a/2$) and is of the form

$$w_{max} = w'_{max} (1 + C + Sm^2\pi^2) \tag{10.54}$$

where w'_{max} is the maximum deflection of a laminated plate in which both bending-extensional coupling and shear deformation are neglected, C is the bending-extensional coupling term, and S represents a shear deformation term. These expressions are given as follows:

$$w'_{max} = \frac{q_0 a^4}{m^4 \pi^4 D_{11}}, \quad C = \frac{B_{11}^2}{(A_{11}D_{11} - B_{11}^2)},$$

$$\tag{10.55}$$

$$S = \frac{D_{11}}{kA_{55}a^2}$$

Thus shear deformation increases the maximum deflection. Let us examine the shear deformation term, S, in more detail. The bending stiffness D_{11} can be writ-

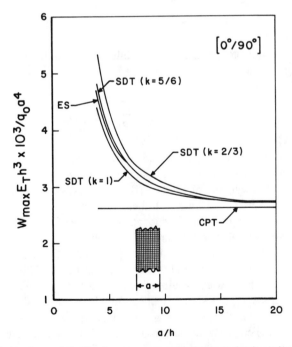

Figure 10.1. *Maximum deflection for an unsymmetric laminated strip under transverse load.*

ten in terms of an effective bending stiffness, $\overline{Q}_{11}$, having the same units as modulus, and A_{55} can be written in terms of an average interlaminar shear modulus, $\overline{G}_{13}$, i.e.,

$$\overline{Q}_{11} = \frac{12}{h^3} D_{11}, \ \overline{G}_{13} = \frac{A_{55}}{h} \tag{10.56}$$

The shear factor S in Equation (10.57) can now be written in the form

$$S = \frac{1}{12k} \left(\frac{\overline{Q}_{11}}{\overline{G}_{13}} \right) \left(\frac{h}{a} \right)^2 \tag{10.57}$$

Thus the magnitude of the shear deformation term depends on the modulus ratio $\overline{Q}_{11}/\overline{G}_{13}$ and on the span-to-depth ratio a/h. Force and moment resultants can be determined from Equations (10.5) and (10.6). For symmetrical laminates, $B_{11} = 0$ and u^0 vanishes.

The maximum deflection as a function of a/h is illustrated in Figures 10.1 and 10.2 [7] for $[0°/90°]$ unsymmetric laminates and for $[0°/\overline{90}°]_s$ laminates, respectively. The following ply properties are utilized in these figures:

$$E_L/E_T = 25, \ G_{LT}/E_T = 0.5, \ G_{TT}/E_T = 0.2, \ \nu_{LT} = 0.25 \tag{10.58}$$

where G_{TT} denotes the interlaminar shear modulus relative to the x_T–z plane. The surface loading is of the form

$$q = q_0 \sin \frac{\pi x}{a} \tag{10.59}$$

In Figures 10.1 and 10.2 comparison is made between the present shear deformation theory (SDT) for several values of k, and exact elasticity solutions (ES) for laminates subjected to cylindrical bending [1]. Solutions obtained from classical laminated plate (CPT) are also shown in these figures.

The stresses calculated from SDT are identical to those of CPT. Thus the inaccuracies in stresses as calculated from CPT for low span-to-depth ratios, a/h, which has been discussed in detail by Pagano [1], are not alleviated by the present theory. This is due to the fact that the stresses are linear in z within each ply for SDT. Improvement in stresses requires higher order displacement terms. It should be noted, however, that significant departure between classical theory and exact elasticity occurs at values of $a/h \leq 10$, which represents a relatively thick plate. In addition, the emphasis in the current theory is on the accurate prediction of global response.

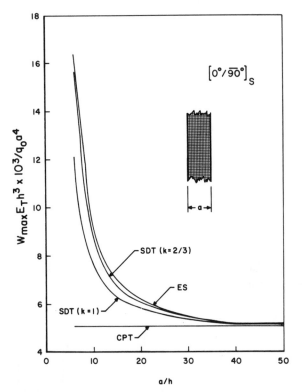

Figure 10.2. *Maximum deflection for a symmetric laminated strip under transverse load.*

Case II: Stability of Symmetric Laminates Under Uniform Compression Loading (q = 0)

We now consider the effect of shear deformation on critical buckling loads. For symmetric, specially orthotropic laminates under static inplane loading, Equations (10.47)–(10.49) reduce to the following:

$$D_{11} \frac{d^2\psi_x}{dx^2} - kA_{55}(\psi_x + \frac{dw}{dx}) = 0$$

$$kA_{55}\left(\frac{d\psi_x}{dx} + \frac{d^2w}{dx^2}\right) + N_x^i \frac{d^2w}{dx^2} = 0$$

(10.60)

We consider the initial inplane load $N_x^i = -N_0 = $ constant, where $N_0 > 0$.

For simple supports the boundary conditions (10.50) reduce to the following:

$$w = M_x = D_{11} \frac{d\psi_x}{dx} = 0 \qquad (10.61)$$

A solution to Equations (10.60) which satisfies the boundary conditions (10.61) is of the same form as Equations (10.52), i.e.,

$$\psi_x = A_m \cos \frac{m\pi x}{a}$$

$$(10.62)$$

$$w = B_m \sin \frac{m\pi x}{a}$$

Substituting these functions into Equations (10.60), we obtain the following homogeneous, algebraic equations:

$$\begin{bmatrix} (D_{11} + \frac{kA_{55}a^2}{m^2\pi^2}) & \frac{kA_{55}a}{m\pi} \\ \frac{kA_{55}a}{m\pi} & (kA_{55} - N_0) \end{bmatrix} \begin{bmatrix} A_m \\ B_m \end{bmatrix} = \begin{bmatrix} 0 \\ 0 \end{bmatrix} \qquad (10.63)$$

A nontrivial solution can be obtained by choosing N_0 such that the determinant of the coefficient matrix (10.63) vanishes, with the result

$$N_0 = \frac{kA_{55}D_{11}m^2\pi^2}{(D_{11}m^2\pi^2 + kA_{55}a^2)} \qquad (10.64)$$

The lowest value of N_0, and hence the critical buckling load, is obtained for $m = 1$. We now write this result in the following form:

$$N_{cr} = N'_{cr} (1 - \frac{S\pi^2}{1 + S\pi^2}) \qquad (10.65)$$

where S is the shear deformation factor defined in Equation (10.55), and N'_{cr} is the

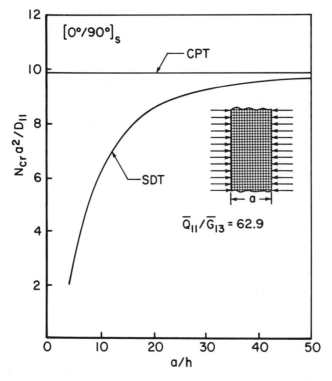

Figure 10.3. *Critical buckling load for a laminated strip under compression loading.*

critical buckling load for the same laminate with shear deformation neglected and is given by Equation (4.32), i.e.,

$$N'_{cr} = D_{11} \frac{\pi^2}{a^2} \tag{10.66}$$

It can be seen from Equation (10.65) that shear deformation decreases critical buckling loads. As in the case of bending under lateral loads, the magnitude of this reduction depends on the ratios $\overline{Q}_{11}/\overline{G}_{13}$ and a/h.

Numerical results are illustrated in Figure 10.3 for a $[0°/90°]_s$ laminate with the ply properties given by equation (10.58) and $k = 5/6$.

Case III: Free Vibration of Symmetric Laminates ($N_x^i = q = 0$)

We now consider the effect of shear deformation on free vibration frequencies.

For symmetric, orthotropic laminates in the absence of lateral and inplane loads, Equations (10.47)–(10.49) reduce to the following:

$$D_{11} \frac{\partial^2 \psi_x}{\partial x^2} - kA_{55}(\psi_x + \frac{\partial w}{\partial x}) = I \frac{\partial^2 \psi_x}{\partial t^2}$$

$$kA_{55}\left(\frac{\partial \psi_x}{\partial x} + \frac{\partial^2 w}{\partial x^2}\right) = \varrho \frac{\partial^2 w}{\partial t^2}$$

(10.67)

For simple supports, the boundary conditions are those given in Equation (10.63). Displacements which satisfy Equations (10.67) and the boundary conditions are of the form

$$\psi_x = A_m e^{-i\omega_m t} \cos \frac{m\pi x}{a}$$

$$\psi = B_m e^{-i\omega_m t} \sin \frac{m\pi x}{a}$$

(10.68)

Substituting Equations (10.68) into Equations (10.67), we obtain the following homogeneous algebraic equations:

$$\begin{bmatrix} (D_{11} \frac{m^2\pi^2}{a^2} + kA_{55} - I\omega_m^2) & kA_{55}\frac{m\pi}{a} \\ kA_{55}\frac{m\pi}{a} & (kA_{55}\frac{m^2\pi^2}{a^2} - \varrho\omega_m^2) \end{bmatrix} \begin{bmatrix} A_m \\ B_m \end{bmatrix} = \begin{bmatrix} 0 \\ 0 \end{bmatrix}$$

(10.69)

For homogeneous plies, i.e., laminates in which each ply is constructed of the same material, ϱ_0 is a constant and

$$\varrho = \varrho_0 h, \quad I = \varrho_0 \frac{h^3}{12}$$

(10.70)

A nontrivial solution to Equations (10.69) can be obtained by choosing ω_m^2 such that the determinant of the coefficient matrix vanishes, with the result

$$\omega_m^2 = \frac{6}{\varrho_0 a^2 h^3}\left\{ D_{11}m^2\pi^2 + kA_{55}(a^2 + \frac{m^2\pi^2 h^2}{12}) \right\}$$

(10.71)

$$\pm \sqrt{[D_{11}m^2\pi^2 + kA_{55}(a^2 + \frac{m^2\pi^2h^2}{12})]^2 - kA_{55}D_{11}\frac{m^2\pi^2h^2}{3}}\Bigg\}$$

If we neglect rotatory inertia terms, Equation (10.71) reduces to the form

$$\omega_m = \omega_m' \sqrt{1 - \frac{Sm^2\pi^2}{1 + Sm^2\pi^2}} \tag{10.72}$$

where S is the shear deformation factor defined in Equation (10.55), and ω_m' is the free vibration frequency for the same laminate with shear deformation neglected and is given by Equation (4.31), i.e.,

$$\omega_m' = \frac{m^2\pi^2}{a^2}\sqrt{\frac{D_{11}}{\varrho_0 h}} \tag{10.73}$$

As in the case of critical buckling loads, shear deformation decreases vibration frequencies. The magnitude of the reduction depends on the ratios $\overline{Q}_{11}/\overline{G}_{13}$ and a/h. It is also obvious from Equation (10.73) that the fundamental vibration frequency occurs for $m = 1$.

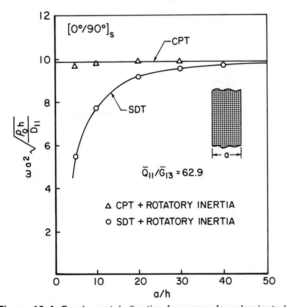

Figure 10.4. *Fundamental vibration frequency for a laminated strip.*

Numerical results are shown in Figure 10.4 for a $[0°/90°]_s$ laminate with the ply properties given by Equation (10.58) and $k = 5/6$. We note that rotatory inertia has little effect compared to shear deformation.

10.6 BENDING OF LAMINATED BEAMS

We now consider the effect of shear deformation on the bending of laminated beams. As in Section 4.5, we limit our discussion to symmetric laminates.

For present purposes it is useful to consider Equation (10.6) along with the bending portion of Equation (10.5) in the inverted forms

$$[\varkappa] = [D^*][M] \tag{10.74}$$

$$
\begin{bmatrix} \epsilon_{yz} \\ \\ \epsilon_{xz} \end{bmatrix} = \frac{1}{k} \begin{bmatrix} A_{44}^* & A_{45}^* \\ \\ A_{45}^* & A_{55}^* \end{bmatrix} \begin{bmatrix} Q_y \\ \\ Q_x \end{bmatrix} \tag{10.75}
$$

where D_{ij}^* are elements of the inverse matrix of D_{ij} and

$$A_{44}^* = \frac{A_{55}}{A} \, , \, A_{45}^* = -\frac{A_{45}}{A} \, , \, A_{55}^* = \frac{A_{44}}{A}$$

and

$$A = (A_{44}A_{55} - A_{45}^2)$$

Equation (10.74) is of the same form as Equation (4.39) with $\varkappa$ defined by Equation (10.3).

We assume that both ψ_x and w are independent of y, i.e.,

$$\psi_x = \psi_x(x) = \psi, \, w = w(x) \tag{10.76}$$

Combining Equations (10.76) with Equations (10.2)–(10.4), we obtain the strain-displacement relations

$$\epsilon_x = z \frac{d\psi}{dx}$$

$$\tag{10.77}$$

$$\epsilon_{xz} = \psi + \frac{dw}{dx}$$

From the constitutive Equations (10.74) we obtain the relationship

$$\varkappa_x = \frac{d\psi}{dx} = D_{11}^* \, M_x \tag{10.78}$$

Since we are discussing beam theory, the assumption given by Equation (4.40) is still applicable, i.e., we assume

$$M_y = M_{xy} = 0 \tag{10.79}$$

Substituting Equation (10.79) into the equilibrium Equation (2.20), it is easily seen that Q_y vanishes. The second of Equations (10.75) can now be written in the form

$$\epsilon_{xz} = \psi + \frac{dw}{dx} = \frac{A_{55}^*}{k} \, Q_x \tag{10.80}$$

Substituting Equations (10.78) and (10.80) into the equilibrium Equation (2.19), we obtain the relationship

$$\frac{d^2\psi}{dx^2} - k(\psi + \frac{dw}{dx}) \frac{G_{xz}bh}{E_x^b I} \tag{10.81}$$

where

$$E_x^b = \frac{12}{h^3 D_{11}^*} \,, \; G_{xz} = \frac{1}{h A_{55}^*} \tag{10.82}$$

Since Q_y vanishes, the static form of Equation (2.16) in the absence of inplane force effects reduces to

$$\frac{dQ_x}{dx} + q = 0 \tag{10.83}$$

Substituting Equation (10.80) into Equation (10.83) and taking into account Equation (10.82), we obtain the differential equation

$$\frac{d^2w}{dx^2} + \frac{d\psi}{dx} + \frac{p}{kG_{xz}h} = 0 \tag{10.84}$$

where, as previously,

$$p = bq$$

Equations (10.81) and (10.84) constitute two equations in the two unknowns ψ and w. Thus these are the governing bending equations for symmetric, laminated, anisotropic beams with shear deformation effects included. These equations are of the same form as Timoshenko beam theory [9].

As in the case of cylindrical bending, the stress distributions will not change for the transverse shear deformation theory compared to the original classical beam theory. Again, this is due to the assumption of linear displacements with respect to the z coordinate.

In order to illustrate the effect of transverse shear deformation on beam bending, the higher order theory will now be applied to beams subjected to 3-point and 4-point loading.

Since the moment distribution is known for these problems, Equation (10.78) can be utilized in the form

$$\frac{d\psi}{dx} = \frac{M}{E_x^b I} \tag{10.85}$$

where, as in Equation (4.44),

$$M = b \, M_x$$

A second equation can be obtained by combining Equation (4.49) and Equation (10.80), with the result

$$\frac{dM}{dx} = kG_{xz}bh(\psi + \frac{dw}{dx}) \tag{10.86}$$

Case I: Bending Under 3-Point Loading

We consider a beam under 3-point bending as illustrated in Figure 4.4. The bending moment in the left half of the beam is given by the relationship

$$M = -\frac{Px}{2}, \quad 0 \le x \le L/2 \tag{10.87}$$

Substituting this relationship into Equation (10.85), and integrating, we obtain the result

$$\psi = -\frac{Px^2}{4E_x^b I} + c_1, \quad 0 \le x \le L/2 \tag{10.88}$$

Symmetry of the loading requires that $u(L/2) = 0$. It is easily seen from Equation (10.1) that this condition leads to the result

$$\psi(L/2) = 0 \tag{10.89}$$

Using this condition in conjunction with Equation (10.88), we obtain the solution

$$\psi = \frac{PL^2}{16E_x^b I} \left[1 - 4 \left(\frac{x}{L} \right)^2 \right], \quad 0 \le x \le L/2 \tag{10.90}$$

Substituting Equation (10.87) into Equation (10.86), and rearranging terms, we obtain the result

$$\frac{dw}{dx} = - (\psi + \frac{P}{2kG_{xz}bh}), \quad 0 \le x \le L/2 \tag{10.91}$$

Note that Equation (10.91) yields a nonvanishing slope at the center of the beam. In particular, since $\psi(L/2) = 0$, then

$$\frac{dw}{dx}(L/2) = \frac{-P}{2kG_{xz}bh}, \quad 0 \le x \le L/2 \tag{10.92}$$

Integrating Equation (10.92) after substituting Equation (10.90) and applying the condition that $w(0) = 0$ leads to the deflection equation

$$w = \frac{PL^2 x}{4E_x^b bh^3} \, [4 \left(\frac{x}{L} \right)^2 - 3 - 2s], \quad 0 \le x \le L/2 \tag{10.93}$$

where S is a shear correction factor given by

$$S = \frac{1}{k} \left(\frac{E_x^b}{G_{xz}} \right) \left(\frac{h}{L} \right)^2 \tag{10.94}$$

Note that this shear deformation factor is of the same form as Equation (10.57) for plates. Thus the effect of shear deformation depends on the span-to-depth ratio L/h and the modulus ratio E_x^b/G_{xz}. If the modulus is expressed in terms of the center deflection w_c, Equation (10.93) becomes

$$E_x^b = \frac{PL^3}{4w_c bh^3} (1 + S) \tag{10.95}$$

Often the exact value of G_{xz} may not be known. Often, however, an approximate value of E_x^b/G_{xz} is known, and Equation (10.95) is useful under such circumstances. When G_{xz} is known, Equation (10.93) can be solved for E_x^b with $x = L/2$. In this case,

$$E_x^b = \frac{PL^3}{4bh^3 \left[w_c - \dfrac{PL}{4kbh\ G_{xz}} \right]} \tag{10.96}$$

It can be easily seen that neglect of shear deformation leads to an underestimated value of E_x^b. Dividing Equation (4.61) by Equation (10.95), we obtain the result

$$\frac{E_x^b(0)}{E_x^b(S)} = \frac{1}{(1 + S)} \tag{10.97}$$

where $E_x^b(S)$ denotes bending modulus with shear deformation effects included, while $E_x^b(0)$ denotes bending modulus without the effect of shear deformation.

Equation (10.97) is plotted as a function of span-to-depth ratio, L/h, in Figure 10.5 for two ratios of E_x^b/G_{xz}. The smaller ratio corresponds to unidirectional S-glass/epoxy material, while the larger corresponds to graphite/epoxy unidirectional composites.

Case II: Bending Under 4-Point Loading

We now consider a beam under 4-point loading at quarter points, as illustrated in Figure 4.4. For the left half of the beam, the bending moment is given by the relationships

$$M = -\frac{Px}{2}, \quad 0 \leq x \leq L/4 \tag{10.98}$$

$$M = -\frac{PL}{8}, \quad L/4 \leq x \leq L/2 \tag{10.99}$$

Substituting Equations (10.98) and (10.99) into Equation (10.85) and integrating, we obtain the following results:

$$\psi = \psi_1 = -\frac{Px^2}{4E_x^b I} + c_1, \quad 0 < x < L/2 \tag{10.100}$$

$$\psi = \psi_2 = -\frac{PLx}{8E_x^b I} + c_2, \quad L/4 \leq x \leq L/2 \tag{10.101}$$

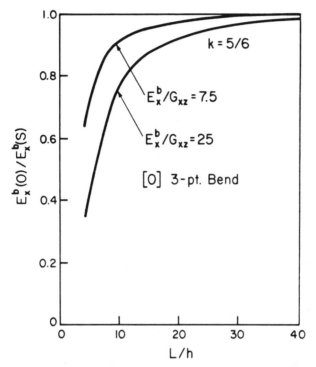

Figure 10.5. *Effect of shear deformation on bending modulus, 3-point bending.*

As in the case of 3-point bending, symmetry requires that ψ vanish at the center of the beam. Thus,

$$\psi_2(L/2) = 0 \tag{10.102}$$

and

$$\psi_2 = \frac{PL^2}{16E_x^b I}\left[1 - 2\left(\frac{x}{L}\right)\right] \tag{10.103}$$

Continuity of displacements at $x = L/4$ requires that

$$\psi_1(L/4) = \psi_2(L/4) \tag{10.104}$$

Combining Equations (10.100), (10.101), and (10.104), we obtain the relationship

$$\psi_1 = \frac{PL^2}{64E_x^b I}\left[3 - 16\left(\frac{x}{L}\right)^2\right] \tag{10.105}$$

Substituting Equations (10.98) and (10.99) into Equation (10.86) and rearranging terms, we obtain the relations

$$\frac{dw_1}{dx} = -(\psi_1 + \frac{P}{2kG_{xz}\mathrm{bh}}), \quad 0 \le x \le L/4 \tag{10.106}$$

$$\frac{dw_2}{dx} = -\psi_2, \quad L/4 \le x \le L/2 \tag{10.107}$$

Integrating Equation (10.107) after substituting Equation (10.103), we obtain the following deflection

$$w_2 = -\frac{PL^2 x}{16 E_x^b I}(1 - \frac{x}{L}) + c_3 \tag{10.108}$$

A similar procedure with Equations (10.105) and (10.106), along with the condition $w_1(0) = 0$, leads to the relationship

$$w_1 = \frac{PL^2 x}{192 E_x^b I}[16\left(\frac{x}{L}\right)^2 - 9 - 8S] \tag{10.109}$$

Continuity of displacements requires that

$$w_1(L/4) = w_2(L/4) \tag{10.110}$$

Combining Equations (10.108) and (10.109), we obtain the result

$$w_2 = \frac{PL^3}{768 E_x^b I}[1 - 48\left(\frac{x}{L}\right) + 48\left(\frac{x}{L}\right)^2 - 8S] \tag{10.111}$$

where S is defined by Equation (10.94). Expressing the modulus in terms of the center deflection, we obtain the result

$$E_x^b = \frac{PL^3}{64 w_c bh^3}(11 + 8S) \tag{10.112}$$

If Equation (10.111) is solved completely in terms of flexural modulus, then

$$E_x^b = \frac{11 PL^3}{64 bh^3 \left[w_c - \dfrac{PL}{8 kbh \, G_{xz}} \right]} \tag{10.113}$$

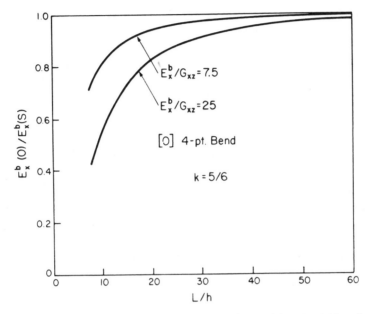

Figure 10.6. *Effect of shear deformation on bending modulus, 4-point bending.*

If modulus is determined at the quarter point, then Equation (10.111) yields

$$E_x^b = \frac{PL^3}{8w_q bh^3}(1 + S) \qquad (10.114)$$

where w_q denotes $w(L/4)$. If Equation (10.111) is solved completely in terms of modulus, then

$$E_x^b = \frac{PL^3}{8bh^3 \left[w_q - \dfrac{PL}{8kbh\, G_{xz}} \right]} \qquad (10.115)$$

If we divide Equation (4.75) by Equation (10.114), we obtain Equation (10.95). Thus shear deformation has the same effect on the determination of flexural modulus for 3-point bending with deflection measured at the center as it does on 4-point bending with deflection measured at the quarter point. Dividing Equation (4.76) by Equation (10.112), we obtain the modulus ratio

$$\frac{E_x^b(0)}{E^b(S)} = \frac{1}{(1 + 8S/11)} \qquad (10.116)$$

Thus shear deformation has less effect on the determination of flexural modulus if a 4-point bend test is utilized in conjunction with a center deflection measurement.

The modulus ratio in Equation (10.116) is plotted as a function of span-to-depth ratio, L/h, in Figure 10.6 for the same two ratios of E_x^b/G_{xz} as considered in Figure 10.5.

10.7 BENDING AND FREE VIBRATION OF ANGLE-PLY RECTANGULAR PLATES

We now consider bending under lateral loads and free vibration of rectangular laminates of the class $[\pm \theta]_n$. For these angle-ply plates, $A_{16} = A_{26} = B_{11} = B_{22} = B_{12} = B_{66} = D_{16} = D_{26} = 0$. Thus B_{16} and B_{26} are the only nonvanishing bending-extensional coupling terms.

In the absence of inplane force effects, Equations (10.14)–(10.18) reduce to the following:

$$A_{11} \frac{\partial^2 u^0}{\partial x^2} + A_{66} \frac{\partial^2 u^0}{\partial y^2} + (A_{12} + A_{66}) \frac{\partial^2 v^0}{\partial x \partial y} + 2B_{16} \frac{\partial^2 \psi_x}{\partial x \partial y}$$

$$+ B_{16} \frac{\partial^2 \psi_y}{\partial x^2} + B_{26} \frac{\partial^2 \psi_y}{\partial y^2} = \varrho \frac{\partial^2 u^0}{\partial t^2} + H \frac{\partial^2 \psi_x}{\partial x^2} \tag{10.117}$$

$$(A_{12} + A_{66}) \frac{\partial^2 u^0}{\partial x \partial y} + A_{66} \frac{\partial^2 v^0}{\partial x^2} + A_{22} \frac{\partial^2 v^0}{\partial y^2} + B_{16} \frac{\partial^2 \psi_x}{\partial x^2}$$

$$+ B_{26} \frac{\partial^2 \psi_x}{\partial y^2} + 2B_{26} \frac{\partial^2 \psi_y}{\partial x \partial y} = \varrho \frac{\partial^2 v^0}{\partial t^2} + H \frac{\partial^2 \psi_y}{\partial t^2} \tag{10.118}$$

$$2B_{16} \frac{\partial^2 u^0}{\partial x \partial y} + B_{16} \frac{\partial^2 v^0}{\partial x^2} + B_{26} \frac{\partial^2 v^0}{\partial y^2} + D_{11} \frac{\partial^2 \psi_x}{\partial x^2}$$

$$+ D_{66} \frac{\partial^2 \psi_x}{\partial y^2} + (D_{12} + D_{66}) \frac{\partial^2 \psi_y}{\partial x \partial y} - kA_{55}(\psi_x + \frac{\partial w}{\partial x}) \tag{10.119}$$

$$= H \frac{\partial^2 u^0}{\partial t^2} + I \frac{\partial^2 \psi_x}{\partial t^2}$$

$$B_{16} \frac{\partial^2 u^0}{\partial x^2} + B_{26} \frac{\partial^2 u^0}{\partial y^2} + 2B_{26} \frac{\partial^2 v^0}{\partial x \partial y} + (D_{12} + D_{66}) \frac{\partial^2 \psi_x}{\partial x \partial y}$$

$$+ D_{66} \frac{\partial^2 \psi_y}{\partial x^2} + D_{22} \frac{\partial^2 \psi_y}{\partial y^2} - kA_{44}(\psi_y + \frac{\partial w}{\partial y}) \qquad (10.120)$$

$$= H \frac{\partial^2 v^0}{\partial t^2} + I \frac{\partial^2 \psi_y}{\partial t^2}$$

$$k[A_{55} \left(\frac{\partial \psi_x}{\partial x} + \frac{\partial^2 w}{\partial x^2} \right) + A_{44} \left(\frac{\partial \psi_y}{\partial y} + \frac{\partial^2 w}{\partial y^2} \right)] + q = \varrho \frac{\partial^2 w}{\partial t^2} \qquad (10.121)$$

For hinged edges, free in the tangential direction, we have the following boundary conditions:
at $x = 0$ and a

$$u^0 = N_{xy} = A_{66} \left(\frac{\partial u^0}{\partial y} + \frac{\partial v^0}{\partial x} \right) + B_{16} \frac{\partial \psi_x}{\partial x} + B_{26} \frac{\partial \psi_y}{\partial y} = 0 \qquad (10.122)$$

$$w = \psi_y = M_x = B_{16} \left(\frac{\partial u^0}{\partial y} + \frac{\partial v^0}{\partial x} \right) + D_{11} \frac{\partial \psi_x}{\partial x} + D_{12} \frac{\partial \psi_y}{\partial y} = 0$$

at $y = 0$ and b

$$v^0 = N_{xy} = A_{66} \left(\frac{\partial u^0}{\partial y} + \frac{\partial v^0}{\partial x} \right) + B_{16} \frac{\partial \psi_x}{\partial x} + B_{26} \frac{\partial \psi_y}{\partial y} = 0 \qquad (10.123)$$

$$w = \psi_x = M_y = B_{26} \left(\frac{\partial u^0}{\partial x} + \frac{\partial v^0}{\partial y} \right) + D_{12} \frac{\partial \psi_x}{\partial x} + D_{22} \frac{\partial \psi_y}{\partial y}$$

If we consider a transverse load of the form

$$q = q_0 \sin \frac{m\pi x}{a} \sin \frac{n\pi y}{b} \qquad (10.124)$$

then a solution to Equations (10.115)–(10.119) which satisfies the boundary conditions (10.120) and (10.121) is found in the following:

$$u^0 = A_{mn} e^{i\omega_{mn} t} \sin \frac{m\pi x}{a} \cos \frac{n\pi y}{b} \qquad (10.125)$$

$$v^0 = B_{mn} \, e^{i\omega mnt} \cos\frac{m\pi x}{a} \sin\frac{n\pi y}{b}$$

$$\psi_x = C_{mn} \, e^{i\omega mnt} \cos\frac{m\pi x}{a} \sin\frac{n\pi y}{b}$$

$$\psi_y = D_{mn} \, e^{i\omega mnt} \sin\frac{m\pi x}{a} \cos\frac{n\pi y}{b}$$

$$w = E_{mn} \, e^{i\omega mnt} \sin\frac{m\pi x}{a} \sin\frac{n\pi y}{b}$$

Substituting Equations (10.125) into Equations (10.117)–(10.121), we arrive at a set of five algebraic equations.

$$
\begin{bmatrix}
(H_{11mn} - \lambda_{mn}) & H_{12mn} & H_{13mn} & H_{14mn} & 0 \\[2mm]
H_{12mn} & (H_{22mn} - \lambda_{mn}) & H_{14mn} & H_{24mn} & 0 \\[2mm]
H_{13mn} & H_{14mn} & (H_{33mn} - \dfrac{h^2}{12}\lambda_{mn}) & H_{34mn} & H_{35mn} \\[2mm]
H_{14mn} & H_{24mn} & H_{34mn} & (H_{44mn} - \dfrac{h^2}{12}\lambda_{mn}) & H_{45mn} \\[2mm]
0 & 0 & H_{35mn} & H_{45mn} & (H_{55mn} - \lambda_{mn})
\end{bmatrix}
$$

$$
\times
\begin{bmatrix}
A_{mn} \\[1mm]
B_{mn} \\[1mm]
C_{mn} \\[1mm]
D_{mn} \\[1mm]
E_{mn}
\end{bmatrix}
=
\begin{bmatrix}
0 \\[1mm]
0 \\[1mm]
0 \\[1mm]
0 \\[1mm]
\bar{q}_0
\end{bmatrix}
\qquad (10.126)
$$

where

$$H_{11mn} = A_{11}m^2 + A_{66}n^2R^2$$

$$H_{12mn} = (A_{12} + A_{66})mnR$$

$$H_{13mn} = 2B_{16}mnR$$

$$H_{14mn} = B_{16}m^2 + B_{26}n^2R^2$$

$$H_{22mn} = A_{66}m^2 + A_{22}n^2R^2$$

$$H_{24mn} = 2B_{26}mnR$$

$$H_{33mn} = D_{11}m^2 + D_{66}n^2R^2 + k\,\frac{A_{55}a^2}{\pi^2}$$

$$H_{34mn} = (D_{12} + D_{66})mnR$$

$$H_{35mn} = kA_{55}\,\frac{ma}{\pi}$$

$$H_{44mn} = D_{66}m^2 + D_{22}n^2R^2 + k\,\frac{A_{55}a^2}{\pi^2}$$

$$H_{45mn} = kA_{44}\,\frac{nRa}{\pi}$$

$$H_{55mn} = k(A_{55}m^2 + A_{44}n^2)$$

$$\lambda_{mn} = \frac{\varrho\omega^2}{\pi^2}\,,\ \bar{q}_0 = q_0\,\frac{a^2}{\pi^2}\,,\ R = a/b$$

In Equations (10.126) it is assumed that each ply is constructed of the same uni-directional material, i.e., $H = 0$. We now consider two classes of problems.

Case I: Static Bending with m = n = 1

In the static case, $\lambda_{mn} = 0$ and Equations (10.126) can be solved for the coefficients A_{mn}, B_{mn}, C_{mn}, D_{mn}, and E_{mn}. Numerical results are illustrated in Figure 10.7 for a $[\pm 45°]$ laminate subjected to the bending load $q = q_0 \sin(\pi x/a) \sin(\pi y/b)$. The maximum deflection, which occurs at $x = a/2$ and $y = b/2$, is plotted as a function of a/h for a square plate. The following ply properties are utilized:

$$E_L/E_T = 40,\ G_{LT}/E_T = 0.6,\ G_{TT}/E_T = 0.5,\ \nu_{LT} = 0.25 \qquad (10.127)$$

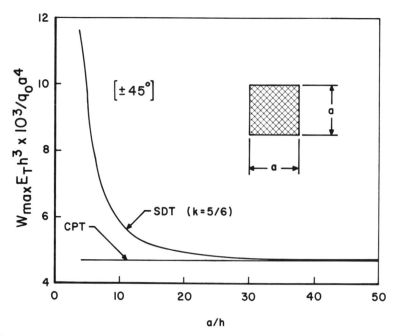

Figure 10.7. *Maximum deflection of an unsymmetric angle-ply plate under transverse load as a function of width-to-thickness ratio.*

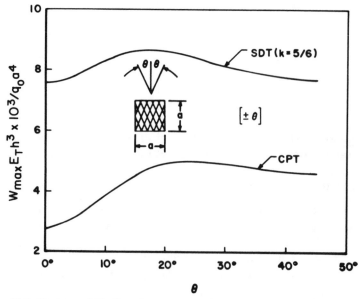

Figure 10.8. *Maximum deflection of an unsymmetric angle-ply plate under transverse load as a function of ply orientation.*

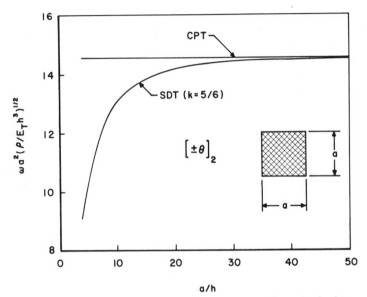

Figure 10.9. *Fundamental vibration frequency for an unsymmetric angle-ply plate as a function of width-to-thickness ratio.*

Numerical results are also shown from classical laminated plate theory (CPT) in which shear deformation is neglected.

The maximum deflection for $[\pm\theta]$ square laminates are illustrated in Figure 10.8, with $a/h = 6$ and the ply properties as given by (10.127). The bending load is the same as in Figure 10.7. As a basis for comparison, CPT results are also illustrated in Figure 10.8. Note that $k = 5/6$ in both Figures 10.7 and 10.8.

Case II: Free Vibration

In the case of free vibration, $q = 0$ and (10.126) reduces to a set of homogeneous equations. In order to obtain a nontrivial solution, λ_{mn} is chosen such that the determinant of the coefficient matrix in (10.126) vanishes.

The lowest frequency associated with $[\pm 45]_2$ square laminates is shown in Figure 10.9 as a function of a/h. The ply properties are those given in (10.127). Again the value $k = 5/6$ is chosen as the shear correction factor.

10.8 ANALYSIS OF SANDWICH PLATES

In this section the governing equations for sandwich plates are derived. Bending under lateral loads, buckling under inplane loads, and flexural vibration frequencies are discussed.

We now consider a sandwich plate consisting of a core material of thickness h, having face sheets bonded to the top and bottom surfaces. The face sheets are assumed to be laminates with the top plate having thickness h_1 and the bottom surface having thickness h_2. The core material is assumed to be of low density, and its primary function is to stabilize the face sheets. The coordinate system is placed at the center of the core, as shown in Figure 10.10. Specific assumptions are as follows:

1. The core material is transversely isotropic and of much greater thickness than the face sheets, i.e., $h >> h_1, h_2$.
2. The inplane stresses σ_x, σ_y, σ_{xy} of the core are assumed to be negligible.
3. The inplane core displacements u_c and v_c are assumed to be linear functions of the z coordinate.
4. The face sheets are constructed of an arbitrary number of layers of orthotropic material bonded together. However, the orthotropic axes of material symmetry of an individual ply need not coincide with the x-y axes of the plate.

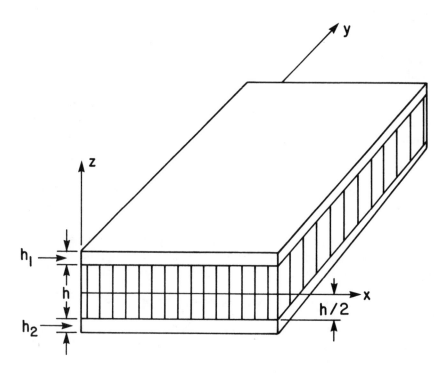

Figure 10.10. Nomenclature for sandwich plates.

5. Inplane displacements u and v are uniform through-the-thickness of the face sheets.
6. The transverse shear stresses σ_{xz} and σ_{yz} are negligible in the face sheets.
7. The plate displacements are small compared to the plate thickness.
8. The plate inplane strains are small compared to unity.
9. The transverse displacement w is independent of the z coordinate, i.e., ϵ_z is negligible.
10. The core and each ply of the face sheets obey Hooke's law.
11. The core and face sheets have constant thickness.

Based on assumption 2, the core displacements are of the form

$$u_c = u^0(x,y) + z\,\psi_x(x,y)$$
$$v_c = v^0(x,y) + z\,\psi_y(x,y)$$

$$(10.128)$$

From assumption 5 and the requirement that the displacements at the interfaces between the core and the face sheets be continuous, we obtain the face sheet displacements

$$u_1 = u^0 + \frac{h}{2}\psi_x \qquad (10.129)$$

$$v_1 = v^0 + \frac{h}{2}\psi_y \qquad (10.130)$$

$$u_2 = v^0 - \frac{h}{2}\psi_x \qquad (10.131)$$

$$v_2 = v^0 - \frac{h}{2}\psi_y \qquad (10.132)$$

From assumption 8,

$$w = w(x,y) \qquad (10.133)$$

Thus sandwich plate theory is based on the same five displacement functions as the shear deformation plate theory previously developed.

Strains are of the form

$$\epsilon_{xz}^{(c)} = \psi_x + \frac{\partial w}{\partial x} \qquad (10.134)$$

$$\epsilon_{yz}^{(c)} = \psi_y + \frac{\partial w}{\partial y} \tag{10.135}$$

$$\epsilon_x^{(1)} = \epsilon_x^0 + \frac{h}{2} \varkappa_x \tag{10.136}$$

$$\epsilon_y^{(1)} = \epsilon_y^0 + \frac{h}{2} \varkappa_y \tag{10.137}$$

$$\epsilon_{xy}^{(1)} = \epsilon_{xy}^0 + \frac{h}{2} \varkappa_{xy} \tag{10.138}$$

$$\epsilon_x^{(2)} = \epsilon_x^0 - \frac{h}{2} \varkappa_x \tag{10.139}$$

$$\epsilon_y^{(2)} = \epsilon_y^0 - \frac{h}{2} \varkappa_y \tag{10.140}$$

$$\epsilon_{xy}^{(2)} = \epsilon_{xy}^0 - \frac{h}{2} \varkappa_{xy} \tag{10.141}$$

where ϵ_x^0, ϵ_y^0, and ϵ_{xy}^0 are defined in Equation (2.6), and $\varkappa_x$, $\varkappa_y$, and $\varkappa_{xy}$ are defined in Equation (10.3). The superscripts c, 1, and 2 refer to the core, top face sheet, and lower face sheet, respectively.

Plate constitutive relations are derived by considering force and moment resultants similar to the ones defined for classical plate theory, i.e.,

$$(N_x, N_y, N_{xy}) = \int_{-(h/2 + h_2)}^{-h/2} (\sigma_x, \sigma_y, \sigma_{xy}) dz$$
$$+ \int_{h/2}^{(h/2 + h_1)} (\sigma_x, \sigma_y, \sigma_{xy}) dz \tag{10.142}$$

$$(M_x, M_y, M_{xy}) = \int_{-(h/2 + h_2)}^{-h/2} (\sigma_x, \sigma_y, \sigma_{xy}) z \, dz$$
$$+ \int_{h/2}^{(h/2 + h_1)} (\sigma_x, \sigma_y, \sigma_{xy}) z \, dz \tag{10.143}$$

$$(Q_x, Q_y) = \int_{-h/2}^{h/2} (\sigma_{xz}, \sigma_{yz}) dz \qquad (10.144)$$

For the face sheets a state of plane stress is assumed to exist in each ply. Thus Equation (2.24) is applicable to the face sheets, i.e., for the kth layer of the ith face sheet,

$$\begin{bmatrix} \sigma_x^{(k)} \\ \sigma_y^{(k)} \\ \sigma_{xy}^{(k)} \end{bmatrix} = \begin{bmatrix} Q_{11}^{(k)} & Q_{12}^{(k)} & Q_{16}^{(k)} \\ Q_{12}^{(k)} & Q_{22}^{(k)} & Q_{26}^{(k)} \\ Q_{16}^{(k)} & Q_{26}^{(k)} & Q_{66}^{(k)} \end{bmatrix} \begin{bmatrix} \epsilon_x^{(i)} \\ \epsilon_y^{(i)} \\ \epsilon_{xy}^{(i)} \end{bmatrix} \quad (i = 1,2) \quad (10.145)$$

where $i = 1$ and 2 refers to the upper and lower face sheets, respectively. For the core,

$$\begin{bmatrix} \sigma_{yz} \\ \sigma_{xz} \end{bmatrix} = \begin{bmatrix} G_{23} & 0 \\ 0 & G_{13} \end{bmatrix} \begin{bmatrix} \epsilon_{yz}^{(c)} \\ \epsilon_{xz}^{(c)} \end{bmatrix} \qquad (10.146)$$

Substituting Equations (10.145) and (10.146) into the force and moment resultants, Equations (10.142) and (10.143), we obtain the constitutive relations which are of a form similar to (2.26):

$$\begin{bmatrix} N_x \\ N_y \\ N_{xy} \\ \hline M_x \\ M_y \\ M_{xy} \end{bmatrix} = \left[\begin{array}{ccc|ccc} A_{11} & A_{12} & A_{16} & B_{11} & B_{12} & B_{16} \\ A_{12} & A_{22} & A_{26} & B_{12} & B_{22} & B_{26} \\ A_{16} & A_{26} & A_{66} & B_{16} & B_{26} & B_{66} \\ \hline C_{11} & C_{12} & C_{16} & D_{11} & D_{12} & D_{16} \\ C_{12} & C_{22} & C_{26} & D_{12} & D_{22} & D_{26} \\ C_{16} & C_{26} & C_{66} & D_{16} & D_{26} & D_{66} \end{array} \right] \begin{bmatrix} \epsilon_x^0 \\ \epsilon_y^0 \\ \epsilon_{xy}^0 \\ \hline \varkappa_x \\ \varkappa_y \\ \varkappa_{xy} \end{bmatrix}$$

$$(10.147)$$

where

$$A_{ij} = A_{ij}^{(1)} + A_{ij}^{(2)}$$

$$B_{ij} = \frac{h}{2}[A_{ij}^{(1)} - A_{ij}^{(2)}]$$

$$C_{ij} = C_{ij}^{(1)} + C_{ij}^{(2)}$$

$$D_{ij} = \frac{h}{2}[C_{ij}^{(1)} - C_{ij}^{(2)}]$$

(10.148)

and

$$(A_{ij}^{(1)}, C_{ij}^{(1)}) = \int_{h/2}^{(h/2 + h_1)} Q_{ij}^{(k)} (1,z) \, dz$$

$$(A_{ij}^{(2)}, C_{ij}^{(2)}) = \int_{-(h/2 + h_2)}^{-h/2} Q_{ij}^{(k)} (1,z) \, dz$$

(10.149)

The transverse shear relations are of the form

$$\begin{bmatrix} Q_y \\ \\ Q_x \end{bmatrix} = h \begin{bmatrix} G_{23} & 0 \\ \\ 0 & G_{13} \end{bmatrix} \begin{bmatrix} \psi_y + \dfrac{\partial w}{\partial y} \\ \\ \psi_x + \dfrac{\partial w}{\partial x} \end{bmatrix}$$

(10.150)

A cursory examination of Equations (10.149) reveals that if the top and bottom face sheets are identical then

$$A_{ij}^{(2)} = A_{ij}^{(1)}, \quad C_{ij}^{(2)} = - C_{ij}^{(1)} \quad \text{(IDENTICAL FACE SHEETS)} \quad (10.151)$$

and as a result

$$\left. \begin{matrix} A_{ij} = 2A_{ij}^{(1)}, \; D_{ij} = h \, C_{ij}^{(1)} \\ \\ B_{ij} = C_{ij} = 0 \end{matrix} \right\} \quad \text{(IDENTICAL FACE SHEETS)} \quad (10.152)$$

Thus for such cases the inplane and bending problems uncouple.

The stiffness coefficients C_{ij} and D_{ij} can also be written in terms of $A_{ij}^{(1)}$ and $A_{ij}^{(2)}$ by treating the face sheets as homogeneous materials with effective stiffnesses $A_{ij}^{(k)}/h_k$ $(k = 1,2)$. This may be a reasonable approximation since the face sheets are thin compared to the core. For this case we have

$$C_{ij} = \frac{h}{2}[(1 + h_1/h)A_{ij}^{(1)} - (1 + h_2/h)A_{ij}^{(2)}]$$

$$D_{ij} = \frac{h^2}{4}[(1 + h_1/h)A_{ij}^{(1)} + (1 + h_2/h)A_{ij}^{(2)}]$$

(10.153)

To illustrate the order of the error introduced by this approximation, let us consider two ply face sheets. Using Equations (10.148) and (10.149), we find that

$$C_{ij} = \frac{h}{8}(h_1 F_{ij}^{(1)} - h_2 F_{ij}^{(2)})$$

$$D_{ij} = \frac{h^2}{16}(h_1 F_{ij}^{(1)} - h_2 F_{ij}^{(2)})$$

(10.154)

where

$$F_{ij}^{(k)} = 2[(1 + h_k/2h)Q_{ij}^{(k)}(\theta_1) + (1 + 3h_k/2h)Q_{ij}^{(k)}(\theta_2)]$$

and θ_1 and θ_2 denote inner ply and outer ply orientations, respectively. If Equations (10.153) are utilized, the results can be written in the same form as Equation (10.154) with

$$F_{ij}^{(k)} = 2(1 + h_k/h)[Q_{ij}^{(k)}(\theta_1) + Q_{ij}^{(k)}(\theta_2)]$$

Thus the ratio h_k/h determines the order of the error introduced by assuming homogeneous face sheets. The two-ply laminate represents a "worst" case situation. Less error is introduced for face sheets with more layers.

If we neglect inplane inertia and rotatory inertia terms, the static form of Equations (10.8)–(10.11) and the dynamic form of Equation (10.12) are applicable to the sandwich plates under consideration. Substituting the constitutive relations (10.147) and (10.150), and taking into account the strain-displacement relations (10.134)–(10.141), we obtain the governing equations in terms of displacements. The first two equations are identical to the static form of Equations (10.14) and (10.15), with A_{ij} and B_{ij} defined by Equation (10.148). The last three equations are as follows:

$$C_{11}\frac{\partial^2 u^o}{\partial x^2} + 2C_{16}\frac{\partial^2 u^o}{\partial x \partial y} + C_{66}\frac{\partial^2 u^o}{\partial y^2} + C_{16}\frac{\partial^2 v^o}{\partial x^2} + (C_{12} + C_{66})\frac{\partial^2 v^o}{\partial x \partial y} \quad (10.155)$$

$$+ C_{26} \frac{\partial^2 v^0}{\partial y^2} + D_{11} \frac{\partial^2 \psi_x}{\partial x^2} + 2D_{16} \frac{\partial^2 \psi_x}{\partial x \partial y} + D_{66} \frac{\partial^2 \psi_x}{\partial y^2} + D_{16} \frac{\partial^2 \psi_y}{\partial x^2}$$

$$+ (D_{12} + D_{66}) \frac{\partial^2 \psi_y}{\partial x \partial y} + D_{26} \frac{\partial^2 \psi_y}{\partial y^2} - G_{13}h(\psi_x + \frac{\partial w}{\partial x}) = 0$$

$$C_{16} \frac{\partial^2 u^0}{\partial x^2} + (C_{12} + C_{66}) \frac{\partial^2 u^0}{\partial x \partial y} + C_{26} \frac{\partial^2 u^0}{\partial y^2} + C_{66} \frac{\partial^2 v^0}{\partial x^2} + 2C_{26} \frac{\partial^2 v^0}{\partial x \partial y}$$

$$+ C_{22} \frac{\partial^2 v^0}{\partial y^2} + D_{16} \frac{\partial^2 \psi_x}{\partial x^2} + (D_{12} + D_{66}) \frac{\partial^2 \psi_x}{\partial x \partial y} + D_{26} \frac{\partial^2 \psi_x}{\partial y^2} + D_{66} \frac{\partial^2 \psi_y}{\partial x^2}$$

$$+ 2D_{26} \frac{\partial^2 \psi_y}{\partial x \partial y} + D_{22} \frac{\partial^2 \psi_y}{\partial y^2} - G_{23}h(\psi_y + \frac{\partial w}{\partial y}) = 0 \tag{10.156}$$

$$G_{13}h \left(\frac{\partial \psi_x}{\partial x} + \frac{\partial^2 w}{\partial x^2} \right) + G_{23}h \left(\frac{\partial \psi_y}{\partial y} + \frac{\partial^2 w}{\partial y^2} \right) + N_x^i \frac{\partial^2 w}{\partial x^2} + 2N_{xy}^i \frac{\partial^2 w}{\partial x \partial y}$$

$$+ N_y^i \frac{\partial^2 w}{\partial y^2} + q = \varrho \frac{\partial^2 w}{\partial t^2} \tag{10.157}$$

where ϱ is defined by Equation (2.12).

The boundary conditions for this theory are identical to those prescribed in Equation (10.23).

10.9 CYLINDRICAL BENDING OF SANDWICH PLATES

We now consider cylindrical bending of a sandwich plate. For simplicity the top and bottom face sheets are assumed to be identical. As previously, cylindrical bending implies the laminate is of infinite extent in the y-direction and uniformly supported along the edges $x = 0,a$. In addition let us assume the face sheets contain only layers which are orthotropic with respect to the x-y axes of the plate. Under these conditions, $B_{ij} = C_{ij} = A_{16} = A_{26} = D_{16} = D_{26} = 0$ and

$$u^0 = v^0 = \psi_y = 0, \quad \psi_x = \psi_x(x,t), \quad w = w(x,t) \tag{10.158}$$

Since the face sheets are identical, Equations (10.155)–(10.157) are applicable to bending. For the present case they reduce to the following:

$$D_{11} \frac{\partial^2 \psi_x}{\partial x^2} - G_{13}h(\psi_x + \frac{\partial w}{\partial x}) = 0 \tag{10.159}$$

$$G_{13}h \left(\frac{\partial \psi_x}{\partial x} + \frac{\partial^2 w}{\partial x^2} \right) + N_x^i \frac{\partial^2 w}{\partial x^2} + q = \varrho \frac{\partial^2 w}{\partial t^2} \qquad (10.160)$$

Bending under lateral load, buckling under inplane load, and free vibration in conjunction with Equations (10.159) and (10.160) will now be considered.

Case I: Bending Under Uniform Transverse Load ($N_x^i = 0$)

For a uniform transverse load $q = q_0 = $ constant, in the absence of inplane load, Equations (10.159) and (10.160) reduce to the following differential equations:

$$D_{11} \frac{d^2 \psi_x}{dx^2} - G_{13}h(\psi_x + \frac{dw}{dx}) = 0 \qquad (10.161)$$

$$G_{13}h \left(\frac{d \psi_x}{dx} + \frac{d^2 w}{dx^2} \right) + q_0 = 0 \qquad (10.162)$$

For simply-supported boundary conditions, we require that
at $x = 0$ and a

$$w = M_x = \frac{d \psi_x}{dx} = 0 \qquad (10.163)$$

Equations (10.161) and (10.162) can be uncoupled by integrating Equation (10.162) with the result

$$G_{13}h(\psi_x + \frac{dw}{dx}) - q_0 x + c_1 \qquad (10.164)$$

Substituting (10.164) into Equation (10.161), we obtain the result

$$D_{11} \frac{d^2 \psi_x}{dx^2} + q_0 x + c_1 = 0 \qquad (10.165)$$

Integrating (10.165) in conjunction with the boundary condition (10.163), we find that

$$c_1 = - \frac{q_0 a}{2} \qquad (10.166)$$

and

$$\frac{d\psi_x}{dx} = -\frac{q_0 x}{2D_{11}} (x - a) \tag{10.167}$$

Substituting this expression into Equation (10.162), we obtain the following differential equation in w:

$$\frac{d^2 w}{dx^2} = q_0 \left[\frac{x}{2D_{11}} (x - a) - \frac{1}{G_{13}h} \right] \tag{10.168}$$

Integrating this equation twice and applying the condition that w vanish on the edges, we obtain the result

$$w = \frac{q_0 x}{24D_{11}} \left[x^3 - 2ax^2 + a^3 + \frac{12D_{11}}{G_{13}h} (a - x) \right] \tag{10.169}$$

Equation (10.161) can now be solved directly for ψ_x with the result

$$\psi_x = -\frac{q_0}{24D_{11}} (4x^3 - 6ax^2 + a^3) \tag{10.170}$$

Just as in the case of classical laminated plate theory and in the shear deformation plate theory, the interlaminar shear stress distribution in a sandwich plate can be determined by integrating the equilibrium equations of classical theory of elasticity. The first of Equations (2.8) can be utilized for present purposes, i.e., for the kth layer,

$$\frac{\partial \sigma_x^{(k)}}{\partial x} = \frac{\partial \sigma_{xz}^{(k)}}{\partial z} = 0 \tag{10.171}$$

For the core material, $\sigma_x = 0$ and Equation (10.171) becomes

$$\frac{\partial \sigma_{xz}}{\partial z} = 0 \tag{10.172}$$

which implies $\sigma_{xz} = f(x,y)$ in the core. For the sandwich plate under consideration, the inplane stresses are given by

$$\sigma_x^{(k)} = \pm Q_{11}^{(k)} \frac{h}{2} \frac{d\psi_x}{dx} \tag{10.173}$$

$$\sigma_y^{(k)} = \pm Q_{12}^{(k)} \frac{h}{2} \frac{d\psi_x}{dx} \tag{10.174}$$

$$\sigma_{xy}^{(k)} = 0 \tag{10.175}$$

where a plus sign is associated with the upper face sheet and a minus sign with the lower face sheet. Since the face sheets are identical and the core interlaminar shear stress is constant, it is necessary only to calculate the shear distribution through the bottom face sheet to obtain the entire profile. Let us consider face sheets constructed of two-ply laminates with stacking sequences [0°/90°]. using Equation (10.173) in conjunction with Equation (10.171), and integrating with respect to z, we obtain the following results:

$$\sigma_{xz}(0°) = \frac{q_0 E_L h (2x - a)(h + 2h_1 + 2z)}{8D_{11}(1 - \nu_{LT}^2 E_T/E_L)} \tag{10.176}$$

$$\sigma_{xz}(90°) = \frac{q_0 h (2x - a)[2h_1 E_L + (h + h_1 + 2z)E_T]}{16D_{11}(1 - \nu_{LT}^2 E_T/E_L)} \tag{10.177}$$

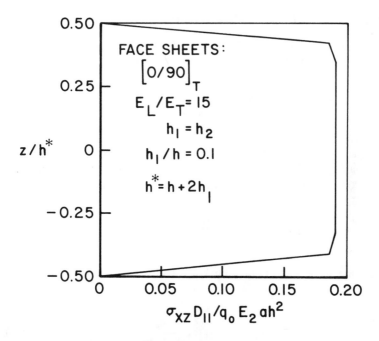

Figure 10.11. *Interlaminar shear stress distribution through-the-thickness of a sandwich plate.*

Since the core interlaminar shear stress is constant and must be continuous at the face sheet/core interface

$$\sigma_{xz}(\text{core}) = \sigma_{xz}(90°), \ z = -h/2 \tag{10.178}$$

Substituting Equation (10.177) into Equation (10.178), we obtain the result

$$\sigma_{xz}(\text{core}) = \frac{q_0 h h_1 (2x - a)(2E_L + E_T)}{16 D_{11}(1 - \nu_{LT}^2 E_T/E_L)} \tag{10.179}$$

Numerical results are shown in Figure 10.11 for $E_L/E_T = 15$ and $h_1/h = 0.1$. Note that the assumption of uniform strain in the face sheets leads to a linear interlaminar shear stress distribution within each ply rather than the parabolic distribution found in classical laminated plate theory and in the shear deformation plate theory. Also note that the interlaminar shear stress distribution is independent of core properties. The maximum interlaminar shear stress occurs in the core, including the face sheet core interfaces. This is an important observation when considering interlaminar shear failure of sandwich plates.

Case II: Buckling Under Uniform Compression Loading (q = 0)

For the case of an initial inplane compression load $N_x^i = -N_0 = $ constant where $N_0 > 0$, Equation (10.160) becomes

$$G_{13} h \left(\frac{d\psi_x}{dx} + \frac{d^2 w}{dx^2} \right) - N_0 \frac{d^2 w}{dx^2} = 0 \tag{10.180}$$

Equations (10.161) and (10.180) are of the same form as Equations (10.60). By direct analogy, a solution for the simply-supported boundary conditions of Equation (10.163) is of the same form as Equation (10.64). Thus,

$$N_0 = \frac{G_{13} D_{11} h m^2 \pi^2}{(D_{11} m^2 \pi^2 + G_{13} h a^2)} \tag{10.181}$$

In a similar manner the critical buckling load is of the same form as Equation (10.65), i.e.,

$$N_{cr} = N_{cr}' \left(1 - \frac{S\pi^2}{1 + S\pi^2} \right) \tag{10.182}$$

where

$$N'_{cr} = D_{11} \frac{\pi^2}{a^2}, \quad S = \frac{D_{11}}{G_{13}ha^2} \tag{10.183}$$

Case III: Free Vibration ($N_x^i = q = 0$)

In the absence of lateral and inplane loads, Equation (10.160) reduces to the following:

$$G_{13}h \left(\frac{\partial \psi_x}{\partial x} + \frac{\partial^2 w}{\partial x^2} \right) = \varrho \frac{\partial^2 w}{\partial t^2} \tag{10.184}$$

Equations (10.159) and (10.184) are of the same form as Equations (10.67) in the absence of rotatory inertia. By direct analogy, a solution for simply-supported boundary conditions, Equation (10.163), is of the same form as Equation (10.72) with

$$\omega_m = \omega'_m \sqrt{1 - \frac{Sm^2\pi^2}{1 + Sm^2\pi^2}} \tag{10.185}$$

where ω'_m and S are defined by Equations (10.73) and (10.183), respectively.

10.10 COMPARISON OF SANDWICH PLATE ANALYSIS TO EXACT THEORY

Consider a rectangular sandwich plate subjected to the surface load

$$q = q_0 \sin \frac{\pi x}{a} \sin \frac{\pi y}{b} \tag{10.186}$$

For the case of identical top and bottom face sheets with orthotropic properties ($D_{16} = D_{26} = 0$), Equations (10.155)–(10.157) are applicable and reduce to the following:

$$D_{11} \frac{\partial^2 \psi_x}{\partial x^2} + D_{66} \frac{\partial^2 \psi_x}{\partial y^2} + (D_{12} + D_{66}) \frac{\partial^2 \psi_y}{\partial x \partial y} - G_{13}h(\psi_x + \frac{\partial w}{\partial x}) = 0 \tag{10.187}$$

$$(D_{12} + D_{66}) \frac{\partial^2 \psi_x}{\partial x \partial y} + D_{66} \frac{\partial^2 \psi_y}{\partial x^2} + D_{22} \frac{\partial^2 \psi_y}{\partial y^2} - G_{23}h(\psi_y + \frac{\partial w}{\partial y}) = 0$$

(10.188)

$$G_{13}h \left(\frac{\partial^2 \psi_x}{\partial x^2} + \frac{\partial^2 w}{\partial x^2} \right) + G_{23}h \left(\frac{\partial^2 \psi_y}{\partial y^2} + \frac{\partial^2 w}{\partial y^2} \right) + q = 0 \quad (10.189)$$

For simple-supports we have the following boundary conditions:
at $x = 0$ and a

$$w = \psi_y = M_x = D_{11} \frac{\partial \psi_x}{\partial x} + D_{12} \frac{\partial \psi_y}{\partial y} = 0 \qquad (10.190)$$

at $y = 0$ and b

$$w = \psi_x = M_y = D_{12} \frac{\partial \psi_x}{\partial x} + D_{22} \frac{\partial \psi_y}{\partial y} \qquad (10.191)$$

A solution to Equations (10.187)–(10.189) which satisfies the boundary conditions (10.190) and (10.191) is of the form

$$\psi_x = A \cos \frac{\pi x}{a} \sin \frac{\pi y}{b}$$

$$\psi_y = B \sin \frac{\pi x}{a} \cos \frac{\pi y}{b} \qquad (10.192)$$

$$w = C \sin \frac{\pi x}{a} \sin \frac{\pi y}{b}$$

Substituting (10.192) into Equations (10.187)–(10.189), and solving the resulting algebraic equations leads to the following results:

$$A = \frac{q_0 a^2}{\pi^2 D} (A_2 A_5 - A_3 A_4) \qquad (10.193)$$

$$B = \frac{q_0 a^2}{\pi^2 D} (A_3 A_2 - A_1 A_5) \qquad (10.194)$$

$$C = \frac{q_0 a^2}{\pi^2 D} (A_1 A_4 - A_2^2) \qquad (10.195)$$

where

$$A_1 = (D_{11} + D_{66}R^2 + G_{13}\frac{ha^2}{\pi^2})$$

$$A_2 = (D_{12} + D_{66})$$

$$A_3 = G_{13}\frac{ha}{\pi}$$

$$A_4 = (D_{66} + D_{22}R^2 + G_{23}\frac{ha^2}{\pi^2})$$

$$A_5 = G_{23}\frac{haR}{\pi}$$

$$A_6 = (G_{13} + G_{23}R^2)$$

$$D = A_1A_4A_6 + 2A_2A_3A_5 - A_1A_5^2 - A_4A_3^2 - A_6A_2^2$$

and R is the aspect ratio a/b.

We now consider the case of homogeneous face sheets composed of unidirectional composites in which the fibers are parallel to the x-axis. Since the face sheets are homogeneous, unidirectional composites, Equation (10.152) is applicable for calculating the plate bending sitffnesses, i.e.,

$$D_{11} = \frac{hh_1(h + h_1)E_L}{4(1 - \nu_{LT}^2 E_T/E_L)}$$

$$D_{22} = \frac{hh_1(h + h_1)E_T}{4(1 - \nu_{LT}^2 E_T/E_L)}$$

$$D_{12} = \nu_{LT}D_{22}$$

(10.196)

$$D_{66} = \frac{hh_1}{4}(h + h_1)G_{LT}$$

The face sheet properties are as follows:

$$E_L/E_T = 25, \; G_{LT}/E_T = 0.5, \; G_{TT}/E_T = 2.5,$$

(10.197)

$$\nu_{LT} = \nu_{TT} = 0.25$$

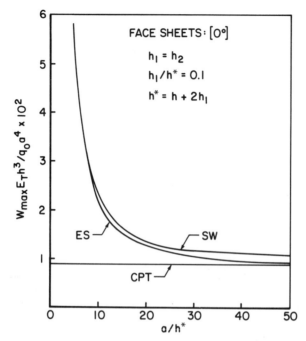

Figure 10.12. *Maximum deflection for a sandwich plate subjected to a transverse bending load.*

where, as previously, G_{TT} denotes the interlaminar shear modulus relative to the x_T-z plane and ν_{TT} is the Poisson ratio determined by measuring the contraction through-the-thickness during a uniaxial tensile test in the x_T direction. The following core properties are also considered:

$$E_{11}/E_T = E_{22}/E_T = 0.04, \quad E_{33}/E_T = 0.5,$$

$$G_{13}/E_T = G_{23}/E_T = 0.06, \quad G_{12}/E_T = 0.016, \tag{10.198}$$

$$\nu_{12} = \nu_{13} = \nu_{23} = 0.25$$

Utilizing classical theory of elasticity, an exact solution to this problem has been obtained by Pagano [2]. Numerical results are shown in Figure 10.12 where the maximum deflection of a square plate as determined from sandwich plate theory (SW) is compared to the exact elasticity solution (ES). The maximum deflection occurs at the center of the plate ($x = a/2$, $y = b/2$). Good agreement between sandwich plate theory and exact elasticity solutions is noted over a wide range of values of a/h. As a point of reference, results obtained from classical laminated plate theory (CPT) are also shown in Figure 10.12.

It should also be noted that almost perfect agreement can be obtained between shear deformation theory and exact elasticity solutions for maximum deflection in the case of the plate considered in Figure 10.12 [13]. Such close correlation, however, requires that two k factors, one associated with A_{55} and a second associated with A_{44}, be utilized in conjunction with the procedure outlined by Chou [11] for determining their value. For the sandwich plate considered in Figure 10.12, $k_1 = 0.4098$ and $k_2 = 0.6915$. Despite the correlation obtained with the shear deformation theory, the sandwich plate theory is less complex.

REFERENCES

1. Pagano, N. J. "Exact Solutions for Composite Laminates in Cylindrical Bending," *Journal of Composite Materials*, 3:398–411 (1969).

2. Pagano, N. J. "Exact Solutions for Rectangular Bidirectional Composites and Sandwich Plates," *Journal of Composite Materials*, 4:20–35 (1970).

3. Pagano, N. J. "Influence of Shear Coupling in Cylindrical Bending of Anisotropic Laminates," *Journal of Composite Materials*, 4:330–343 (1970).

4. Reissner, E. "The Effect of Transverse Shear Deformation on the Bending of Elastic Plates," *Journal of Applied Mechanics*, 12:69–77 (1945).

5. Mindlin, R. D. "Influence of Rotatory Inertia and Shear on Flexural Motions of Isotropic, Elastic Plates," *Journal of Applied Mechanics*, 18:336–343 (1951).

6. Yang, P. C., C. H. Horris and Y. Stavsky. "Elastic Wave Propagation in Heterogeneous Plates," *International Journal of Solids and Structures*, 2:665–684 (1966).

7. Whitney, J. M. and N. J. Pagano. "Shear Deformation in Heterogeneous Anisotropic Plates," *Journal of Applied Mechanics*, 37:1031–1036 (1970).

8. Uflyand, Ya. S. "The Propagation of Waves in the Transverse Vibrations of Bars and Plates," Akad. Nauk SSSR. *Prikl Mat. Meh.*, 12:287–300 (Russian) (1948).

9. Timoshenko, S. P. *Strength of Materials, Vol. 1.* Third Edition. Princeton:D. Von Nostrand Co., Inc. (1955).

10. Medwadowski, S. J. "A Refined Theory for Elastic Orthotropic Plates," *Journal of Applied Mechanics*, 25:437–443 (1958).

11. Chou, T. S. "On the Propagation of Flexural Waves in an Orthotropic Laminated Plate and Its Response to an Impulsive Load," *Journal of Composite Materials*, 5:306–319 (1971).

12. Whitney, J. M. "Shear Correction Factors for Orthotropic Laminates Under Static Loading," *Journal of Applied Mechanics*, 40:302–304 (1973).

13. Whitney, J. M., "Stress Analysis of Thick Laminated Composite and Sandwich Plates," *Journal of Composite Materials*, 6:426–440 (1972).

HAPTER 11

Free-Edge Effects and Higher Order Laminated Plate Theory

11.1 INTRODUCTION

IN CHAPTER X we discussed some limitations on classical laminated plate theory which involved the neglect of transverse shear deformation. Another classical problem which represents a limitation on laminated plate theory is the so-called "free-edge effect." An illustration of how a higher order laminated plate theory can be utilized to address this issue is presented in this chapter. The approach requires the derivation of a higher order plate theory which includes a thickness-stretch mode in addition to shear deformation.

11.2 FREE-EDGE EFFECTS IN LAMINATED PLATES

It was briefly pointed out in Section 4.5 that the stress distribution near the free-edge of a laminate is three-dimensional in nature and not predictable by classical laminated plate theory. This is due to the fact that force and moment resultants are prescribed rather than ply-by-ply stresses. A classical example is a $[0°/90°]_{ns}$ laminate with a uniaxial tensile load applied parallel to the $0°$ plies. According to classical laminated plate theory, the transverse stress, σ_y in the $0°$ and $90°$ plies is equal in magnitude but opposite in sign (see Figure 11.1). In particular, if we consider the inplane load

$$N_y = N_{xy} = 0, \; N_x = N_0 = \text{constant}, \; N_0 > 0 \qquad (11.1)$$

then the ply stresses can be determined by combining the laminate constitutive relations (2.26) with the ply stress-strain relations (2.24). This procedure leads to the result

$$\sigma_y(0°) = \frac{2\nu_{LT}E_T(E_L - E_T)\sigma_0}{[(E_L + E_T)^2 - 4\nu_{LT}^2 E_T^2]}$$

$$\sigma_y(90°) = -\sigma_y(0°) \tag{11.2}$$

$$\sigma_{xy}(0°) = \sigma_{xy}(90°) = 0$$

where σ_0 denotes the average laminate stress N_0/h. The results in Equation (11.2) assure that the boundary conditions $N_y = N_{xy} = 0$ are exactly satisfied. With the existence of free edges, however, there is no physical mechanism for applying $\sigma_y(0°)$ and $\sigma_y(90°)$. Thus we conclude that the stresses must be three-dimensional in nature.

It has been suggested by Pipes and Pagano [1] that the three-dimensional state-of-stress is confined to a boundary region approximately equal to one laminate thickness, h, wide. Classical lamination theory is recovered outside of this boundary.

For the case of general laminates with free-edges, both interlaminar shear and normal stresses will be present in a free-edge boundary zone. Of particular concern, however, is the interlaminar normal stress σ_z, which can lead to delamination. The qualitative nature of free-edge stresses as well as documented failure associated with interlaminar peel stresses have been presented by Pagano and Pipes [2] and [3]. The work in Reference 3 includes a procedure for approximating the distribution of σ_z associated with free-edge problems in laboratory type tensile coupons. The assumed distribution, however, is based on considera-

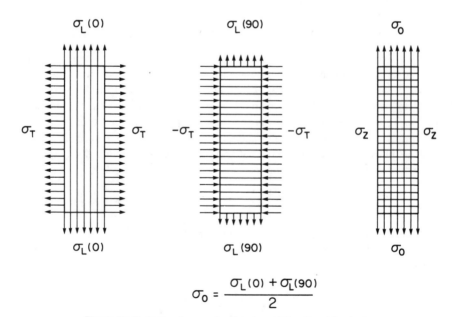

$$\sigma_0 = \frac{\sigma_L(0) + \sigma_L(90)}{2}$$

Figure 11.1. Free-edge mechanism in a bidirectional laminate.

tions of statics and contains no description of the influence of material and geometric parameters on the interlaminar state-of-stress. In this chapter a procedure is discussed for determining the distribution of σ_z along the central plane of a symmetric, finite-width laminate. For simplicity the discussion is limited to laminates constructed of plies in which the orthotropic axes of material symmetry are parallel to the x-y axes of the plate. The procedure requires the use of a higher order laminated plate theory which includes a thickness-stretch displacement mode in addition to transverse shear deformation.

11.3 A THICKNESS-STRETCH DEFORMATION MODE IN LAMINATED PLATES

Let us consider a laminated plate theory in which the transverse displacement w, as well as the inplane displacements u and v, are linear functions of the z coordinate. As previously, we utilize a coordinate system located at the midplane of the plate. Thus for static loading the displacements are of the form

$$u = u^0(x,y) + z\,\psi_x(x,y)$$

$$v = v^0(x,y) + z\,\psi_y(x,y) \tag{11.3}$$

$$w = w^0(x,y) + z\,\phi(x,y)$$

The static form of Equations (10.8)–(10.12) in the absence of inplane force effects are applicable to the current theory. Thus,

$$\frac{\partial N_x}{\partial x} + \frac{\partial N_{xy}}{\partial y} = 0 \tag{11.4}$$

$$\frac{\partial N_{xy}}{\partial x} + \frac{\partial N_y}{\partial y} = 0 \tag{11.5}$$

$$\frac{\partial M_x}{\partial x} + \frac{\partial M_{xy}}{\partial y} - Q_x = 0 \tag{11.6}$$

$$\frac{\partial M_{xy}}{\partial x} + \frac{\partial M_y}{\partial y} - Q_y = 0 \tag{11.7}$$

$$\frac{\partial M_x}{\partial x} + \frac{\partial Q_y}{\partial y} + q = 0 \tag{11.8}$$

Since we are dealing with six displacement variables, we need one additional equilibrium equation. In order to accomplish this, it is necessary to define the following force and moment resultants:

$$N_z = \int_{-h/2}^{h/2} \sigma_z^{(k)} \, dz$$

$$(11.9)$$

$$(R_x, R_y) = \int_{-h/2}^{h/2} (\sigma_{xz}^{(k)}, \sigma_{yz}^{(k)}) z \, dz$$

Multiplying the static form of the third equation in (2.8) by z and integrating with respect to z, we obtain the result

$$\frac{\partial R_x}{\partial x} + \frac{\partial R_y}{\partial y} - N_z + m = 0 \qquad (11.10)$$

where

$$m = \frac{h}{2} [\sigma_z(h/2) + \sigma_z(-h/2)] \qquad (11.11)$$

In order to arrive at a set of displacement equations, it is necessary to derive constitutive relations for the laminate.

Substituting Equations (11.3) into the strain-displacement relations (1.10), we obtain Equations (10.2) and (10.3). In addition,

$$\epsilon_z = \phi(x,y) \qquad (11.12)$$

$$\epsilon_{xz} = (\psi_x + \frac{\partial w}{\partial x}) + z \frac{\partial \phi}{\partial x} \qquad (11.13)$$

$$\epsilon_{yz} = (\psi_y + \frac{\partial w}{\partial y}) + z \frac{\partial \phi}{\partial y} \qquad (11.14)$$

Substituting Equations (10.2) and (11.12) into the orthotropic constitutive relations

(1.27), we obtain the following ply stresses:

$$
\begin{bmatrix} \sigma_x^{(k)} \\ \sigma_y^{(k)} \\ \sigma_z^{(k)} \\ \sigma_{xy}^{(k)} \end{bmatrix} = \begin{bmatrix} c_{11}^{(k)} & c_{12}^{(k)} & c_{13}^{(k)} & 0 \\ c_{12}^{(k)} & c_{22}^{(k)} & c_{23}^{(k)} & 0 \\ c_{13}^{(k)} & c_{23}^{(k)} & c_{33}^{(k)} & 0 \\ 0 & 0 & 0 & c_{66}^{(k)} \end{bmatrix} \begin{bmatrix} \epsilon_x^0 \\ \epsilon_y^0 \\ \phi \\ \epsilon_{xy}^0 \end{bmatrix}
$$

$$
+ z \begin{bmatrix} c_{11}^{(k)} & c_{12}^{(k)} & c_{13}^{(k)} & 0 \\ c_{12}^{(k)} & c_{22}^{(k)} & c_{23}^{(k)} & 0 \\ c_{13}^{(k)} & c_{23}^{(k)} & c_{33}^{(k)} & 0 \\ 0 & 0 & 0 & c_{66}^{(k)} \end{bmatrix} \begin{bmatrix} \varkappa_x \\ \varkappa_y \\ 0 \\ \varkappa_{xy} \end{bmatrix} \tag{11.15}
$$

where $\varkappa_x$, $\varkappa_y$, and $\varkappa_{xy}$ are defined in Equation (10.3). Note that in the derivation of Equation (11.15), a state-of-plane-stress is not assumed.

Substituting Equations (11.13) and (11.14) into Equations (1.27), we obtain the following results:

$$
\begin{bmatrix} \sigma_{yz}^{(k)} \\ \sigma_{xz}^{(k)} \end{bmatrix} = \begin{bmatrix} c_{44}^{(k)} & 0 \\ 0 & c_{55}^{(k)} \end{bmatrix} \begin{bmatrix} \psi_y + \dfrac{\partial w}{\partial y} \\ \psi_x + \dfrac{\partial w}{\partial x} \end{bmatrix}
$$

$$
+ z \begin{bmatrix} c_{44}^{(k)} & 0 \\ 0 & c_{55}^{(k)} \end{bmatrix} \begin{bmatrix} \dfrac{\partial \phi}{\partial y} \\ \dfrac{\partial \phi}{\partial x} \end{bmatrix} \tag{11.16}
$$

As in the case of the shear deformation theory, Equation (10.22), the interlaminar stresses generated from Equation (11.16) are discontinuous at the ply interfaces. Thus the procedure outlined in Chapter 2 for calculating interlaminar stresses should again be utilized in conjunction with the higher order theory currently under discussion.

Substituting Equations (11.15) and (11.16) into the definitions of force and moment resultants, we arrive at the following constitutive relations:

$$
\begin{bmatrix}
N_x \\
N_y \\
N_z/k_1 \\
N_{xy} \\
M_x \\
M_y \\
M_{xy}
\end{bmatrix}
=
\begin{bmatrix}
A'_{11} & A'_{12} & A'_{13} & 0 & B'_{11} & B'_{12} & 0 & 0 \\
A'_{12} & A'_{22} & A'_{23} & 0 & B'_{12} & B'_{22} & 0 & 0 \\
A'_{13} & A'_{23} & A'_{33} & 0 & B'_{13} & B'_{23} & 0 & 0 \\
0 & 0 & 0 & A'_{66} & 0 & 0 & 0 & B'_{66} \\
B'_{11} & B'_{12} & B'_{13} & 0 & D'_{11} & D'_{12} & 0 & 0 \\
B'_{12} & B'_{22} & B'_{23} & 0 & D'_{12} & D'_{22} & 0 & 0 \\
0 & 0 & 0 & B'_{66} & 0 & 0 & 0 & D'_{66}
\end{bmatrix}
\begin{bmatrix}
\epsilon^0_x \\
\epsilon^0_y \\
k_1\phi \\
\epsilon^0_{xy} \\
\varkappa_x \\
\varkappa_y \\
\varkappa_{xy}
\end{bmatrix}
\quad (11.17)
$$

$$
\begin{bmatrix}
Q_y/k_1 \\
Q_x/k_1 \\
R_y/k_2 \\
R_x/k_2
\end{bmatrix}
=
\begin{bmatrix}
A'_{44} & 0 & B'_{44} & 0 \\
0 & A'_{55} & 0 & B'_{55} \\
B'_{44} & 0 & D'_{44} & 0 \\
0 & B'_{55} & 0 & D'_{55}
\end{bmatrix}
\begin{bmatrix}
k_1\left(\psi_y + \dfrac{\partial w}{\partial y}\right) \\
k_1\left(\psi_x + \dfrac{\partial w}{\partial x}\right) \\
k_2 \dfrac{\partial \phi}{\partial y} \\
k_2 \dfrac{\partial \phi}{\partial x}
\end{bmatrix}
\quad (11.18)
$$

where

$$
(A'_{ij}, B'_{ij}, D'_{ij}) = \int_{-h/2}^{h/2} c_{ij}(1, z, z^2)\, dz
$$

The quantities k_1 and k_2 are parameters introduced by Whitney and Sun [4] to improve the theory. These parameters have also been utilized by Pagano [5] in conjunction with free-edge analysis. For the results presented in this chapter,

$$
k_1^2 = \frac{\pi^2}{12}, \quad k_2^2 = \frac{\pi^2}{15}
$$

which are the values used in both References [4] and [5]. These values were originally derived by Mindlin and Medick [6] from a dynamic analysis of homogeneous, isotropic plates.

Governing equations in terms of displacements can be obtained by substituting Equations (11.17) and (11.18) into the equilibrium Equations (11.4)–(11.8) and (11.10). The complete equations will not be required for our purposes and are not presented here.

11.4 CALCULATION OF INTERLAMINAR NORMAL STRESS IN A BIDIRECTIONAL LAMINATE

We now illustrate the application of the higher order laminated plate theory developed in Section 11.3 to the determination of an approximate distribution of interlaminar normal stress along the midplane of orthotropic laminates. The procedure outlined here was first suggested by Pagano [5].

Consider a laminated plate of the class $[0°/90°]_{ns}$ subjected to the uniform axial strain $\epsilon_x = \epsilon = $ constant. Thus the stress field is independent of x. Since the laminate is symmetric, the vertical displacement w and the interlaminar shear stresses σ_{xz} and σ_{yz} must vanish along the midplane $z = 0$. In addition most laboratory tensile coupons are wide, i.e., $b/h >> 2$. As a result we analyze the plate response by considering the upper half of the laminate, $z \geq 0$, as a plate under cylindrical bending, where all force and moment resultants are independent of x.

We now adopt a new coordinate system in which the origin is located at the center of the upper half of the laminate. Because of symmetry with respect to the plane $y = 0$, we consider the region of $y \geq 0$, $-h/2 \leq z \leq h/2$ as shown in Figure 11.2. In this approach the normal surface traction at $z = -h/2$ is treated as a dependent variable, i.e., the problem is analogous to that of a plate fastened to a rigid foundation.

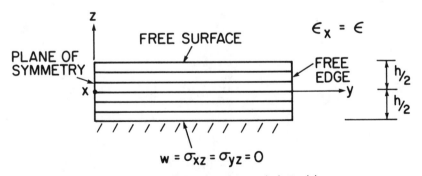

Figure 11.2. Free-edge plate analysis model.

As discussed by Pagano [5], there are certain minimum requirements that must be met by the plate theory chosen to attack the problem in Figure 11.2. In particular, if we choose classical laminated plate theory to solve the problem, we find that q vanishes identically. This difficulty can be overcome by considering the shear deformation theory developed in Chapter 10. In this case, however, the transverse displacement w is independent of z, and by prescribing $w = 0$ at the surface $z = -h/2$, we force w to vanish throughout the plate. Such constraint does not allow the transverse shear force resultant, Q_y, to be prescribed at both ends ($y = 0, -b$), leading to a distribution of q which differs from exact analysis. Thus the simplest appropriate theory must include both transverse shear deformation and a thickness-stretch deformation mode.

For cylindrical bending of orthotropic laminates, the following displacements are utilized:

$$u^0 = x\epsilon, \quad \psi_x = 0 \tag{11.19}$$

$$v^0 = v_p^0(y) + V(y), \quad \psi_y = \Psi(y) \tag{11.20}$$

$$w^0 = w_p^0(y) + W(y), \quad \phi = \phi_p(y) + \Phi(y) \tag{11.21}$$

where ϵ is a uniform axial strain and the subscript p denotes a particular solution. The equilibrium Equations (11.4)–(11.8) and (11.10) reduce to the following:

$$\frac{dN_y}{dy} = 0 \tag{11.22}$$

$$\frac{dM_y}{dy} - Q_y = 0 \tag{11.23}$$

$$\frac{dQ_y}{dy} - \sigma_z = 0 \tag{11.24}$$

$$\frac{dR_y}{dy} - N_z + \sigma_z \frac{h}{2} = 0 \tag{11.25}$$

where

$$\sigma_z = \sigma_z(y, -h/2)$$

Equation (11.22) in conjunction with the free-edge boundary condition at $y = b$ implies that

$$N_y = 0 \tag{11.26}$$

while vanishing of the transverse displacement, w, at $z = -h/2$ in conjunction with Equation (11.3) and (11.21) yields

$$w_p^0(y) = \frac{h}{2} \phi_p(y), \quad W(y) = \frac{h}{2} \Phi(y) \tag{11.27}$$

In order to improve the accuracy of the results, the particular solutions are developed in terms of classical laminated plate stiffnesses, i.e., a state-of-plane-stress is assumed to exist within each ply. In particular, if we consider small values of y, we find that $\sigma_z \rightarrow 0$ and the force and moment resultants all become constant, assuming that $b > 2h$. Therefore the solution in this region represents the response of a laminate of infinite width, i.e., having no free edges. As discussed by Pagano [7], the exact solution for this problem is represented by classical laminated plate theory in which each layer is assumed to be in a state-of-plane-stress.

Combining Equation (11.26) with the constitutive relations from classical laminated plate theory, Equations (2.26), we find that

$$v_p^0 = - \frac{A_{12}}{A_{22}} \epsilon y \tag{11.28}$$

The condition of plane stress in conjunction with the constitutive relations (11.17) leads to the result

$$\phi_p(y) = \left(\frac{A_{12}A_{23}' - A_{22}A_{13}'}{k_1 A_{22} A_{33}'} \right) \epsilon \tag{11.29}$$

It should be noted from Equation (2.26) that the particular solution leads to a bending moment M_y, which we denote here by M_0, which is given by the relationship

$$M_0 = - \frac{A_{12}B_{22}}{A_{22}} \epsilon \tag{11.30}$$

This expression is derived by denoting that $B_{12} = 0$ for the $[0°/90°]_n$ class of laminates under consideration with the present model.

Substituting the constitutive relations (11.17) and (11.18) into the equilibrium Equations (11.26) and (11.23)–(11.25), we obtain the following homogeneous differential equations:

$$A'_{22} \frac{dV}{dy} + k_1 A'_{23} \Phi + B'_{22} \frac{d\Psi}{dy} = 0 \tag{11.31}$$

$$B'_{22} \frac{dV}{dy} + k_1 (B'_{23} - k_2 B'_{44} - k_1 A'_{44} \frac{h}{2}) \frac{d\Phi}{dy} + D'_{22} \frac{d^2\Psi}{dy^2} - k_1^2 A'_{44} \Psi = 0 \tag{11.32}$$

$$k_1 (\frac{k_1 h}{2} A'_{44} + k_2 B'_{44}) \frac{d^2\Phi}{dy^2} + k_1^2 A'_{44} \frac{d\Psi}{dy} - \sigma_z = 0 \tag{11.33}$$

$$- k_1 A'_{23} \frac{dV}{dy} + k_2 (k_1 B'_{44} \frac{h}{2} + k_2 D'_{44}) \frac{d^2\Phi}{dy^2} - k_1^2 A'_{33} \Phi$$

$$\tag{11.34}$$

$$- k_1 (B'_{23} - k_2 B'_{44}) \frac{d\psi}{dy} + \sigma_z \frac{h}{2} = 0$$

Solutions to Equations (11.31)–(11.34) are assumed to be of the form

$$(\sigma_z, V, \Phi \Psi) = (A, B, C, D) e^{\lambda y} \tag{11.35}$$

where A, B, C, and D are constants. Substituting (11.35) into Equations (11.31)–(11.34), we obtain the following homogeneous algebraic equations for the complementary solutions:

$$A'_{22} \lambda B + k_1 A'_{23} C + B'_{22} \lambda D = 0 \tag{11.36}$$

$$B'_{22} \lambda B + k_1 (B'_{23} - k_2 B'_{44} - k_1 A'_{44} \frac{h}{2}) \lambda C + (D'_{22} \lambda^2 - k_1^2 A'_{44}) D = 0 \tag{11.37}$$

$$A - k_1 (\frac{k_1 h}{2} A'_{44} + k_2 B'_{44}) \lambda^2 C - k_1^2 A'_{44} \lambda D = 0 \tag{11.38}$$

$$A \frac{h}{2} - k_1 A'_{23} \lambda B + [k_2(k_1 B'_{44} \frac{h}{2} + k_2 D'_{44})\lambda^2 - k_1^2 A'_{33}]C$$

$$- k_1(B'_{23} - k_2 B'_{44})\lambda D = 0 \tag{11.39}$$

A nontrivial solution to Equations (11.36)–(11.39) can be obtained by choosing λ such that the determinant of the coefficient matrix vanishes. This procedure leads to the following polynomial for λ:

$$r_1 \lambda^4 + r_2 \lambda^2 + r_3 = 0 \tag{11.40}$$

where

$$r_1 = - (A'_{22} D'_{22} - B'_{22})(k_1 k_2 h B'_{44} + k_2^2 D'_{44} + \frac{k_1^2 h^2}{4} A'_{44})$$

$$r_2/k_1^2 = - A'_{22}(k_2 B'_{44} - B'_{23})^2 + 2A'_{23}B'_{22}(B'_{23} - k_2 B'_{44} - \frac{k_1 h}{2} A'_{44})$$

$$+ A'_{22}A'_{44}(k_1 h B'_{23} + k_2^2 D'_{44}) - D'_{22}A'^2_{23} + A'_{33}(A'_{22}D'_{22} - B'^2_{22})$$

$$r_3 = - A'_{44}k_1^4(A'_{22}A'_{33} - A'^2_{23})$$

Equation (11.40) yields two pairs of roots which we denote by $\pm\lambda_i$ ($i = 1,2$). We now introduce a subscript to denote the respective constants, A_i, B_i, C_i, and D_i, associated with a particular value λ_i. We can solve Equations (11.36)–(11.38) in the form

$$B_i = \bar{B}_i A_i, \ C_i = \bar{C}_i A_i, \ D_i = \bar{D}_i A_i \tag{11.41}$$

where

$$\bar{B}_i = \frac{k_1}{S_i} \{[B'_{22}(k_2 B'_{44} + \frac{k_1 h}{2}(A'_{44} - B'_{23}) + A'_{23}D'_{22}]\lambda_i^2 - k_1^2 A'_{23}A'_{44}\}$$

$$\bar{C}_i = \frac{\lambda_i}{S_i} [(B'^2_{22} - A'_{22}D'_{22})\lambda_i^2 + k_1^2 A'_{22}A'_{44}]$$

$$\bar{D}_i = \frac{k_1 \lambda_i^2}{S_i} [A'_{23}B'_{22} - A'_{22}(B'_{23} - k_2 B'_{44} - \frac{k_1 h}{2} A'_{44})]$$

and

$$S_i = k_1 \lambda_i^3 [(B_{22}'^2 - A_{22}' D_{22}')(k_2 B_{44}' + \frac{k_1 h}{2} A_{44}')\lambda_i^2 + k_1^2 A_{44}'(A_{22}' B_{23}' - A_{23}' B_{23}')]$$

We now consider the case where the roots to Equation (11.40) are real and we can write the solutions in terms of hyperbolic sine and cosine functions. To accomplish this we define the following relationships:

$$G_i(y) = A_i \cosh \lambda_i y + A_i' \sinh \lambda_i y$$

$$\overline{G}_i(y) = A_i' \cosh \lambda_i y + A_i \sinh \lambda_i y \tag{11.42}$$

where A_i and A_i' are arbitrary constants. We can now write the solution in terms of four arbitrary constants as follows:

$$\sigma_z = \sum_{i=1}^{2} G_i(y)$$

$$v^0 = -\frac{A_{12}}{A_{22}} \epsilon y + \sum_{i=1}^{2} \overline{B}_i \overline{G}_i(y)$$

$$\phi = \left(\frac{A_{12} A_{23}' - A_{22} A_{13}'}{k_1 A_{22} A_{33}'} \right) \epsilon + \sum_{i=1}^{2} \overline{C}_i G_i(y) \tag{11.43}$$

$$\psi_y = \sum_{i=1}^{2} \overline{D}_i \overline{G}_i(y)$$

The boundary conditions are as follows:
   at $y = 0$

$$\psi_y = Q_y = 0 \tag{11.44}$$

   at $y = b$

$$M_y = Q_y = 0 \tag{11.45}$$

Substituting the first of Equations (11.43) into Equation (11.44) and integrating, we obtain the transverse shear stress

$$Q_y = \sum_{i=1}^{2} \frac{\bar{G}_i(y)}{\lambda_i} \tag{11.46}$$

Combining this result with Equation (11.23) and again integrating, we find

$$M_y = M_0 + \sum_{i=1}^{2} \frac{\bar{G}_i(y)}{\lambda_i^2} \tag{11.47}$$

where M_0 is associated with the particular solution as defined by Equation (11.30). Equations (11.46) and (11.47) can also be obtained by substituting Equations (11.43) into the constitutive relations (11.17) and (11.18), and simplifying the results.

The boundary conditions (11.44) are satisfied if

$$A_1' = A_2' = 0 \tag{11.48}$$

Substituting Equations (11.46) and (11.47) into Equation (11.45), we obtain the algebraic equations

$$\frac{A_1}{\lambda_1} \sinh \lambda_1 b + \frac{A_2}{\lambda_2} \sinh \lambda_2 b = 0 \tag{11.49}$$

$$\frac{A_1}{\lambda_1^2} \cosh \lambda_1 b + \frac{A_2}{\lambda_2^2} \cosh \lambda_2 b = -M_0 \tag{11.50}$$

Solving Equations (11.49) and (11.50) for A_1 and A_2, and substituting the results into the first of Equations (11.43), we obtain the result

$$\sigma_z = \frac{(\lambda_1 \sinh \lambda_2 b \cosh \lambda_1 y - \lambda_2 \sinh \lambda_1 b \cosh \lambda_2 y)\lambda_1\lambda_2 M_0}{\lambda_1 \sinh \lambda_1 b \cosh \lambda_2 b - \lambda_2 \cosh \lambda_1 b \sinh \lambda_2 b} \tag{11.51}$$

Numerical results are shown in Table 11.1 [5] for a $[0°/90°]_s$ laminate with the following unidirectional properties in terms of the stiffnesses, C_{ij}:

$$c_{11}(0°)/c_{22}(0°) = R, \; c_{12}(0°)/c_{22}(0°) = c_{13}(0°)/c_{22}(0°) =$$

$$c_{23}(0°)/c_{22}(0°) = 0.25, \; c_{33}(0°)/c_{22}(0°) = 1,$$

$$c_{44}(0°)/c_{22}(0°) = c_{55}(0°)/c_{22}(0°) = c_{66}(0°)/c_{22}(0°) = 0.4 \tag{11.52}$$

Table 11.1. Maximum stress and boundary layer width [5].

R	b/2h	$\sigma_z(b, -h/2)/C_{22}(0°)\epsilon$	b*/2h
	2	0.0760	0.86
2	3	0.0760	0.86
	5	0.0760	0.86
	∞	0.0760	0.86
	2	0.1247	0.92
5	3	0.1247	0.93
	5	0.1247	0.92
	∞	0.1247	0.93
	2	0.1301	0.98
10	3	0.1301	0.98
	5	0.1301	0.98
	∞	0.1301	0.98
	2	0.1179	1.04
20	3	0.1180	1.04
	5	0.1180	1.04
	∞	0.1180	1.03
	2	0.1062	1.10
30	3	0.1063	1.08
	5	0.1063	1.08
	∞	0.1063	1.07

The results in Table 11.1 give the maximum value of σ_z, i.e., $\sigma_z(b, -h/2)$, for various values of anisotropy ratio, R, and aspect ratios, $b/2h$. The results indicate that the maximum value of σ_z is independent of $b/2h$, provided $b/2h > 1$. A "boundary layer zone," denoted by b^*, is also shown in Table 11.1. This is defined as the distance from the free edge at which $|\sigma_z|$ reaches 7% of its maximum value. A cursory examination of Table 11.1 reveals that the boundary layer width closely approximates the plate thickness. It should be noted that the roots of Equation (11.40) are complex for the cases $R = 2, 5$ and real for the remaining cases.

Many practical laminates are wide, i.e., $b/2h > 2$. For this case a very ac-curate approximation to Equation (11.51) can be obtained by letting $b/2h \rightarrow \infty$, with the result

$$\sigma_z(-h/2, y) = \frac{M_0\lambda_1\lambda_2}{(\lambda_1 - \lambda_2)} (\lambda_1 e^{-\lambda_1\xi} - \lambda_2 e^{-\lambda_2\xi}) \qquad (11.53)$$

where

$$\xi = b - y$$

is the distance from the free edge.

11.5 COMPARISON TO EXACT THEORY

Pipes [8] obtained a finite difference solution to the equations of classical theory of elasticity for a $[0°/90°]_s$ laminate under tensile loading. The unidirectional properties in terms of elastic stiffnesses assumed by Pipes are as follows:

$$c_{11}(0°)/c_{22}(0°) = 9.14, \quad c_{33}(0°)/c_{22}(0°) = 1,$$

$$c_{13}(0°)/c_{22}(0°) = c_{12}(0°)/c_{22}(0°) = 0.253,$$

$$c_{23}(0°)/c_{22}(0°) = 0.217, \quad c_{44}(0°)/c_{22}(0°) =$$

$$c_{55}(0°)/c_{22}(0°) = c_{66}(0°)/c_{22}(0°) = 0.385 \tag{11.54}$$

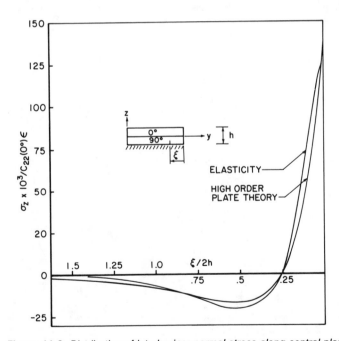

Figure 11.3. *Distribution of interlaminar normal stress along central plane.*

The distribution of σ_z along the central plane $z = -h/2$ given by Pipes [8] is compared in Figure 11.3 to results obtained from Equation (11.51) for a plate aspect ratio $b/2h = 4$. It is easily seen that excellent agreement is obtained between the elasticity solution and the approximate theory.

REFERENCES

1. Pipes, R. B. and N. J. Pagano. "Interlaminar Stresses in Composite Laminates Under Uniform Axial Extension," *Journal of Composite Materials*, 4:538–548 (1970).

2. Pagano, N. J. and R. B. Pipes. "Influence of Stacking Sequence on Laminate Strength," *Journal of Composite Materials*, 5:50–57 (1971).

3. Pagano, N. J. and R. B. Pipes. "Some Observations on the Interlaminar Strength of Composite Laminates," *International Journal of Mechanical Sciences*, 15:679–688 (1973).

4. Whitney, J. M. and C. T. Sun. "A Higher Order Theory for Extensional Motion of Laminated Composites," *Journal of Sound and Vibration*, 30:85–97 (1973).

5. Pagano, N. J. "On the Calculation of Interlaminar Normal Stress in Composite Laminates," *Journal of Composite Materials*, 8:65–82 (1974).

6. Mindlin, R. D. and M. A. Medick. "Extensional Vibrations of Elastic Plates," *Journal of Applied Mechanics*, 26:561–569 (1959).

7. Pagano, N. J. "Exact Moduli of Anisotropic Laminates," *Composite Materials*, edited by G. P. Sendeckyj, Vol. 2 of Series entitled *Composite Materials*, edited by L. J. Broutman and R. H. Krock, Academic Press, pp. 23–44 (1974).

8. Pipes, R. B. "Solution to Certain Problems in the Theory of Elasticity for Laminated Anisotropic Systems," *Ph.D. Dissertation*, University of Texas, Arlington, (March 1972).

Laminated Plate Calculations (LAMPCAL—Users' Guide)

1. LAMPCAL OVERVIEW

L AMPCAL IS A microcomputer program for bending, buckling, and free vibration analysis of laminated plates. The program is written in basic and is available in conjunction with Apple II, McIntosh, and IBM compatible computers.

Calculations include maximum deflection and bending moments under transverse loads, critical buckling loads, and free vibration frequencies for a wide variety of plates, including hybrids and asymmetric laminates. Results can be obtained in conjunction with both classical laminated plate theory and shear deformation plate theory. Calculations are based on theory presented in the book, *Structural Analysis of Laminated Anisotropic Plates*. Specific equations in this book which correspond to calculations under discussion will be cited throughout this manual.

LAMPCAL is delivered to the users on a single-sided floppy disk for use on either single or double sided disk drives. The program requires at least 64k of memory.

2. RUNNING THE PROGRAM

The program LAMPCAL is automatically loaded into computer memory by simply turning the computer on with the disk loaded into the disk drive. A menu displays the program for the Apple II and IBM versions. In the McIntosh version the program appears as an icon with the title LAMPCAL. Thus, the McIntosh version employs pull down menus. After the program has been completely loaded, the following main menu will appear on the screen (icons in the case of McIntosh):

Laminated Plate Calculations (LAMPCAL)

1. One-Dimensional Analysis
2. Rectangular Plate Analysis
3. Shear Deformation Analysis
4. Change Printer Slot Number (Apple II Version Only)

2.1 Units

All results are displayed in normalized non-dimensional form. The following normal-

ized quantities are utilized throughout Lampcal:
For a Uniform Transverse Load

$$W(\text{Max}) = w\,\frac{h^3 E_T}{q_0 a^4}$$

$$MX(\text{Max}) = \frac{M_x}{q_0 a^2}$$

$$MY(\text{Max}) = \frac{M_y}{q_0 a^2}$$

$$MXY(\text{Max}) = \frac{M_{xy}}{q_0 a^2}$$

For Concentrated Loads

$$W(\text{Max}) = w\,\frac{bh^3 E_T}{Pa^3}$$

$$MX(\text{Max}) = \frac{M_x b}{Pa}$$

$$MY(\text{Max}) = \frac{M_y b}{Pa}$$

$$MXY(\text{Max}) = \frac{M_{xy} b}{Pa}$$

Critical Buckling Loads

$$N(\text{Critical}) = N_{cr}\,\frac{a^2}{E_T h^3}$$

Vibration Frequencies

$$\text{Omega} = \omega\,a^2 \sqrt{\frac{\varrho}{h^3 E_T}}$$

2.2 Data Input

The user is prompted for all data input. Since all inputs are dimensionless, the user is usually asked for property ratios. The nomenclature is as follows:

EL/ET = Ratio of longitudinal modulus to transverse modulus.

GLT/ET = Ratio of inplane shear modulus to transverse modulus.

NULT = Poisson's ratio as determined from transverse contraction during a uniaxial tensile test in the longitudinal direction.

A/B = Ratio of length in the *x*-direction to length in the *y*-direction

AN = Angle of fiber orientation within a ply

2.3 Laminate Stacking Geometry

The program will handle laminates up to 30 plies. The plies are numbered beginning at the top of the laminate. The *x*-axis is the reference axis for ply orientation. The ply reference axis is positive for a *counterclockwise* rotation relative to the *x*-axis.

2.4 Hybrid Laminates

The program will handle hybrid laminates containing up to 4 different materials. The user is prompted to enter the ply properties for each material. Data input for geometry includes the material number of each ply (denoted by MAT #) and angle of orientation. The materials are ordered in the sequence they are entered into the program. It is important to note that the normalizing factor for the elastic constants in a hybrid is the transverse modulus of the first material entered, denoted by $ET(1)$. Thus, after entering the properties of the first material, the user will be prompted for the elastic constants of the remaining materials in terms of $ET(1)$. In addition, ply thickness is also normalized by the per ply thickness of the first material entered. Thus, after entering the properties of the first material, the user will be prompted to enter the per ply thickness of the remaining materials in the form of a ratio to the per ply thickness of the first material.

2.5 Hard Copies

The user is given the option of printing results for items 1 through 3 in the main menu. In addition to printing all calculated results, all input parameters are also printed if the print option is chosen.

3. ONE-DIMENSIONAL ANALYSIS

When the user chooses the one-dimensional analysis option from the main menu, the following sub-menu will appear on the screen (icon with the title 1-D Analysis in conjunction with the McIntosh version):

One-Dimensional Analysis

1. Cylindrical Bending
2. Analysis of Laminated Beams

3.1 Cylindrical Bending

With cylindrical bending we consider plates which have a very high length-to-width ratio such that the plate deformation may be considered to be independent of the length coordinate (which in this case is the y coordinate). Such behavior is referred to as cylindrical bending. The theory is covered in Sections 4.2 and 4.3 of *Structural Analysis of Laminated Anisotropic Plates*. When the user chooses this option from the sub-menu the following list of problems will appear (pull down menu in conjunction with icon having the title 1-D Analysis for the McIntosh version):

Cylindrical Bending

1. Bending Under Uniform Transverse Load – SS boundary Conditions
2. Bending Under Uniform Transverse Load – CC Boundary Conditions
3. Bending Under Uniform Transverse Load – CF Boundary Conditions
4. Buckling – SS Boundary Conditions
5. Free Vibration – SS Boundary Conditions

This option considers laminates of arbirary stacking geometry, including antisymmetric laminates.

3.3.1 Bending Under Transverse Load – SS Boundary Conditions

In this option the maximum deflection and bending moments are calculated for laminates under uniform transverse load. Simply-supported (SS) boundary conditions are considered at $x = 0, a$. Equations (4.15) and (4.16) are utilized in obtaining the numerical results.

3.1.2 Bending Under Uniform Transverse Load – CC Boundary Conditions

In this option the maximum deflection and bending moments are calculated for laminates under uniform transverse load. Clamped (CC) boundary conditions are considered at $x = 0, a$. Equations (4.18) and (4.19) are utilized in obtaining the numerical results.

3.1.3 Bending Under Uniform Transverse Load – CF Boundary Conditions

In this option the maximum deflection and bending moments are calculated for laminates under uniform transverse load. Clamped (C) boundary conditions at $x = 0$, and free (F) boundary conditions at $x = a$ are considered. This represents a cantilever plate under uniform load. Although specific results for this case are not presented in section 4.2, the solution can be derived from Equation (4.10) with the constants determined from the boundary conditions.

3.1.4 Buckling – SS Boundary Conditions

In this option the critical buckling load is determined for laminates subjected to uniaxial compression loading in the x-direction. Simply-supported (SS) boundary conditions are considered at $x = 0, a$. Equation (4.32) is utilized in obtaining the critical buckling load.

3.1.5 Free Vibration – SS Boundary Conditions

In this option natural vibration frequencies for laminates are calculated. The user is prompted to enter the mode number, m, for which the vibration frequency is to be determined. Simply-supported (SS) boundary conditions are considered at $x = 0, a$. Equation (4.30) is utilized for obtaining the vibration frequencies.

3.2 Laminated Beam Analysis

Bending, buckling, and free vibration of symmetrically laminated beams are considered. Laminated beam analysis is derived from plate theory by considering laminated plates which are very long in the x-direction and narrow in the y-direction, that is a/b is very large. As in the case of cylindrical bending, all functions are considered to be independent of y. The difference between beam theory and cylindrical bending is analogous to the difference between plane stress and plane strain, respectively, in classical theory of elasticity. The solutions presented in conjunction with the laminated beam analysis option can be found in Sections 4.5–4.7 of the book *Structural Analysis of Laminated Anisotropic Plates*. When the user chooses this option from the sub-menu the following list of problems will appear (pull down menu in conjunction with icon having the title 1-D Analysis for the McIntosh version):

Laminated Beam Analysis

1. 3-Point Bending
2. 4-Point Bending
3. Buckling – SS Boundary Conditions
4. Buckling – CC Boundary Conditions
5. Free Vibration – SS Boundary Conditions

3.2.1 3-Point Bending
In this option a laminated beam under classic 3-point bending is considered (see Figure 4.4). Maximum deflection and bending moment are calculated using Equations (4.62) and (4.58), respectively.

3.2.2 4-Point Bending
In this option a laminated beam under classic 4-point bending is considered (see Figure 4.5). Maximum deflection and bending moment are calculated using Equations (4.74) and (4.67), respectively.

3.2.3 Buckling – SS Boundary Conditions
In this option the critical buckling load is calculated for laminated beams subjected to uniaxial compression loading along the axis of the beam (x-axis). Simply-supported (SS) boundary conditions are considered at $x = 0, a$. Equation (4.92) is utilized for calculating the buckling load.

3.2.4 Buckling – CC Boundary Conditions
In this option the critical buckling load is calculated for laminated beams subjected to uniaxial compression loading along the axis of the beam (x-axis). Clamped (CC) boundary conditions are considered at $x = 0, a$. Equation (4.97) is utilized for calculating the buckling load.

3.2.5 Free Vibration – SS Boundary Conditions
In this option the natural vibration frequencies for laminated beams are calculated. The user is prompted to enter the mode, m, for which the vibration frequency is determined.

Simply-supported (SS) boundary conditions are considered at $x = 0, a$. Equation (4.91) is utilized for obtaining the vibration frequencies.

4. RECTANGULAR LAMINATE ANALYSIS

With rectangular plate analysis we consider laminates which are specially orthotropic, that is laminates which are symmetric and in which the bending-twisting coupling terms vanish. Antisymmetric laminates are handled, however, by using the reduced bending stiffness approximation. In particular, the specially orthotropic plate analysis as discussed in Chapter 5 of *Structural Analysis of Laminated Anisotropic Plates* is utilized with the bending stiffnesses replaced by the reduced bending stiffnesses. The reduced bending stiffness approximation is discussed in Section 7.8. When the user chooses the rectangular laminate analysis option from the main menu, the following sub-menu will appear on the screen (icon with the title Rec Analysis in conjunction with the McIntosh version):

Bending of Simply-Supported Plates
1. Uniform Load
2. Concentrated Load

Buckling
3. Uniaxial Compression—SSSS Boundary Conditions
4. Uniaxial Compression—SSCC Boundary Conditions
5. Biaxial Loading—SSSS Boundary Conditions

Free Vibration
6. SSSS Boundary Conditions
7. CCCC Boundary Conditions
8. SSCC Boundary Conditions

4.1 Bending of Simply-Supported Plates

This option involves bending under either uniform transverse load or a concentrated load. Simply-supported (SSSS) boundary conditions are considered at $x = 0, a$ and $y = 0, b$. Loads are represented by a double Fourier series with 25 terms ($M = N = 5$).

4.1.1 Uniform Load
In this option the maximum deflection and bending moments are calculated for laminates under uniform load. Equations (5.7)–(5.10) and (5.16) are utilized in obtaining numerical results.

4.1.2 Concentrated Load
In this option the maximum deflection and bending moments are calculated for laminates under a concentrated load. The user is prompted to enter the normalized coordinates ($X/A = x/a$, $Y/B = y/b$) of the location of the concentrated load. Equations (5.7)–(5.10) and (5.17) are utilized in obtaining numerical results.

4.2 Buckling

This option involves the calculation of critical buckling loads. In addition to displaying the magnitude of the critical buckling load, the mode numbers m and n corresponding to the critical load are also displayed.

4.2.1 Uniaxial Compression – SSSS Boundary Conditions

In this option critical buckling loads are calculated for laminates subjected to uniform inplane compression loading in the x-direction. Simply-supported (SSSS) boundary conditions are considered at $x = 0$, a and $y = 0$, b. Equation (5.74) is utilized in obtaining the critical buckling load.

4.2.2 Uniaxial Compression – SSCC Boundary Conditions

In this option critical buckling loads are calculated for laminates subjected to uniform inplane compression loading in the x-direction. Simply-supported (SS) boundary conditions are considered at $x = 0$, b and clamped (CC) boundary conditions at $y = 0$, b. Critical buckling loads are obtained from an iteration procedure in conjunction with Equations (5.92), (5.93), and (5.96).

4.2.3 Biaxial Loading – SSSS Boundary Conditions

In this option critical buckling loads are calculated for laminates subjected to uniform compression loading in the x-direction and arbitrary uniform load in the y-direction. The user is prompted to input the parameter k, which is the ratio of the load in the x-direction to the load in the y-direction. A positive value of k indicates compression loading in the y-direction, while a negative value of k indicates tensile loading in the y-direction. Equation (5.72) is utilized in obtaining the critical buckling load.

4.3 Free Vibration

This option involves the calculation of free vibration frequencies. The user is prompted to enter the mode numbers m and n for which the vibration frequency is to be determined. For the case of simply-supported boundary conditions, the exact solution as given by Equation (5.137) is utilized for determining the vibration frequency. The approximation discussed in Section 5.11 is utilized for the other boundary conditions. Thus, these frequencies are determined from Equation (5.157) in conjunction with Table 10.

4.3.1 SSSS Boundary Conditions

In this option the vibration frequency is determined for simply-supported (SSSS) boundary conditions at $x = 0$, a and $y = 0$, b.

4.3.2 CCCC Boundary Conditions

In this option the vibration frequency is determined for clamped (CCCC) boundary conditions at $x = 0$, a and $y = 0$, b.

4.3.3 SSCC Boundary Conditions

In this option the vibration frequency is determined for simply-supported (SS) boundary conditions at $x = 0$, a and clamped (CC) boundary conditions at $y = 0$, b.

5. SHEAR DEFORMATION ANALYSIS

In this option the calculations involve shear deformation plate theory as discussed in Chapter 10 of *Structural Analysis of Laminated Anisotropic Plates*. Three additional input parameters are required with this theory as compared to classical laminated plate theory. The first is *GTT/ET*, which is the ratio of the transverse shear modulus to the transverse modulus. The second parameter is the ratio of the plate length in the x-direction to the thickness ($A/H = a/h$), while the third parameter is the shear correction factor, K, which is usually given the value 5/6. When the user chooses the shear deformation analysis option from the main menu, the following sub-menu will appear on the screen (icon with the title Shear Def in conjunction with the McIntosh version):

Shear Deformation Analysis

1. Cylindrical Bending
2. Analysis of Laminated Beams
3. Analysis of Specially Orthotropic Plates

5.1 Cylindrical Bending

As discussed in Section 3.1, with cylindrical bending we consider plates which have a very high length-to-width ratio such that the plate deformation may be considered to be independent of the length coordinate (which is the y coordinate). When the user chooses this option from the sub-menu the following list of problems will appear (pull down menu in conjunction with icon having the title Shear Def for the McIntosh version):

Cylindrical Bending

1. Bending Under Uniform Transverse Load—SS Boundary Conditions
2. Buckling—SS Boundary Conditions
3. Free Vibration—SS Boundary Conditions

This option is limited to orthotropic lamintes (shear coupling terms vanish). The buckling and free vibration problems are also limited to symmetric laminates.

5.1.1 Bending Under Uniform Transverse Load—SS Boundary Conditions

In this option the maximum deflection and bending moments are calculated for laminates under uniform transverse load. Both symmetric and unsymmetric laminates are considered. Simply-supported (SS) boundary conditions are considered at $x = 0, a$. Solutions are based on Equation (10.53) with Equations (10.51) and (10.52) expanded in a Fourier series in order to represent the uniform load distributed over the plate surface.

5.1.2 Buckling—SS Boundary Conditions

In this option the critical buckling load is determined for laminates subjected to uniaxial compression loading in the x-direction. Simply-supported (SS) boundary conditions are considered at $x = 0, a$. Equation (10.65) is utilized in obtaining the critical buckling load.

5.1.3 Free Vibration—SS Boundary Conditions

In this option natural vibration frequencies for laminates are calculated. The user is prompted to enter the mode number, m, for which the vibration frequency is to be determined. Simply-supported (SS) boundary conditions are considered at $x = 0, a$. Equation (10.72) is utilized for obtaining the vibration frequencies. Thus, rotary inertia is neglected.

5.2 Analysis of Laminated Beams

Bending of laminated beams are considered in this option. As discussed in Section 3.2, laminated beam analysis is derived from plate theory by considering laminated plates which are very long in the x-direction and narrow in the y-direction, that is a/b is very large. As in the case of cylindrical bending, all functions are considered to be independent of y. Solutions presented in conjunction with this beam analysis can be found in Section 10.6 of the book *Structural Analysis of Laminated Anisotropic Plates*. When the user chooses this option from the sub-menu the following list of problems will appear (pull down menu in conjunction with icon having the title Shear Def for the McIntosh version):

Analysis of Laminated Beams

1. 3-Point Bending
2. 4-Point Bending

5.2.1 3-Point Bending

In this option a laminated beam under classic 3-point bending is considered (see Figure 4.4). Maximum deflection and bending moment are calculated using Equations (10.93) and (10.87), respectively. Note that Equation (10.87) is exactly the same as Equation (4.58). Thus, shear deformation has no effect on the maximum bending moment.

5.2.2 4-Point Bending

In this option a laminated beam under classic 4-point bending is considered (see Figure 4.5). Maximum deflection and bending moment are calculated using Equations (10.111) and (10.99), respectively. Note that equation (10.99) is exactly the same as Equation (4.67). Thus, shear deformation has no effect on the maximum bending moment.

5.3 Analysis of Specially Orthotropic Plates

As discussed in Section 5, specially orthotropic plates include symmetric laminates in which the bending-twisting coupling terms vanish. Simply-supported (SSSS) boundary conditions are considered at $x = 0$, a and $y = 0$, b. When the user chooses this option from the sub-menu the following list of problems will appear (pull down menu in conjunction with icon having the title Shear Def with the McIntosh version):

Analysis of Specially Orthotropic Plates

1. Bending Under Uniform Transverse Load
2. Free Vibration

5.3.1 Bending Under Uniform Transverse Load

In this option the maximum deflection and bending moments are calculated for laminates under uniform transverse load. Equation (10.126) is utilized in conjunction with a 25 term ($M = N = 5$) double Fourier series.

5.3.2 Free Vibration

In this option natural vibration frequencies for specially orthotropic plates are calculated. The user is prompted to enter the mode number, m and n, for which the vibration frequency is to be determined. Equation (10.126) utilized for obtaining the vibration frequencies. Rotary inertia is neglected in this calculation.

6. CHANGE PRINTER SLOT NUMBER

This option is applicable to the Apple II program only. The program is set to print output through computer slot number 1. If the user wishes a different slot number, then this option is chosen from the main menu. The user is immediately prompted to enter the slot number of his choice. Unless further changes are made, output will always be printed through the new slot number.

Index

LAMPCAL

☐ YES, send me the new software support package that performs bending, buckling, and free vibration analysis of laminated plates. I understand that I must select the machine compatibility for my needs and have done so on the order form. In addition to performing calculations of maximum deflection under bending loads, critical buckling loads, and free vibration frequencies, LAMPCAL is designed to produce results in conjunction with both classical laminated plate theory and shear deformation plate theory. ONLY **$99.**

ORDER FORM: **LAMPCAL**

Check machine compatibility requirement
☐ IBM PC, PC XT, PC AT, or compatibles (625239)
☐ Apple (625190)

$99.00

Billing to organization & street address only. Add 6% sales tax to PA orders. Please prepay personal orders. $2.00 postage/handling (P/H) added to billing. If prepaid we pay P/H. Offer subject to acceptance.

(Please Print)
ORGANIZATION _____

NAME_____

ADDRESS _____

CITY_____STATE_____ZIP_____

CREDIT CARD ORDERS: ☐ VISA   ☐ Mastercard

CARD NO._____EXP. DATE _____

SIGNATURE _____

| *Toll-Free 800 Number For Fast Delivery on Credit Card Orders* |
| To order with Visa or Mastercard, phone |
| Outside PA: **1-800-233-9936**   In PA: **(717) 291-5609** |
| *This is a direct line to our order desk for all credit card orders.* |
| *From 8:30 A.M. to 5:00 P.M. Eastern Time* |

Please detach and return to:
TECHNOMIC Publishing Company, Inc.
851 New Holland Ave., Box 3535, Lancaster, PA 17604, U.S.A.